2nd RCN Conference on Pan American Biofuels and Bioenergy Sustainability 2016

Buenos Aires, Argentina
13 - 16 September 2016

ISBN: 978-1-5108-4165-9

Printed from e-media with permission by:

Curran Associates, Inc.
57 Morehouse Lane
Red Hook, NY 12571

Some format issues inherent in the e-media version may also appear in this print version.

Printed by Curran Associates, Inc. (2017)

For permission requests, please contact AIChE
at the address below.

AIChE
120 Wall Street, FL 23
New York, NY 10005-4020

Phone: (800) 242-4363
Fax: (203) 775-5177

www.aiche.org

Additional copies of this publication are available from:

Curran Associates, Inc.
57 Morehouse Lane
Red Hook, NY 12571 USA
Phone: 845-758-0400
Fax: 845-758-2633
Email: curran@proceedings.com
Web: www.proceedings.com

TABLE OF CONTENTS

SESSION 6: LARGE SCALE PRODUCTION CASE STUDIES – BIOGEOCHEMICAL ASPECTS/CASOS DE ESTUDIO EN LA PRODUCCION EN GRAN ESCALA – ASPECTOS BIOGEOQUIMICOS

SESSION 7: THE FUTURE OF THE BIOENERGY INDUSTRY/EL FUTURO DE LA INDUSTRIA DE LA BIOENERGIA

SESSION 8: ENVIRONMENTAL LCA & ENERGY BALANCES OF BIOFUEL PATHWAYS/ANALISIS DE CICLO DE VIDA ABIENTALES Y BALANCES DE ENERGIA DE LAS CADENAS DE BIOCOMBUSTIBLES

GREAT LAKES
BIOENERGY
RESEARCH CENTER

Bioenergy and Biofuels: Getting to (Very) Large Scale--Sustainably

Bruce E. Dale, University Distinguished Professor
Michigan State University

Second Conference on Pan-American Biofuels and Bioenergy Sustainability
Buenos Aires, Argentina, September 13, 2016

U.S. DEPARTMENT OF ENERGY
DOE Bioenergy Research Centers

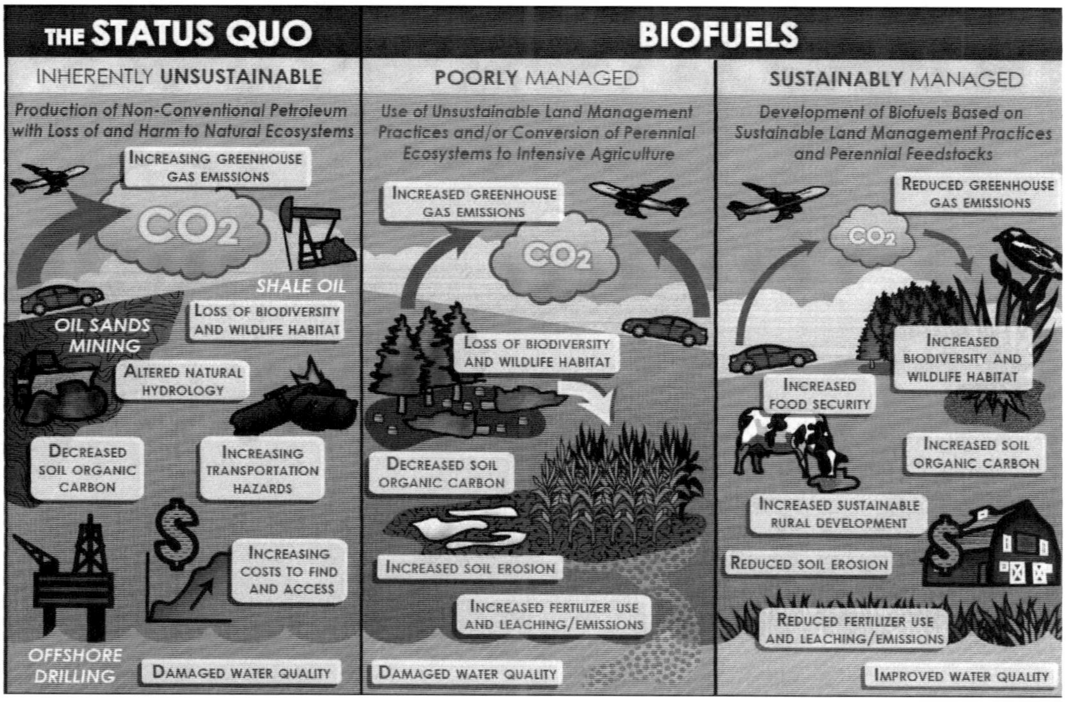

(Renewable) Energy is Critical for Human Well Being

1. Rate of energy use (rate of work done) largely determines national wealth and opportunities for human development
2. Many people seem ignorant of or indifferent to this fact: ***energy use makes us prosperous (lack of energy will make us poor!!)***
3. Current prosperity is based largely on non-renewable fossil fuels—thus our prosperity has an "expiration date"- it will end!
4. Lack of energy access currently condemns billions of human beings to poverty—and will also impoverish future generations of those who are currently wealthy unless....
5. We implement terawatt scale renewable energy systems—in the next few decades
6. Bioenergy is an essential part of a renewable energy future
7. Thus farmers—those who own and manage land--must benefit from bioenergy production
8. **So my question is: "How can we produce terawatts of sustainable biofuels--soon?**

3

ENERGY DOE Bioenergy Research Centers GREAT LAKES BIOENERGY RESEARCH CENTER

Power Consumption and GDP (World Regions)

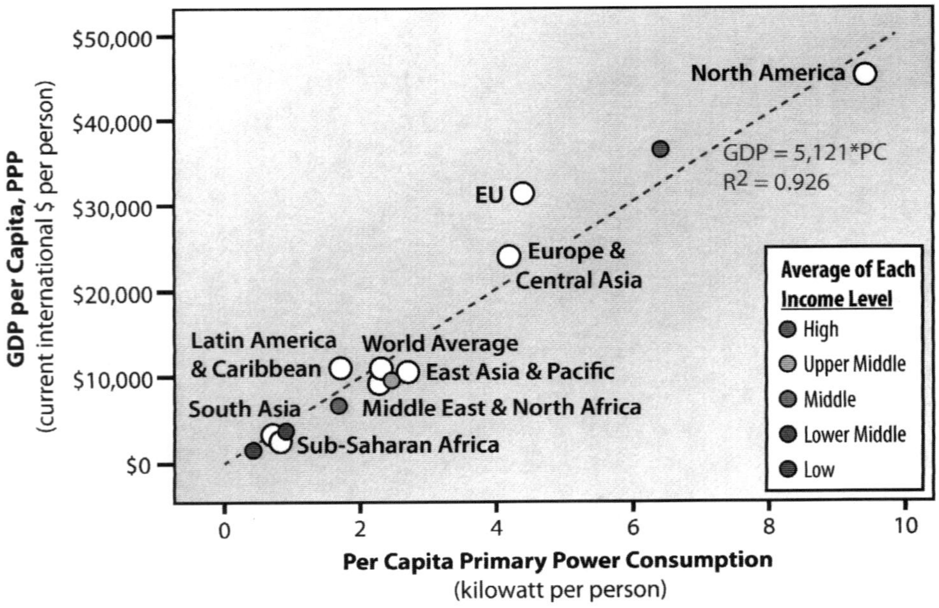

Energy Consumption & Human Well Being are Linked: How Much Energy is "Enough"?

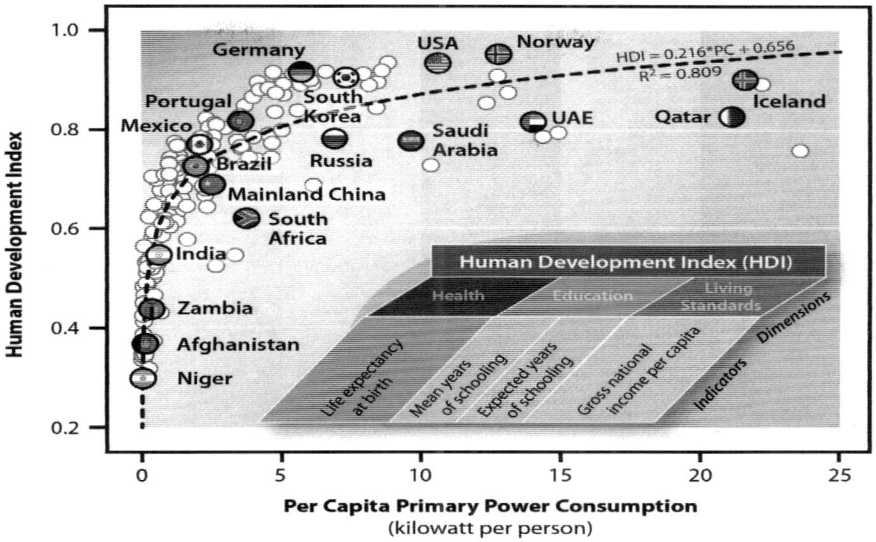

Figure 1. Human Development by Country versus Per Capita Power Consumption in 2010

Some Basic Energy Facts: *Why Liquid Fuels are So Important*

- Services we need from energy (current primary sources of these services: fossil and renewable)
  - Heat (natural gas, coal— solar, wind, geothermal, biomass)
  - Light/electricity (coal, natural gas, hydro/nuclear– solar, wind, geothermal, biomass)
  - Mobility (liquid fuels from oil—96%, some ethanol & biodiesel, & CNG)- **most commerce**
- **All energy services (all BTU, ergs, GJ) are not created equal—we value mobility (=oil) above all other energy carriers**
- Electricity/batteries can never provide more than about half of mobility needs—and they cannot support commerce at all
- Commerce moves by trucks, ocean shipping, rail & jet aircraft
- *Liquid fuels: not "energy" is the key economic security issue—and right now liquid fuels means refined oil products*
- *The <u>only</u> potentially sustainable, very large scale source of renewable liquid fuels is from sustainable plant matter— or "biofuels"*

U.S. DEPARTMENT OF ENERGY DOE Bioenergy Research Centers GREAT LAKES BIOENERGY RESEARCH CENTER

Bioenergy/biofuels are NOT optional

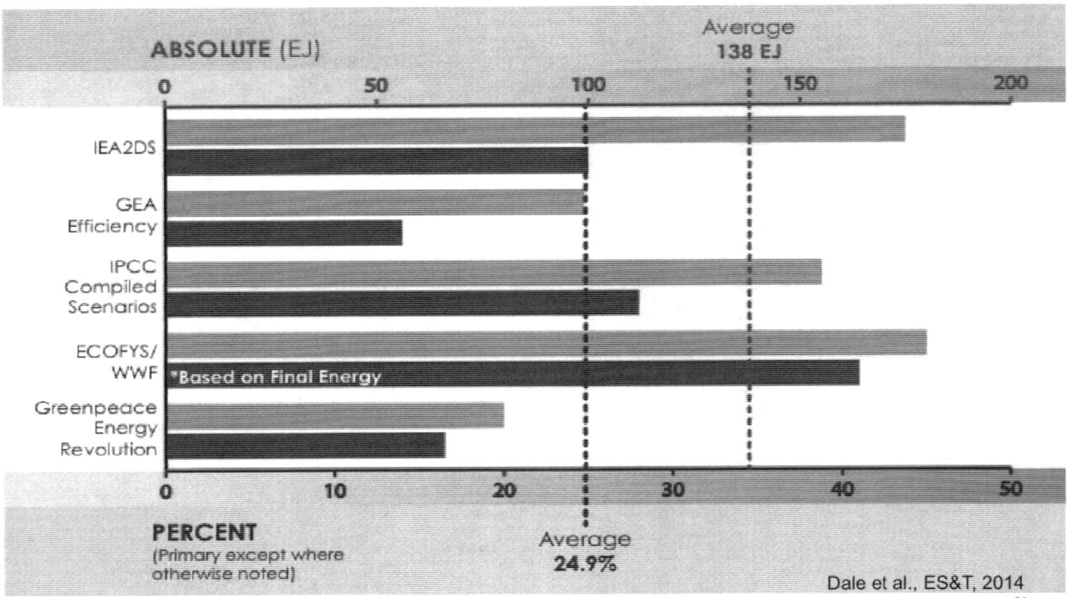

Dale et al., ES&T, 2014

7

ENERGY | DOE Bioenergy Research Centers | GREAT LAKES BIOENERGY RESEARCH CENTER

Agriculture and Biofuels:
we must ask the right questions

- The current approach to biofuel production is to impose a large new demand for biofuels on an **existing agricultural system that otherwise does not change**
- We should be asking: Can we redesign US (world) agriculture to produce biofuels, food/feed & large environmental benefits?
- Agriculture has changed before; *it can, should (must) change again—and farmers are key to that change*
- Examine actual land uses: most land is used for animal feed
- Therefore, how can we sustainably:
  - *integrate biofuel production with animal feed production?*
  - *motivate farmers to participate in biofuel production?*

ENERGY | DOE Bioenergy Research Centers | GREAT LAKES BIOENERGY RESEARCH CENTER

Nutritional Requirements: **Livestock** vs. **Human**

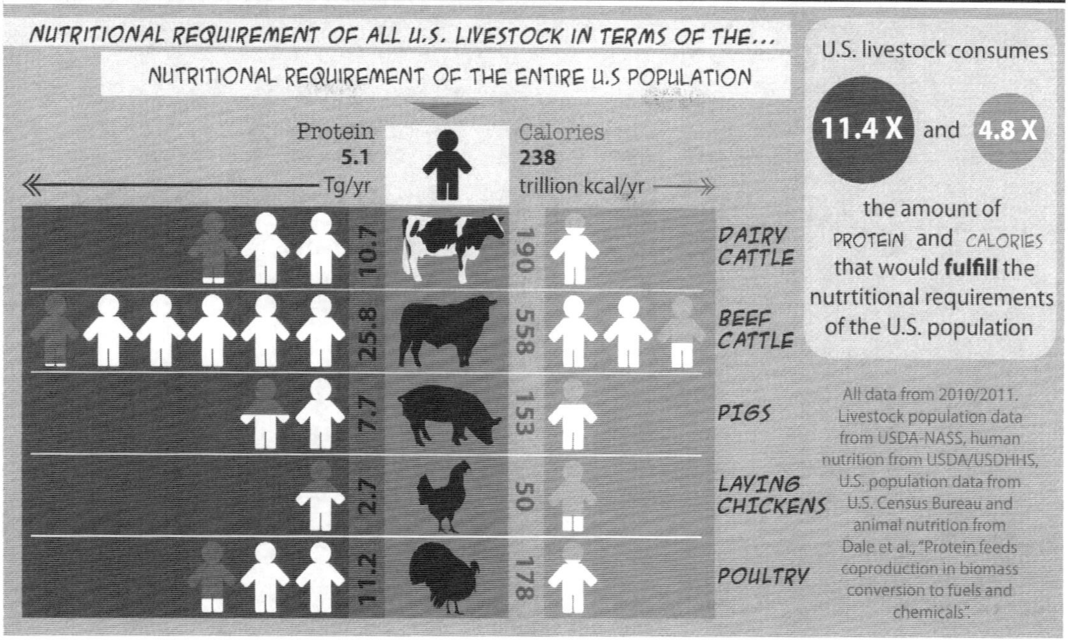

Three Biofuel/Bioenergy Systems: Some Scale and Growth Characteristics

- **Brazilian Sugar Cane to Ethanol System**
  - Very large (~30 billion liters/yr), large refineries, not growing much. Looking for next growth model… by pasture intensification/feeding animals on cane residues?
- Italian Biogas (to Electricity) Consortium
  - Existing, growing, large scale (~1 GW) based on individual farms, developing Biogasdoneright™ model, NOT the German model (apologies to our German attendees ☺)
- Emerging Cellulosic Biofuels System-US
  - Existing, very slow growth, based on custom, expensive supply chains, need a viable large scale model

GeMt ESALQ-USP

Number of formal workers by producing regions and sector 2012

	Region	2000	2012	Δ%
Sugarcane	NNE	81,191	74,484	-8.3↓
	CS	275,795	256,226	-7.1↓
	Total Brazil	356,986	330,710	-7.4↓
Sugar	NNE	143,303	230,443	60.8↑
	CS	74,421	322,431	333.3↑
	Total Brazil	217,724	552,874	153.9↑
Ethanol	NNE	25,730	45,232	75.8↑
	CS	42,408	162,759	283.8↑
	Total Brazil	68,138	207,991	205.2↑
Total for Brazil 3 sectors		642,848	1,091,575	69.8↑

Slide courtesy of Dr. Marcia de Moraes
University of Sao Paulo, Brazil

- Sugarcane, sugar, and ethanol sectors employed over one million people

UNICAMP

Energy from sugar cane

• 1 Mg of cane	• Energy MJ
• 90 litres of ethanol	• 2.300
• 280 kg of bagasse (50% mc)	• 2.300
• 175 kg of trash (20% mc)	• 2.100
• Total	• 6.700 (0.160 TEP)
665 million Mg of sugarcane (2015/2016)	• 106 million TEP
Primary energy production in Brazil - 2014	• 226 million TEP

Slide courtesy of Dr. Paulo Graziano
University of Campinas, Brazil

Leaves

1/3

WASTE!

In São Paulo state in 2015 95% of the area was mechanically green harvested

Slide courtesy of Dr. Paulo Graziano
University of Campinas, Brazil

Why green harvesting ?

- Environmental and legal restriction
- Potential of using trash as energy source
- Soil protection against erosion
- Improve water availability
- Carbon sequestration

Cane Unloading Area-70 tons/minute

15

ENERGY DOE Bioenergy Research Centers www.glbrc.org GREAT LAKES BIOENERGY RESEARCH CENTER

Most of all-Sao Martinhao is big: ~1 million tons in this bagasse pile

16

Three Biofuel/Bioenergy Systems: Some Scale and Growth Characteristics

17

- Brazilian Sugar Cane to Ethanol System
  - Very large (~30 billion liters/yr), large scale model, not growing much. Looking for next growth model... perhaps pasture intensification?
- **Italian Biogas (to Electricity) Consortium**
  - Existing, growing, large scale (~1 GW) based on individual farms, developing Biogasdoneright™ model
- Emerging Cellulosic Biofuels System
  - Existing, slow growth, based on custom supply chains, need a viable large scale model

U.S. DEPARTMENT OF ENERGY   DOE Bioenergy Research Centers   www.glbrc.org   GREAT LAKES BIOENERGY RESEARCH CENTER

Installed biogas plants in Italy

18

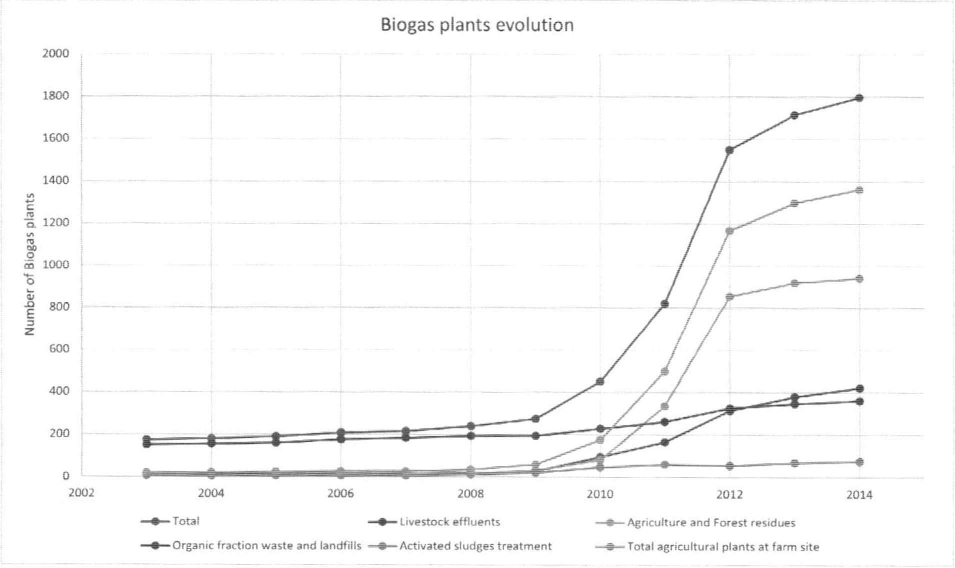

U.S. DEPARTMENT OF ENERGY   DOE Bioenergy Research Centers   www.glbrc.org   GREAT LAKES BIOENERGY RESEARCH CENTER

Double Cropping

- Grow crops (grasses) over winter & spring on corn or soy land while still growing corn/soy
  - Does **NOT** require new land- no expansion of agricultural "frontier"
  - Increases sustainable corn stover harvest rate
  - Biomass can be used for biofuels, animal feed, etc
  - Reframes the "food vs. fuel" debate—now "food AND fuel" (and environmental services!)

Energy crops: environmental benefits

- 20 ***Crop rotation and plant diversity is improved, greater biodiversity,*** *reduced pesticides & herbicides (double cropping, nitrogen fixing crops, etc)*
- ***Year round soil coverage****, reduced erosion, reduced nutrients loss to ground and surface water, reduced nitrous oxide (GHGs), more agricultural residues available*
- *Fertilizer inputs greatly reduced by* ***nutrient recycling via digestate***
- ***Soil carbon levels increase,*** *increased soil fertility, less nutrient loss, carbon sequestration, soil biodiversity increases, more resistant to drought*

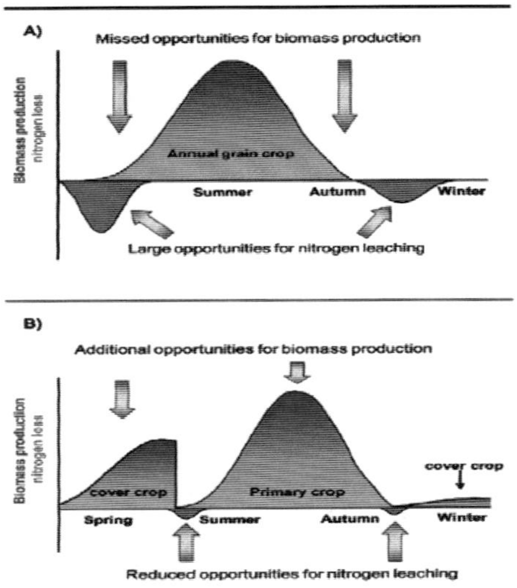

Fig. 1. Hypothesized representation of the seasonal dynamics of dry matter production and NO_3–N leaching (A) in an annual grain cropping system and (B) in a bioenergy double-cropping system.

Cover crops before and after grain and oil crops

21

CURRENT AGRICULTURAL LAND USE

	1st year		2nd year	
monocropping	bare soil	corn grain	bare soil	soy bean
	nov dec jan feb mar	apr may Jun Jul Aug Sept oct	nov dec jan feb mar	apr may Jun Jul Aug Sept oct
	GHG emissions (CO_2, N_2O, etc.) soil C erosion		GHG emissions (CO_2, N_2O, etc.) soil C erosion	

AGRICULTURAL ECOLOGICAL INTENSIFICATION

	1st year		2nd year	
Double cropping	cover crop (triticale)	cash crop (soy bean)	cover crop (rye grass)	cash crop (corn grain)
	nov dec jan feb mar apr	may Jun Jul Aug Sept oct	nov dec jan feb mar apr	may Jun Jul Aug Sept oct
		no tillage seeding		no tillage seeding

cover crop for energy
cash crops food/feed market
bare soil

ENERGY DOE Bioenergy Research Centers consorzio italiano biogas GREAT LAKES BIOENERGY RESEARCH CENTER

Before Biogasdoneright™

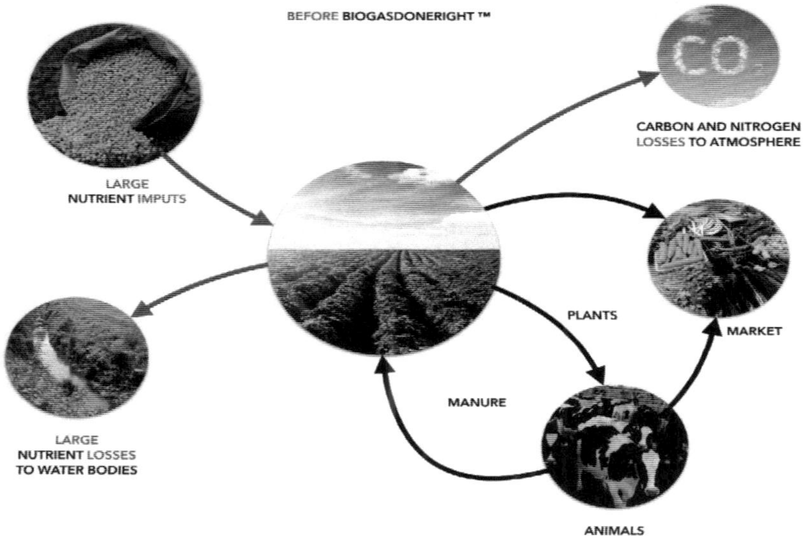

ENERGY DOE Bioenergy Research Centers www.glbrc.org GREAT LAKES BIOENERGY RESEARCH CENTER

After Biogasdoneright

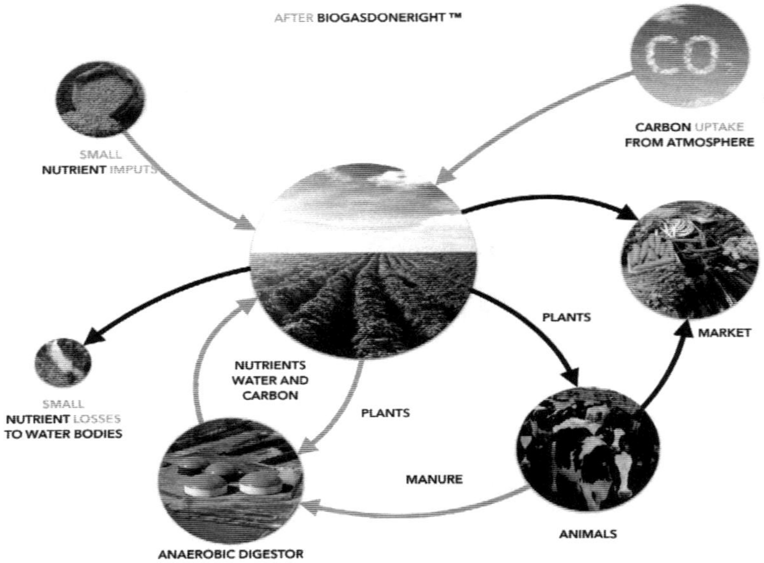

Guiding Principles: "Biogasdoneright"

- Grow regular crop for feed/food market– no "food vs. fuel" conflict
- Grow a double crop to feed the anaerobic digesters (plus manure, other locally-available "wastes")
- Burn biogas on site to generate electricity for power grid or upgrade and export biomethane through natural gas grid
- Fertilize fields with digestate liquid → reduce purchased fertilizers (and associated GHGs)→ reduce irrigation water
- Apply the residual digestate solids → rising soil carbon levels→ increased fertility and farm productivity→ low cost carbon capture & storage
- Pursue opportunities for increased resilience/value added:
  - Biomethane to biodegradable polymers for irrigation piping
  - Biomethane-powered tractors
- Improve farm profitability—increase income and resilience, reduce expenses
- We will be documenting/analyzing the Italian biogas experience in the journal *Biofuels, Bioproducts and Biorefining*

Dual purpose perennial crops: Energy & Forage

Cactus Agronomic Potential

- ✓Potential productivity of **20 Mg of DM/ha/yr**
- ✓Carrying capacity of **4.8 AU/ha/yr**
- ✓This is **57.6 x** higher than the carrying capacity of native rangeland (12 ha/AU/yr)
- ✓In low-input systems, 5 – 6 Mg of DM/ha/yr is easily obtained

consorzio italiano biogas

U.S. DEPARTMENT OF ENERGY DOE Bioenergy Research Centers GREAT LAKES BIOENERGY RESEARCH CENTER

Available at http://www.consorziobiogas.it/Content/public/attachments/527-Biogasdoneright%20No%20VEC%20-%20LowRes.pdf

U.S. DEPARTMENT OF ENERGY DOE Bioenergy Research Centers GREAT LAKES BIOENERGY RESEARCH CENTER

Three Biofuel/Bioenergy Systems: Some Scale and Growth Characteristics

- Brazilian Sugar Cane to Ethanol System
  - Very large (~30 billion liters/yr), large refineries, not growing much. Looking for next growth model... perhaps pasture intensification?
- Italian Biogas (to Electricity) Consortium
  - Existing, growing, large scale (~1 GW) based on individual farms, developing Biogasdoneright™ model
- **Emerging Cellulosic Biofuels System**
  - Existing, slow growth, based on custom, expensive supply chains, need a viable large scale model *(AFEX depots to pellets to integrated biorefineries??)*

Challenges in Building a Sustainable Biobased Economy: *How Will We Supply the Cellulosic Biomass?*

- Discussions of biofuels and biobased chemicals usually assume biomass availability as a given fact
- But cellulosic biomass availability is <u>not</u> a given fact
- Corn availability is a given fact because it is a dense, stable, storable, shippable commodity with <u>multiple existing markets</u>
- *Cellulosic biomass has none of these properties!*
- We have largely ignored this problem until now...
- In fact, biorefinery-oriented research has largely ignored farmers until now....

Oops!! We forgot about them...

- The emerging cellulosic bioeconomy will not grow strongly unless farmers benefit from that growth
- We have already incentivized the formation of the first biorefineries
- We need to get serious about incentivizing farmers
- Farmers will manage land for feed and food, energy and environmental (sustainability) services

ENERGY DOE Bioenergy Research Centers www.glbrc.org 29 GREAT LAKES BIOENERGY RESEARCH CENTER

To Achieve Scale: *Think About the Farmers' Needs*

- Current models of cellulosic biofuel systems require/assume low delivered price of the cellulosic biomass feedstock
- ***But low biomass prices discourage farmer participation in the supply chains. Why would they participate??***
- Biomass transport is expensive. Predicted biorefineries are thus very small, limiting potential economies of scale (~3,000 tons/day vs. ~30,000 tons/day for oil refineries)
- ***But it is biofuel selling price that largely determines market penetration, not just the delivered biomass price.***
- We allowed biomass feedstock prices to rise and explored the effects on biofuel selling price, biofuel volume produced, global warming impact and job generation
- Feedstock supply chain systems were studied as a function of feedstock prices by determining feedstock supply clusters.

ENERGY DOE Bioenergy Research Centers GREAT LAKES BIOENERGY RESEARCH CENTER

Supply Clusters and Biorefinery Sizes

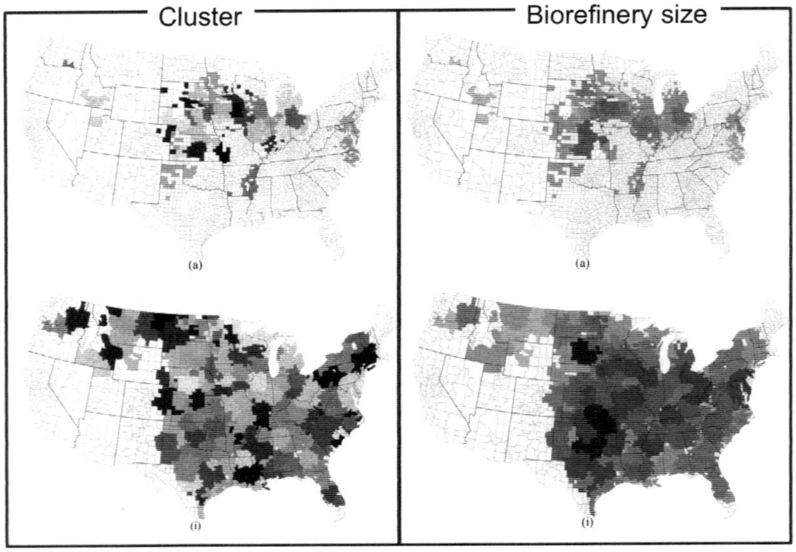

Farmgate price:$40/dry ton
Clusters (biorefineries):43
Ethanol:4.5 Billion gallons
Mostly Corn Belt States
Senators: 20 (20%)
Representatives: 85 (20%)

Farmgate price:$80/dry ton
Clusters (biorefineries):207
Ethanol:42.1 Billion gallons
States with Clusters
Senators: 76 (76%)
Representatives: 334 (77%)

Capacity (dry Mg/day)
(30,000 ≤)
(20,000 to 30,000)
(10,000 to 20,000)
(5,000 to 10,000)
GR (1,000 to 5,000)
(<1,000)

ENERGY DOE Bioenergy Research Centers

Resulting Ethanol Selling Price & Volume

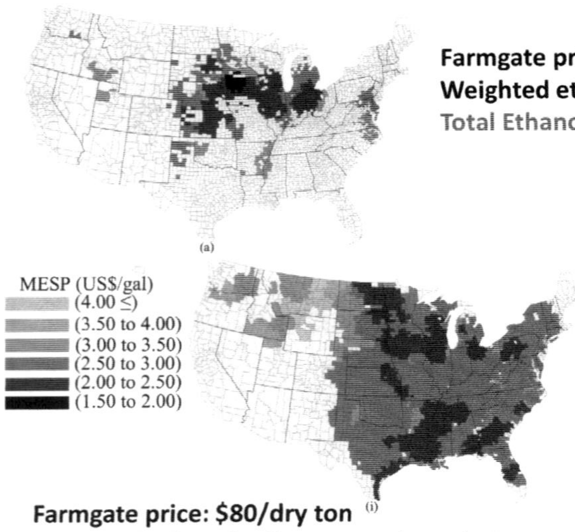

Farmgate price: $40/dry ton
Weighted ethanol selling price: $2.18/gal
Total Ethanol Produced: 4.5 billion gal/yr

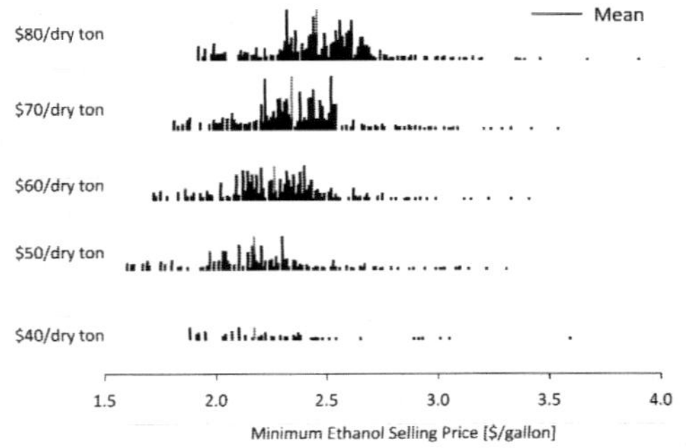

Farmgate price: $80/dry ton
Weighted ethanol selling price: $2.46/gal
Total Ethanol Produced: 42 billion gal/yr

All biomass is local: "average" values do not provide much useful information

ENERGY DOE Bioenergy Research Centers GREAT LAKES BIOENERGY RESEARCH CENTER

Global Warming Impact

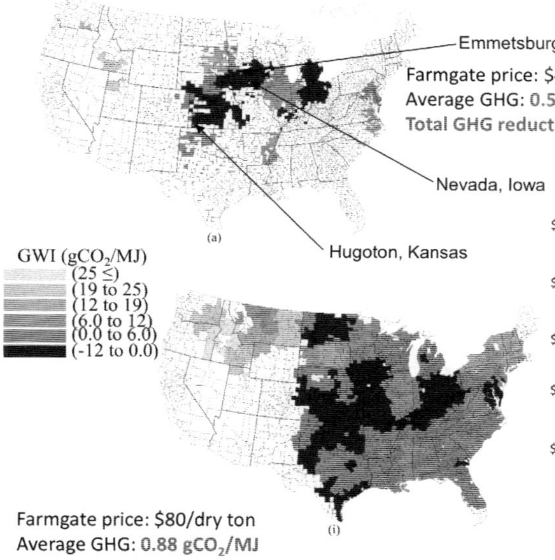

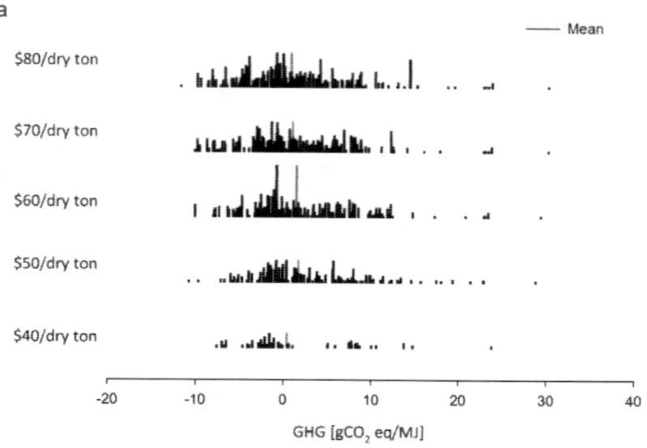

Farmgate price: $80/dry ton
Average GHG: 0.88 gCO_2/MJ
Total GHG reduction vs. gasoline: 313 million metric ton of CO_2 equivalent

U.S. DEPARTMENT OF ENERGY DOE Bioenergy Research Centers **Total GHG from the US transportation in 2013: 1,842 MMT CO_2 eq.** GREAT LAKES BIOENERGY RESEARCH CENTER

Job Generation

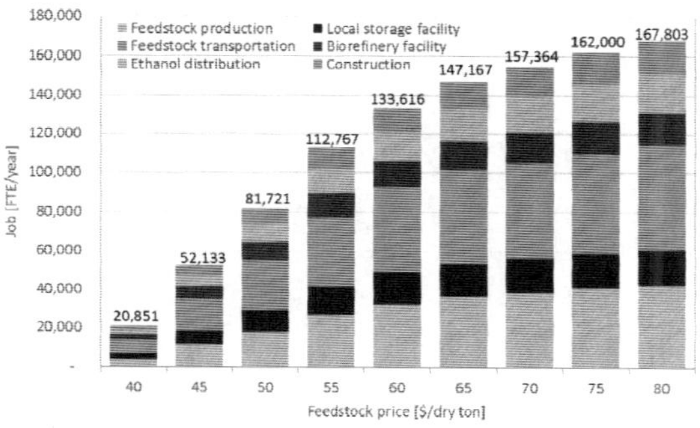

One dollar per ton increase in the farm gate feedstock price generates:

~5,700 FTE per year in the US for feedstock prices below US$ 60/dry ton

~1,400 FTE per year in the US for feedstock prices above US$ 65/ton

Increasing feedstock price paid farmers creates many more jobs without increasing the ethanol selling price very much—in fact, for some specific locations <u>the predicted ethanol selling price declines!</u>

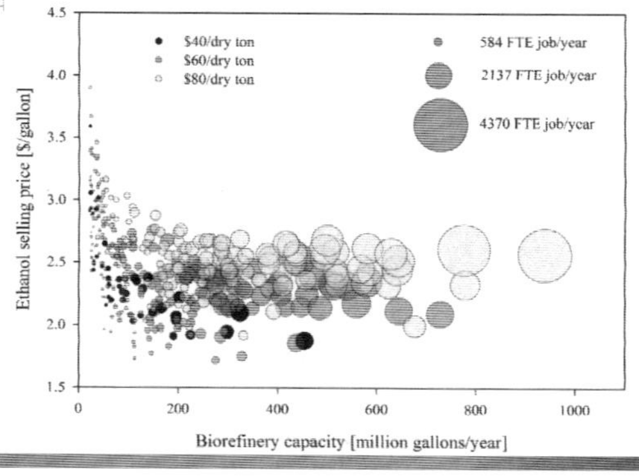

U.S. DEPARTMENT OF ENERGY DOE Bioenergy Research Centers ❖ FTE: Full time equivalent

Findings

- Cellulosic ethanol fuel production systems are highly local (regional) in their key characteristics of scale, cost and volume of fuel produced, GHG reductions and job generation
- Feedstock price greatly affects the formation of supply chains but does not affect ethanol selling price very much
- Higher feedstock prices increase the amount of ethanol produced, generate many more jobs and greatly reduce national GHGs—**and will increase political/social support**
- Farmers supply much more cellulosic biomass at higher feedstock prices → shorter transportation distances → lower transportation costs → larger biorefineries → better economies of scale → thus reducing the ethanol selling price.
- ***Any other factors that increase farmer participation in the supply would have the same effects (eg, ownership in biomass preprocessing systems "depots", payments for environmental services, etc.)***

U.S. DEPARTMENT OF ENERGY DOE Bioenergy Research Centers GREAT LAKES BIOENERGY RESEARCH CENTER

Some Concluding Thoughts

- Energy use (power consumption) is critical to human prosperity and well-being
- Without sustainable, large scale energy sources, we will not have sustainable, large scale prosperity— must act soon
- We need about 4 kW/capita— part of which must be bioenergy- to achieve good levels of wealth and well-bring
- We cannot limit GHGs without involving farms and the farmers who manage land-to reduce emissions and sequester carbon
- Agriculture can provide food, feed, and large sustainability services, *if redesigned to do so.*
- Start incentivizing/involving farmers to produce lots of energy, sustainably—**think about what we are doing!**

U.S. DEPARTMENT OF ENERGY DOE Bioenergy Research Centers GREAT LAKES BIOENERGY RESEARCH CENTER

COMO TRATAR LOS ASPECTOS AMBIENTALES DELOS
BIOCOMBUSTIBLES Y LA BIOENERGÍA
DESDE UNA ERSPECTIVA SISTÉMICA
HOW TO DEAL WITH ENVIRONMENTAL
ASPECTS OF BIOFUELS AND BIOENERGY WITH A SYTEMIC
PERPECTIVE
PAN AMERICAN BIOFUELS AND BIOENERGY SUSTAINABILITY · AN NFS RESEARCH COORDINATION NETWORK
PAN-AMERICAN BIOFUELS & BIOENERGY SUSTAINABILITY
2ª RCN Conferencia Panamericana sobre
Sustentabilidad en Biocombustibles y Bioenergia
Buenos Aires, Argentina - Septiembre 13-16, 2016
Ing.Agr. Jorge Antonio Hilbert
UNIVERSIDAD TECNOLÓGICA NACIONAL
Ministerio de Agroindustria
Presidencia de la Nación
INTA

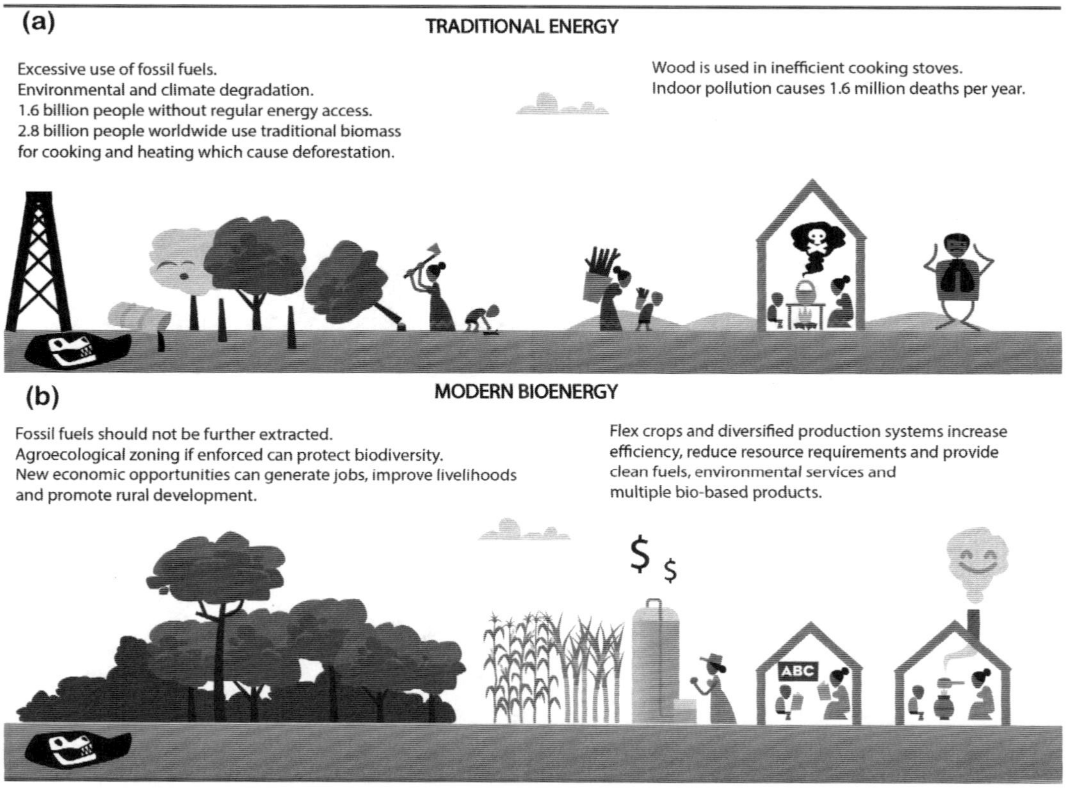
(a)
TRADITIONAL ENERGY
Excessive use of fossil fuels.
Environmental and climate degradation.
1.6 billion people without regular energy access.
2.8 billion people worldwide use traditional biomass for cooking and heating which cause deforestation.
Wood is used in inefficient cooking stoves.
Indoor pollution causes 1.6 million deaths per year.
(b)
MODERN BIOENERGY
Fossil fuels should not be further extracted.
Agroecological zoning if enforced can protect biodiversity.
New economic opportunities can generate jobs, improve livelihoods and promote rural development.
Flex crops and diversified production systems increase efficiency, reduce resource requirements and provide clean fuels, environmental services and multiple bio-based products.
$ $
ABC

Contenido energético
En promedio, un kilogramo de biomasa permite obtener 3.500 kcal
Eficiencia conversión de la energía solar
Promedio general de todas las especies 2 %
Densidad energética
0,6 W/m²
LA ESTUDIOS ESTAN CONDICIONADOS POR
LA BAJA DENSIDAD ENERGETICA
Y
LA ALTA DISPERSION GEOGRÁFICA
LOS SISTEMAS DE PROVISION Y LOGISTICA SE TORNAN
CRITICOS ASI COMO
EL TRATAMIENTO ESPECIFICO

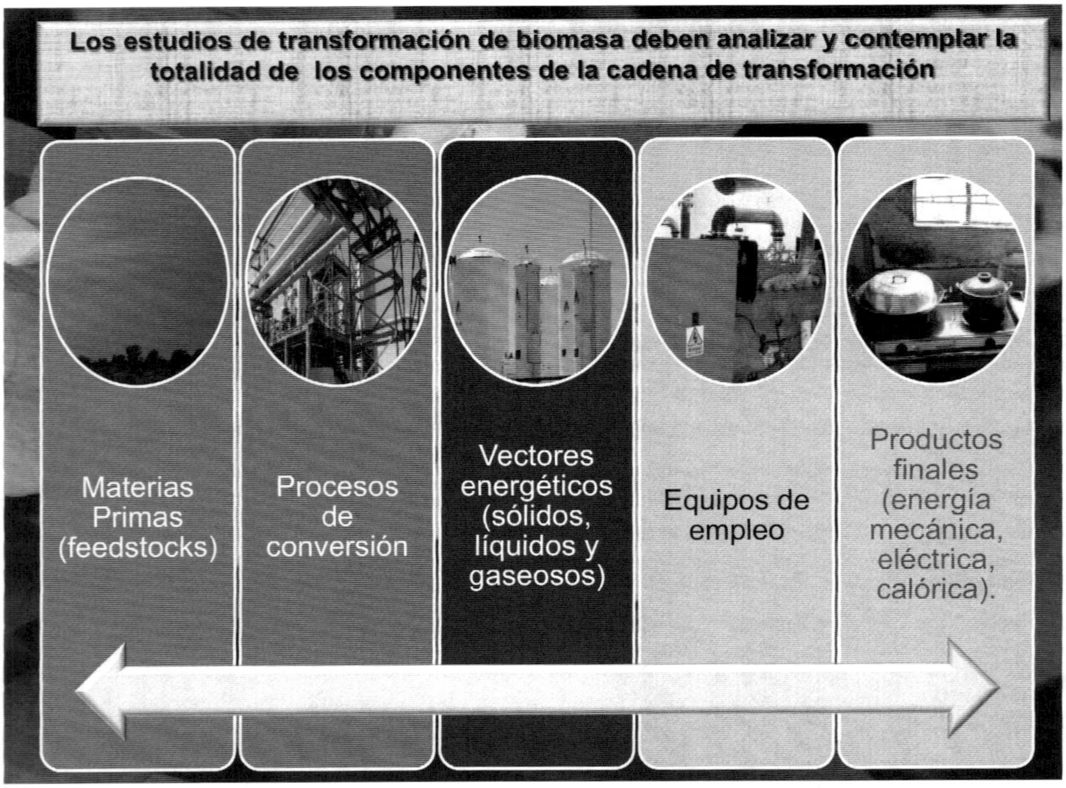
Los estudios de transformación de biomasa deben analizar y contemplar la totalidad de los componentes de la cadena de transformación
Materias Primas (feedstocks)
Procesos de conversión
Vectores energéticos (sólidos, líquidos y gaseosos)
Equipos de empleo
Productos finales (energía mecánica, eléctrica, calórica).

Factores particulares del sector agropecuario

- Fuerte influencia de la temporalidad de los análisis (variación de las condiciones agroecológicas).
- Existencia de una línea de base en los sistemas que se intervienen.
- Multiplicidad de productos, mercados y destinos a partir de una misma materia prima
- Cambios anuales de sistemas productivos (rotaciones).

3/7/2012 5

Factores particulares de la provisión y uso del agua

- El hemisferio sur se caracteriza por una fuerte oscilación de los ciclos climáticos.
- Existe una tendencia al incremento en los ciclos hídricos especialmente en otoño en las zonas productoras.
- Esto implica que los análisis deben ser realizados sitio específicos.
- Los cultivos peremnes o de cobertura podrían ser una salida al incremento de las napas freáticas.
- En estos casos los cultivos bioenergéticos podrían incrementar en lugar de disminuir la producción de alimentos asi como su estabilidad.

6

ANUAL RAIN VARIABILITY AND WATER TABLE EVOLUTION

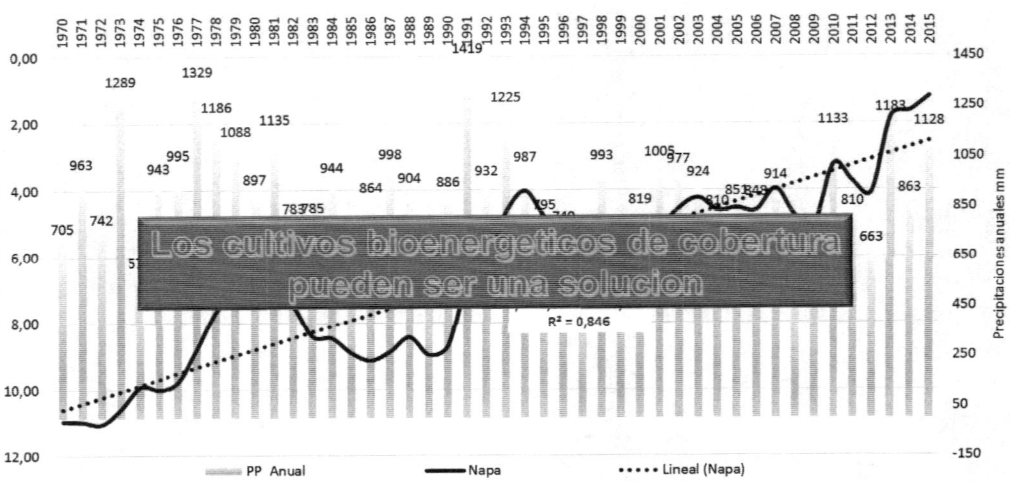

Fuente: Chiachiera y Bertram 2013

Datos: Observatorio meteorológico de la EEA INTA Marcos Juárez.

Gentileza StellaCarballo

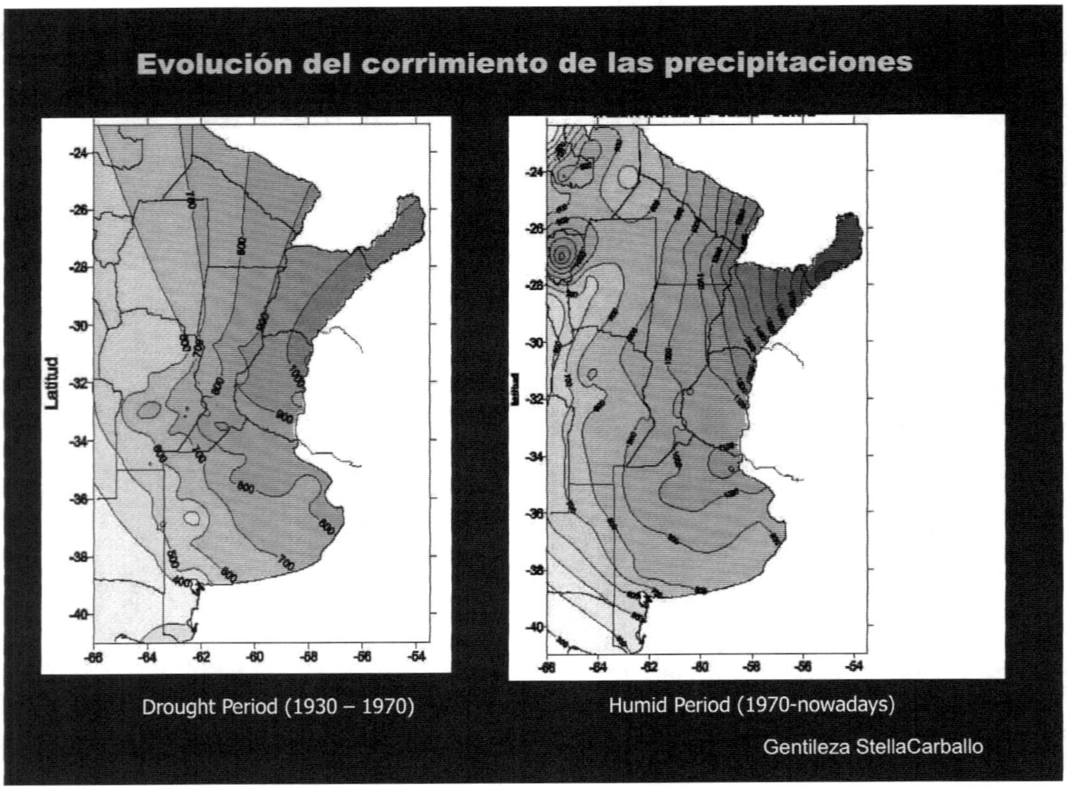

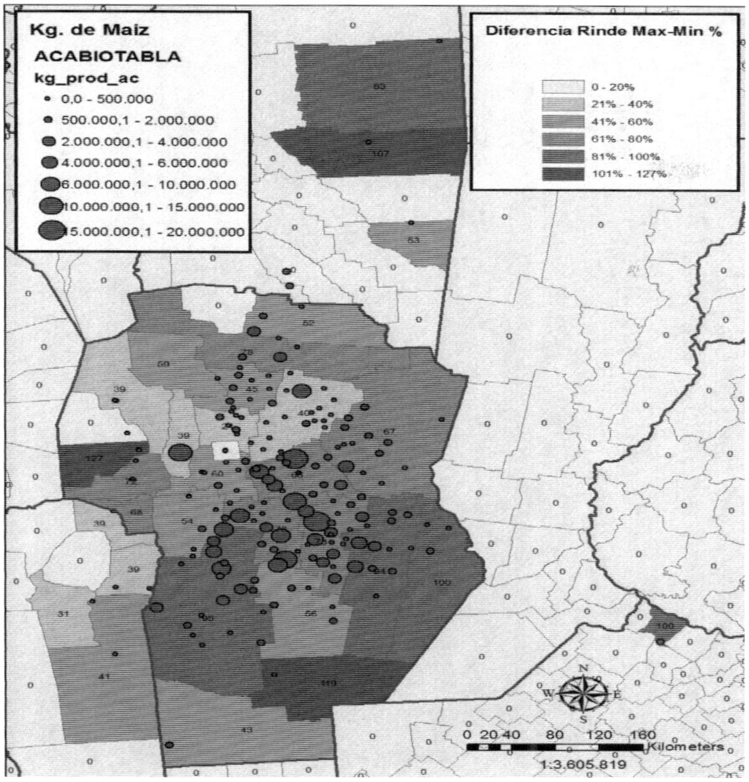

Territorialidad de los análisis y conclusiones.

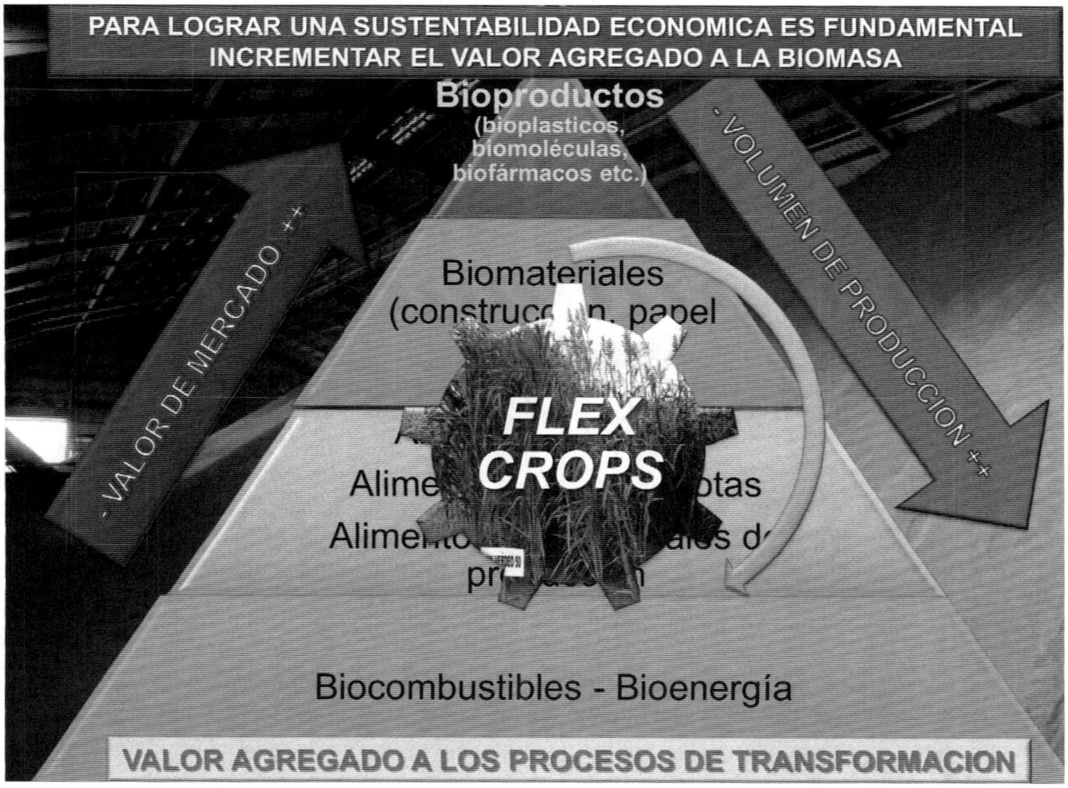

EL TRATAMIENTO DE RESIDUOS CONSTITUYE UNA IMPORTANTE FUENTE DE RECURSOS
Prevención y reduccion
Tecnologías de conversión
Recuperación
Disposición final
- - - VOLUMEN + + +
+ + VALOR - -

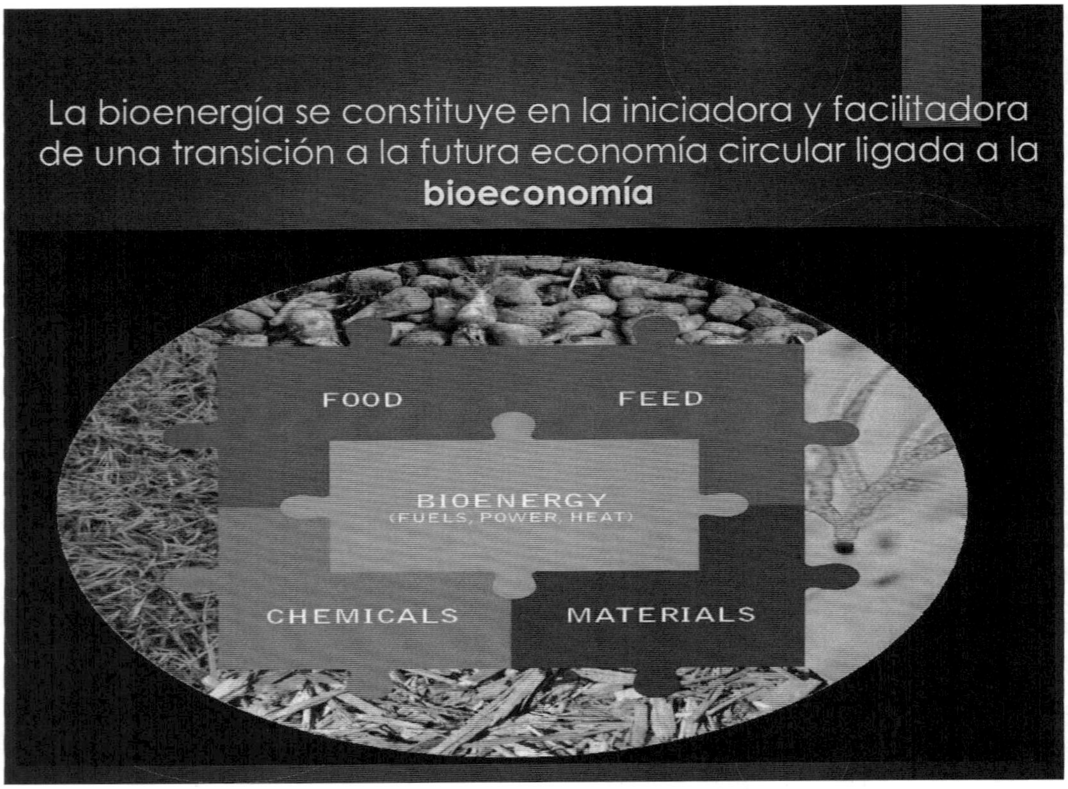
La bioenergía se constituye en la iniciadora y facilitadora de una transición a la futura economía circular ligada a la **bioeconomía**
FOOD
FEED
BIOENERGY
(FUELS, POWER, HEAT)
CHEMICALS
MATERIALS

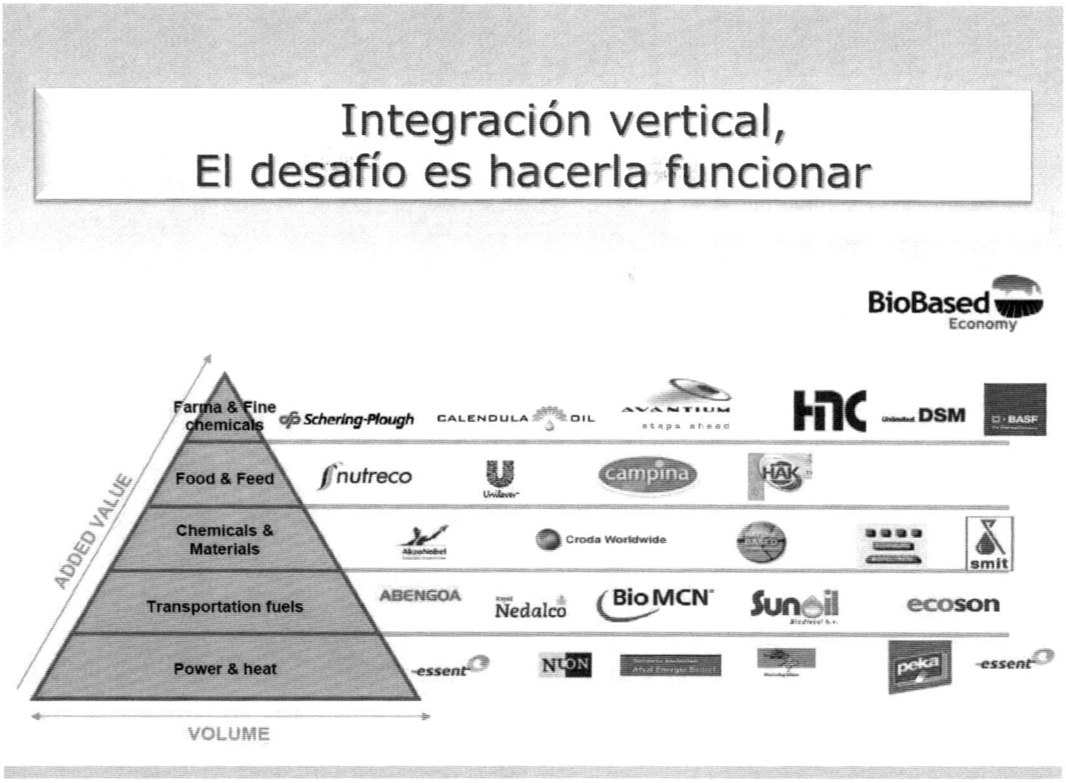
Integración vertical,
El desafío es hacerla funcionar
BioBased Economy
Farma & Fine chemicals
Food & Feed
Chemicals & Materials
Transportation fuels
Power & heat
ADDED VALUE
VOLUME
Schering-Plough
CALENDULA OIL
AVANTIUM
DSM
BASF
nutreco
Unilever
campina
Croda Worldwide
AkzoNobel
smit
ABENGOA
Nedalco
BioMCN
Sunoil
ecoson
essent
NUON
peka

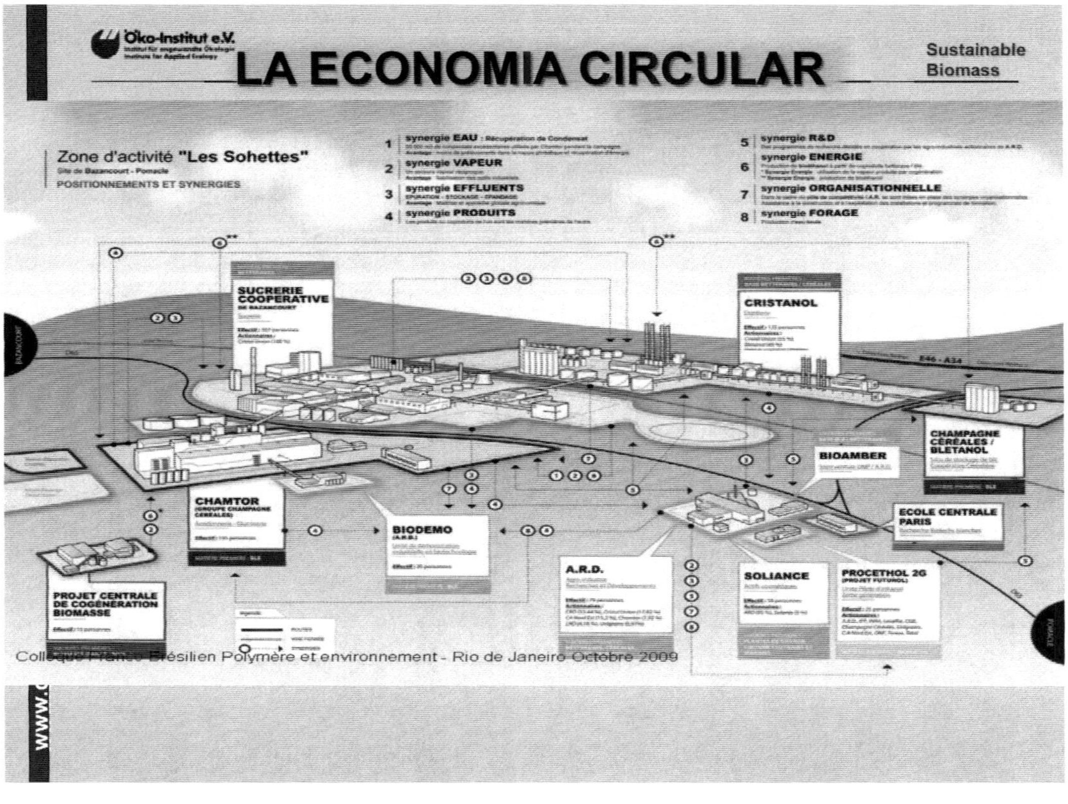
Öko-Institut e.V.
LA ECONOMIA CIRCULAR
Sustainable Biomass
Zone d'activité "Les Sohettes"
POSITIONNEMENTS ET SYNERGIES
1 synergie EAU
2 synergie VAPEUR
3 synergie EFFLUENTS
4 synergie PRODUITS
5 synergie R&D
6 synergie ENERGIE
7 synergie ORGANISATIONNELLE
8 synergie FORAGE
SUCRERIE COOPERATIVE DE BAZANCOURT
CRISTANOL
CHAMPAGNE CÉRÉALES / BLETANOL
BIOAMBER
ECOLE CENTRALE PARIS
CHAMTOR
BIODEMO
A.R.D.
SOLIANCE
PROCETHOL 2G
PROJET CENTRALE DE COGÉNÉRATION BIOMASSE
E46 - A34
Colloque France-Brésilien Polymère et environnement - Rio de Janeiro Octobre 2009

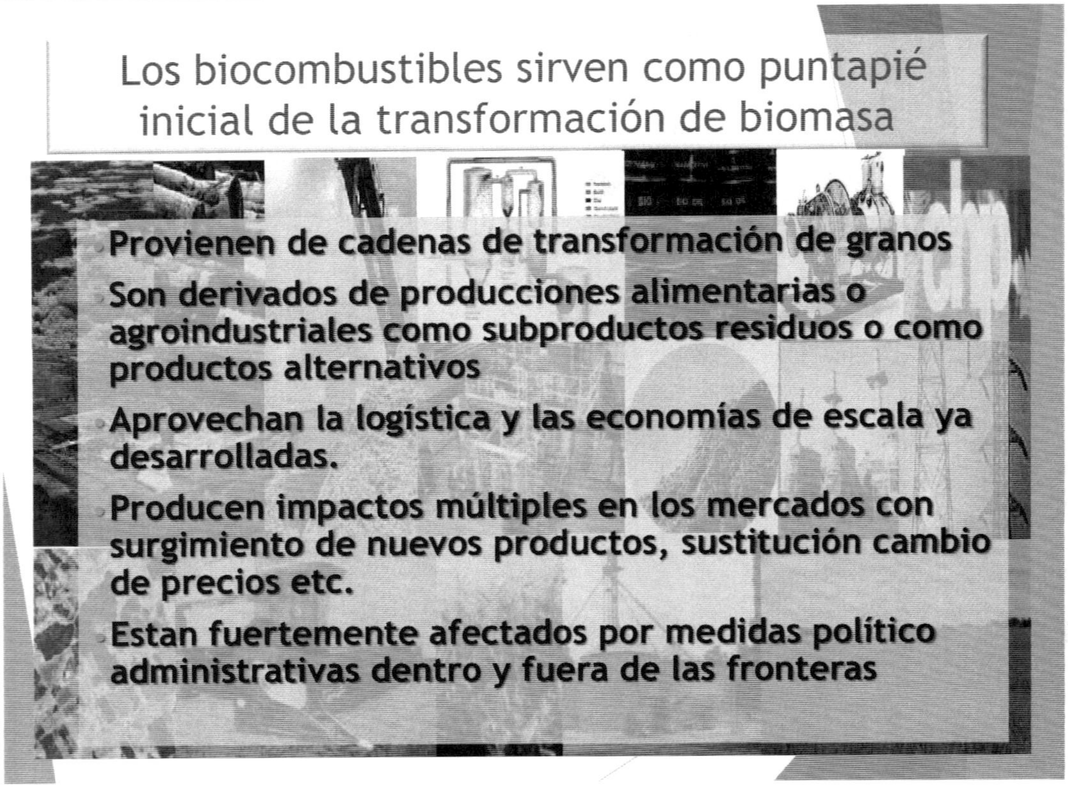
Los biocombustibles sirven como puntapié inicial de la transformación de biomasa
Provienen de cadenas de transformación de granos
Son derivados de producciones alimentarias o agroindustriales como subproductos residuos o como productos alternativos
Aprovechan la logística y las economías de escala ya desarrolladas.
Producen impactos múltiples en los mercados con surgimiento de nuevos productos, sustitución cambio de precios etc.
Estan fuertemente afectados por medidas político administrativas dentro y fuera de las fronteras

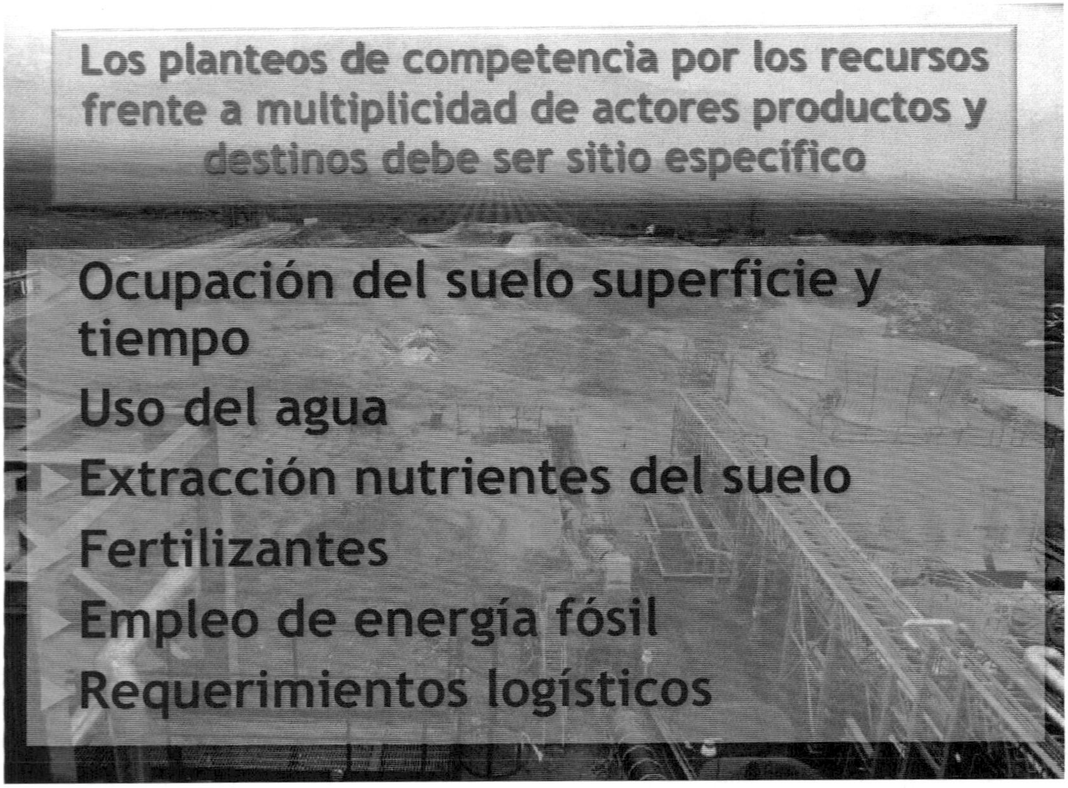
Los planteos de competencia por los recursos frente a multiplicidad de actores productos y destinos debe ser sitio específico
Ocupación del suelo superficie y tiempo
Uso del agua
Extracción nutrientes del suelo
Fertilizantes
Empleo de energía fósil
Requerimientos logísticos

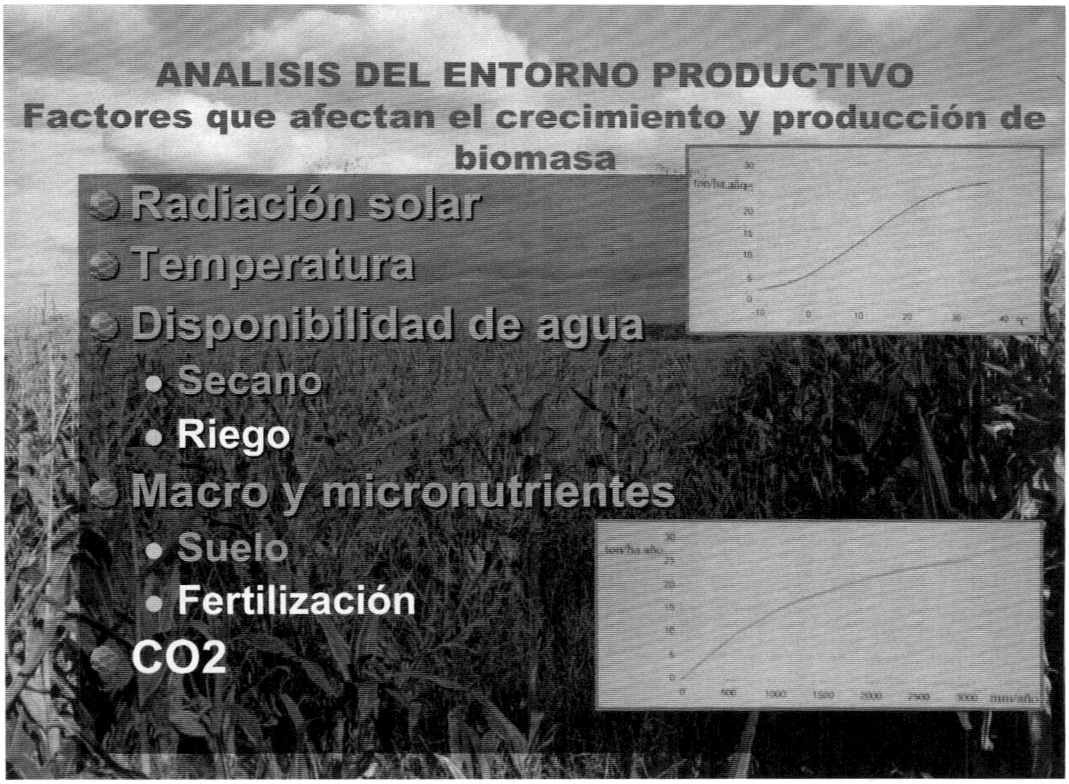
ANALISIS DEL ENTORNO PRODUCTIVO
Factores que afectan el crecimiento y producción de biomasa
Radiación solar
Temperatura
Disponibilidad de agua
Secano
Riego
Macro y micronutrientes
Suelo
Fertilización
CO2

Water Footprint
INTA
El planteo de un retiro de residuos se realiza en una agricultura bajo siembra directa que conserva energía agua y carbono
MAS DEL 80 % DE LA SUPERFICIE
Cuanto retirar ????
COS
AÑOS CON SIEMBRA DIRECTA
1 y 3 : Menor Tasa de Acumulación de COS
2 : Mayor Tasa de Acumulación de COS

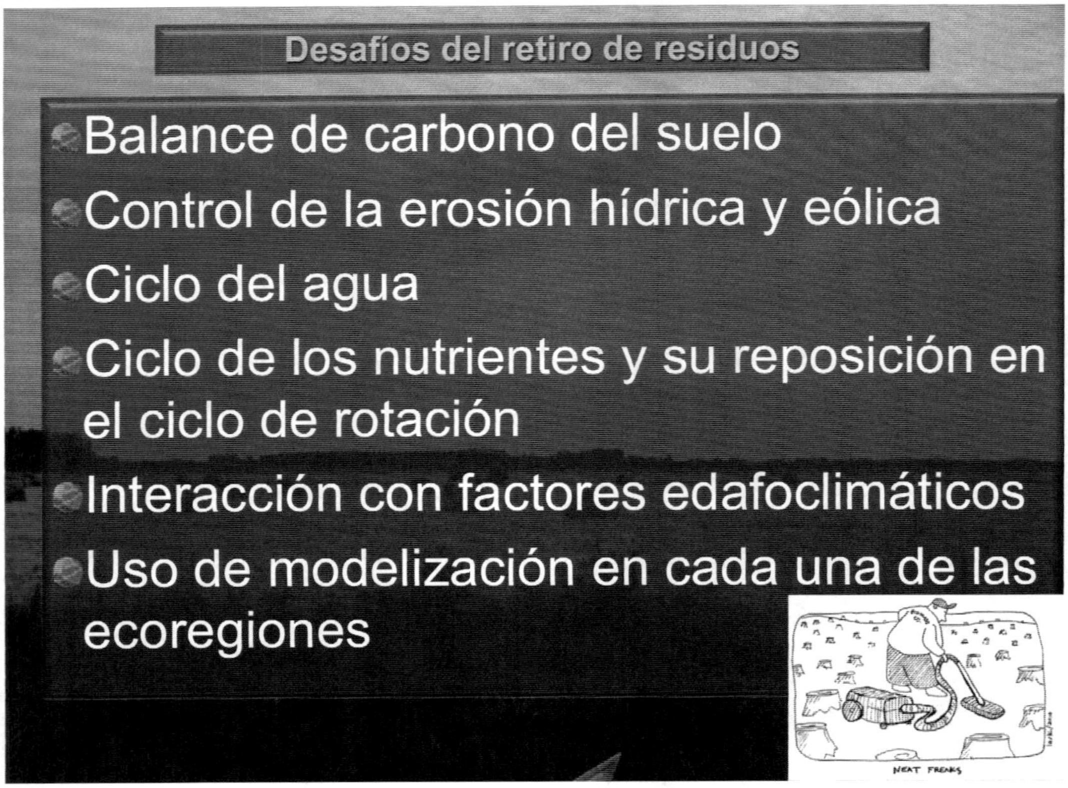

Desafíos del retiro de residuos
Balance de carbono del suelo
Control de la erosión hídrica y eólica
Ciclo del agua
Ciclo de los nutrientes y su reposición en el ciclo de rotación
Interacción con factores edafoclimáticos
Uso de modelización en cada una de las ecoregiones

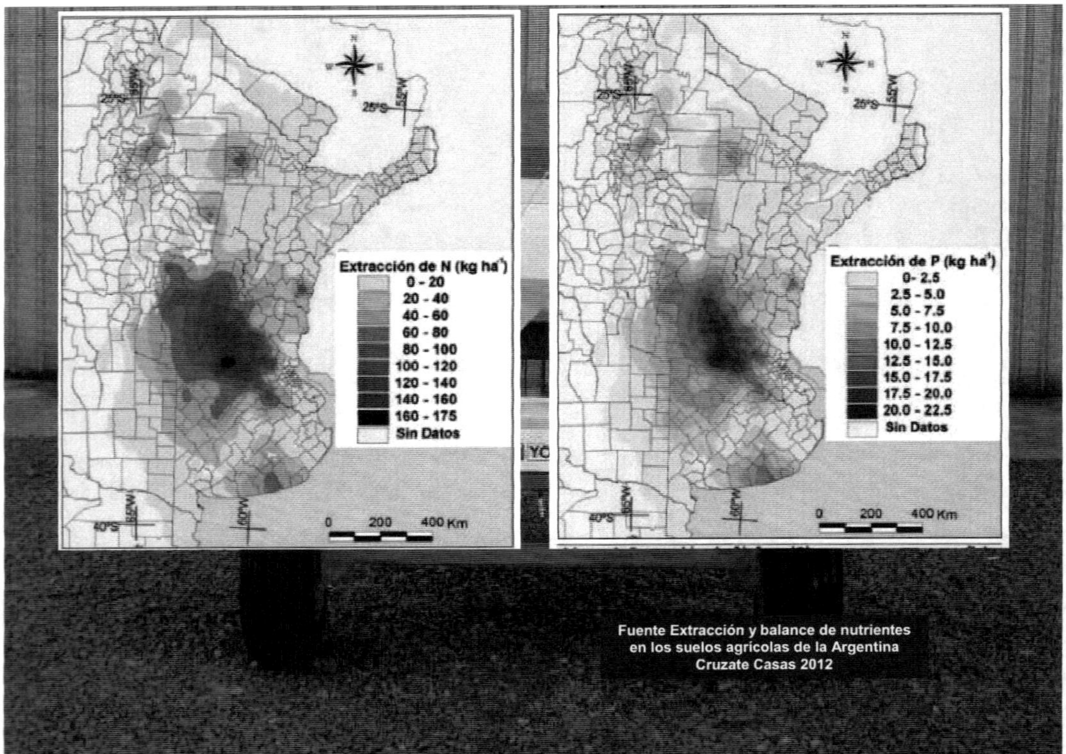

Extracción de N (kg ha-1)
0 - 20
20 - 40
40 - 60
60 - 80
80 - 100
100 - 120
120 - 140
140 - 160
160 - 175
Sin Datos
Extracción de P (kg ha-1)
0 - 2.5
2.5 - 5.0
5.0 - 7.5
7.5 - 10.0
10.0 - 12.5
12.5 - 15.0
15.0 - 17.5
17.5 - 20.0
20.0 - 22.5
Sin Datos
0 200 400 Km
Fuente Extracción y balance de nutrientes en los suelos agrícolas de la Argentina
Cruzate Casas 2012

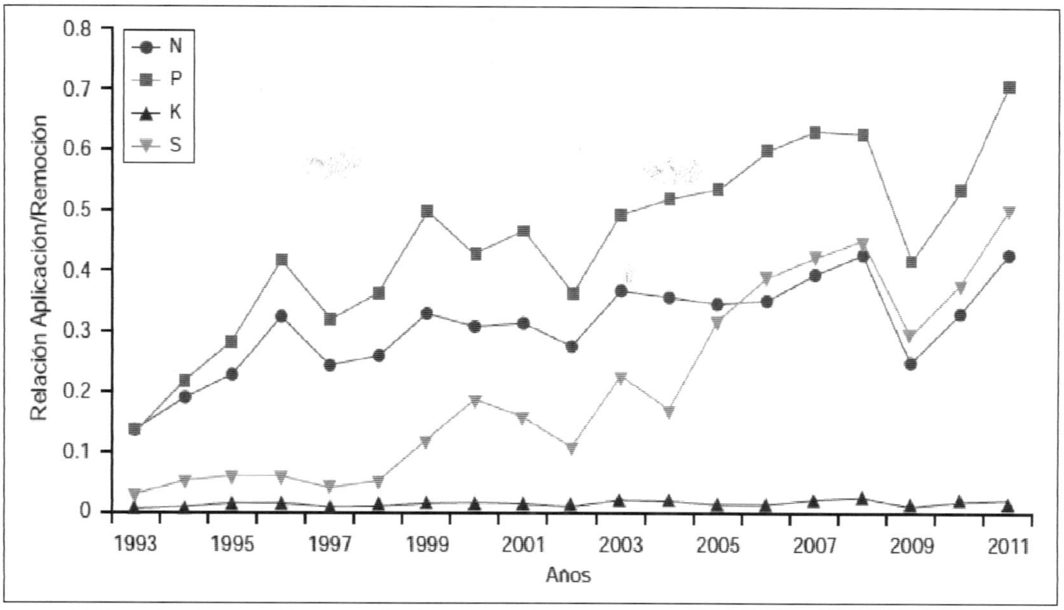

Figura 2. Relaciones Aplicación/Remoción de nitrógeno (N), fósforo (P), potasio (K) y azufre (S) para los cuatro principales cultivos de grano de Argentina (soja, maíz, trigo y girasol) en el periodo 1993-2011. Elaborado a partir de datos de MinAgri y Fertilizar AC.

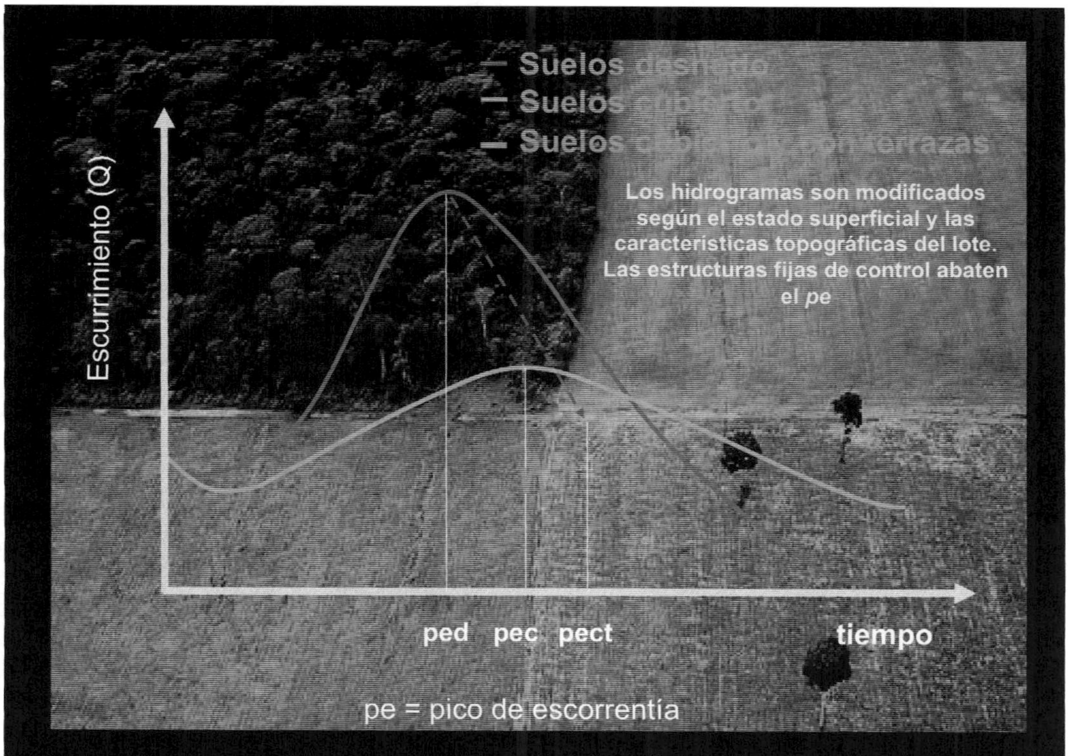

Fuente Sergio Montico UNR

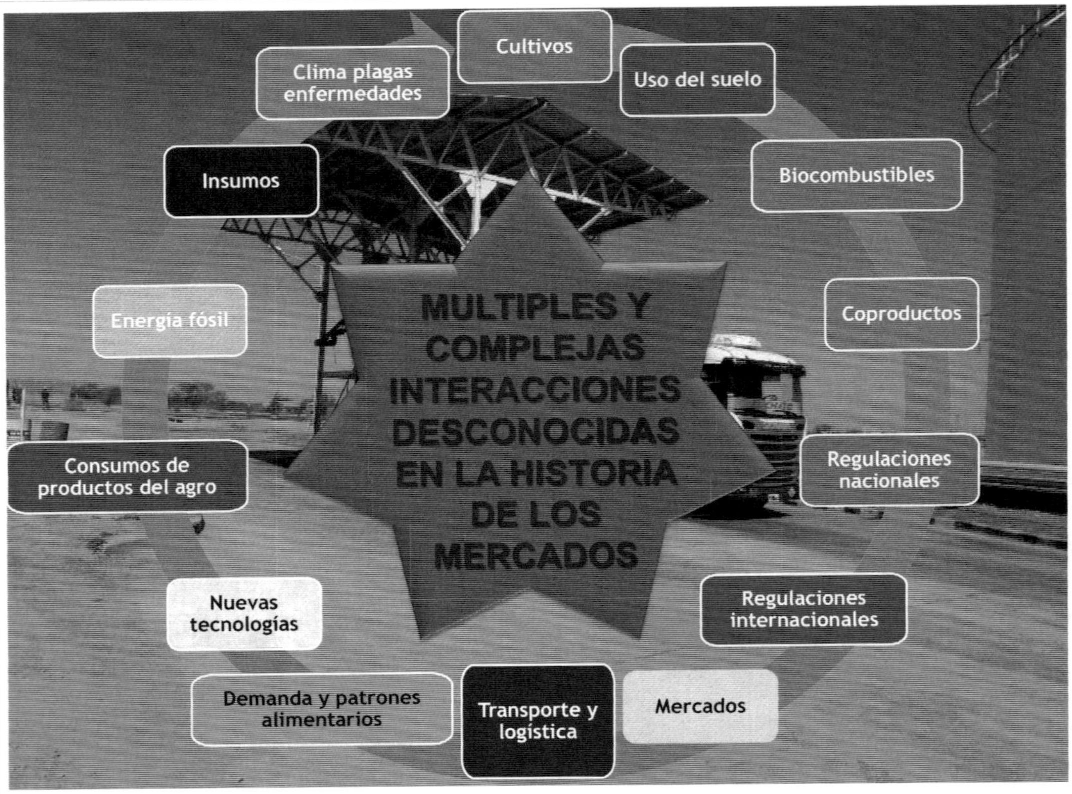

EL DESAFIO QUE ENFRENTAMOS ES COMO NORMALIZAR Y TRANSPARENTAR LOS DIFERENTES TIPOS DE ESTUDIOS QUE SE REALIZAN

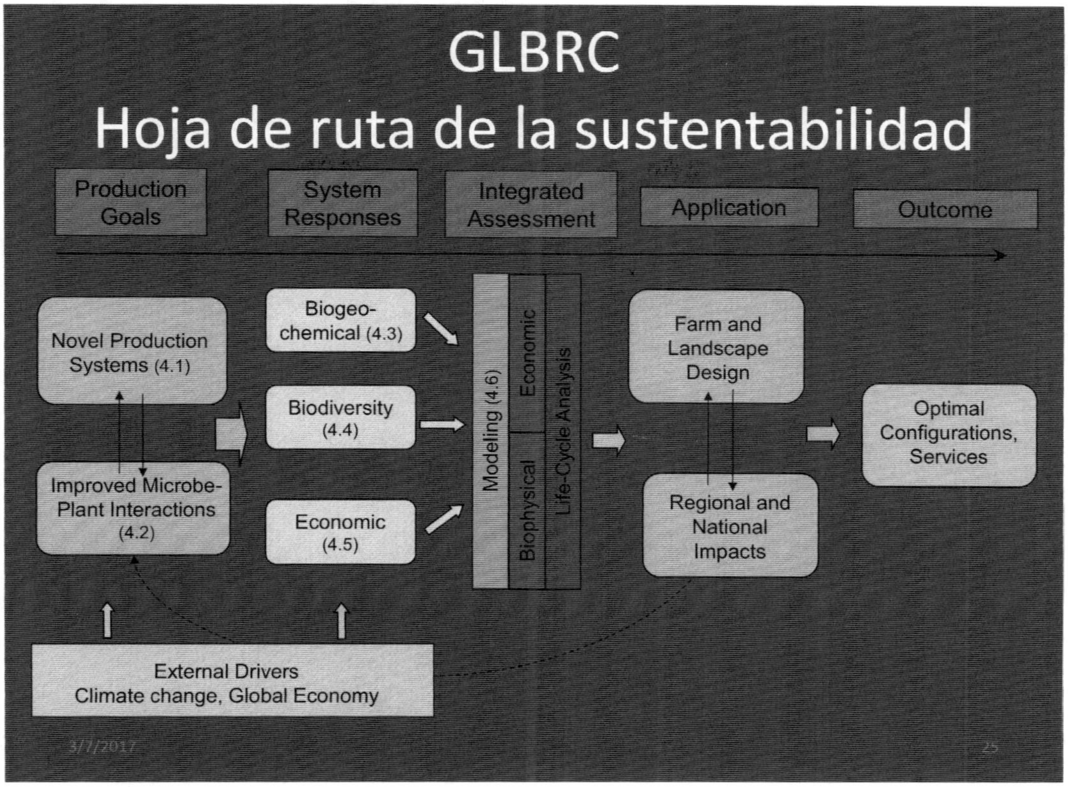
GLBRC
Hoja de ruta de la sustentabilidad
Production Goals
System Responses
Integrated Assessment
Application
Outcome
Novel Production Systems (4.1)
Improved Microbe-Plant Interactions (4.2)
Biogeo-chemical (4.3)
Biodiversity (4.4)
Economic (4.5)
Modeling (4.6)
Biophysical
Economic
Life-Cycle Analysis
Farm and Landscape Design
Regional and National Impacts
Optimal Configurations, Services
External Drivers
Climate change, Global Economy
3/7/2017
25

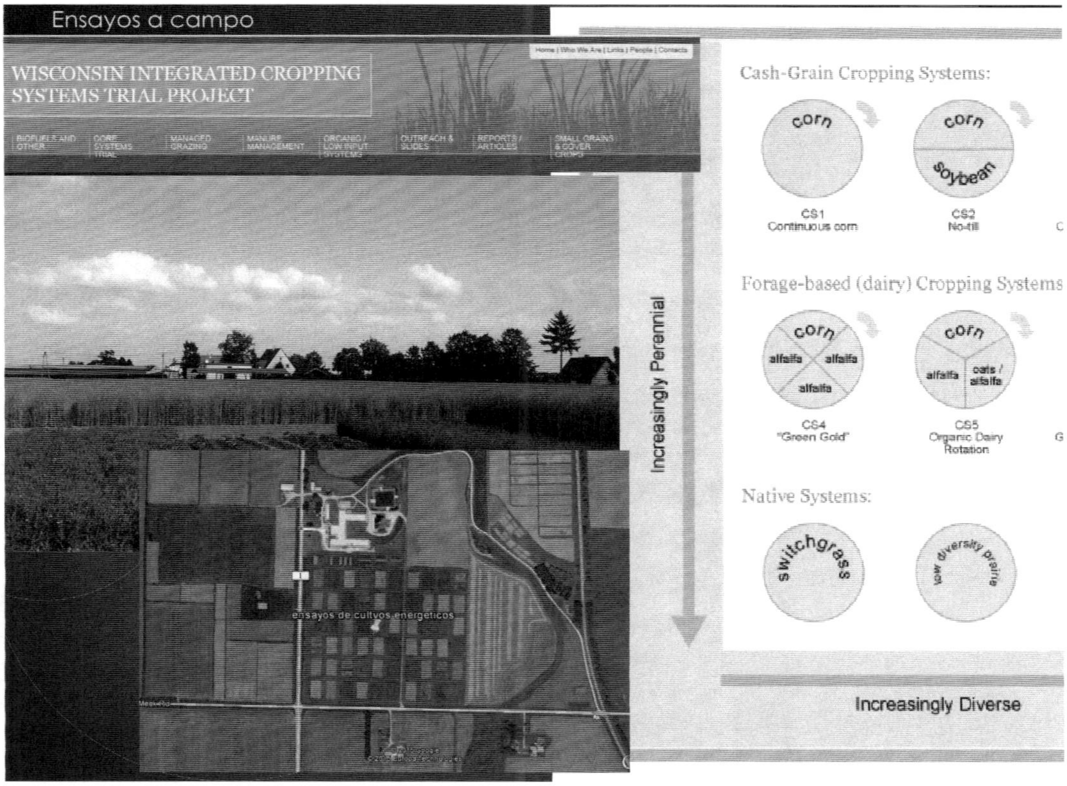
Ensayos a campo
WISCONSIN INTEGRATED CROPPING SYSTEMS TRIAL PROJECT
ensayos de cultivos energéticos
Increasingly Perennial
Cash-Grain Cropping Systems:
corn
corn
soybean
CS1
Continuous corn
CS2
No-till
Forage-based (dairy) Cropping Systems
corn
alfalfa
alfalfa
alfalfa
corn
alfalfa
oats / alfalfa
CS4
"Green Gold"
CS5
Organic Dairy Rotation
Native Systems:
switchgrass
Increasingly Diverse

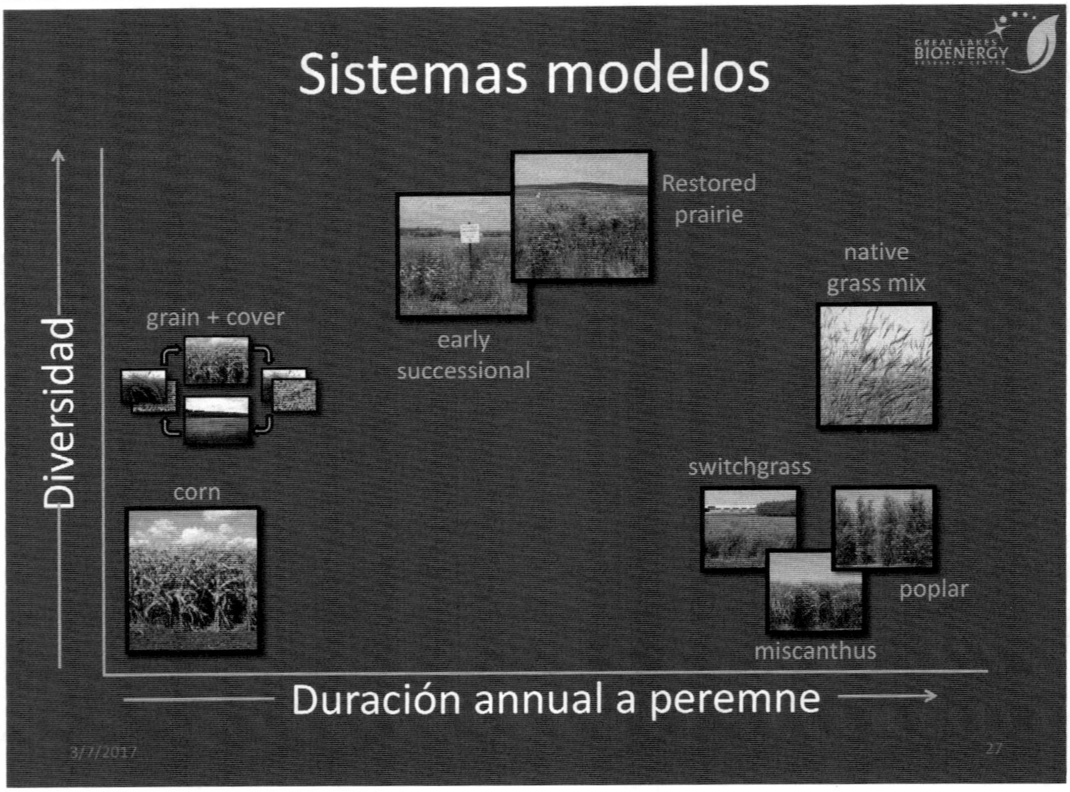

Proyecto BabetrealV Abordaje basado en un tratamiento multicriterio

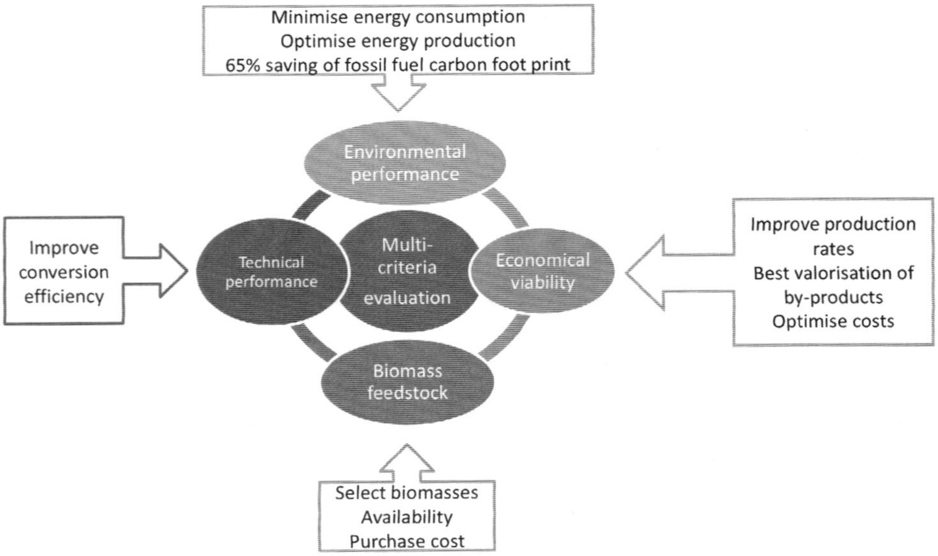

Jorge Antonio Hilbert INTA CIA IIR

28

Desafíos del estudio y evaluación de las nuevas plantas transformadoras de biomasa multipropósito

BALANCES ENERGETICOS
GRAN DESAFÍO PARA LOS BIOCOMBUSTIBLES
DE II GENERACIÓN
• Temporalidad y sitio específicos
• Fronteras de los estudios
• Alocación de los impactos a los diferentes productos y coproductos
• Criterios y metodologías de valoración de cada producto
• Valoración de la producción y logística de la materia prima empleada.

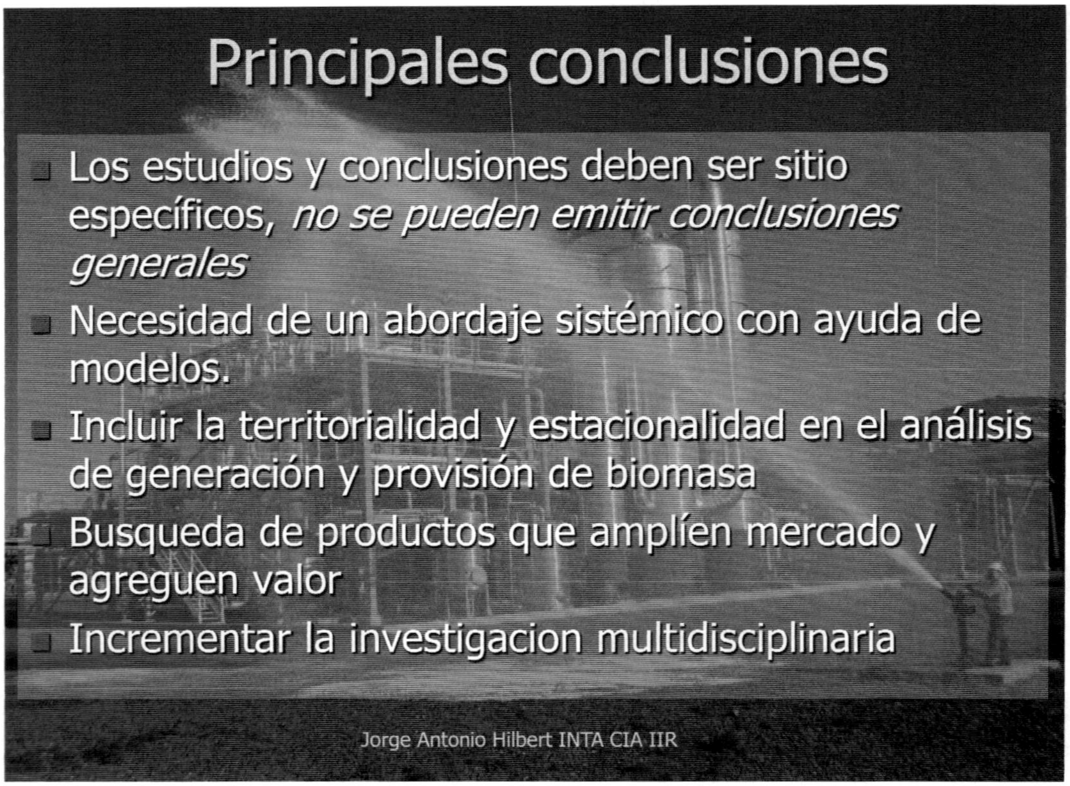
Principales conclusiones
Los estudios y conclusiones deben ser sitio específicos, *no se pueden emitir conclusiones generales*
Necesidad de un abordaje sistémico con ayuda de modelos.
Incluir la territorialidad y estacionalidad en el análisis de generación y provisión de biomasa
Busqueda de productos que amplíen mercado y agreguen valor
Incrementar la investigacion multidisciplinaria
Jorge Antonio Hilbert INTA CIA IIR

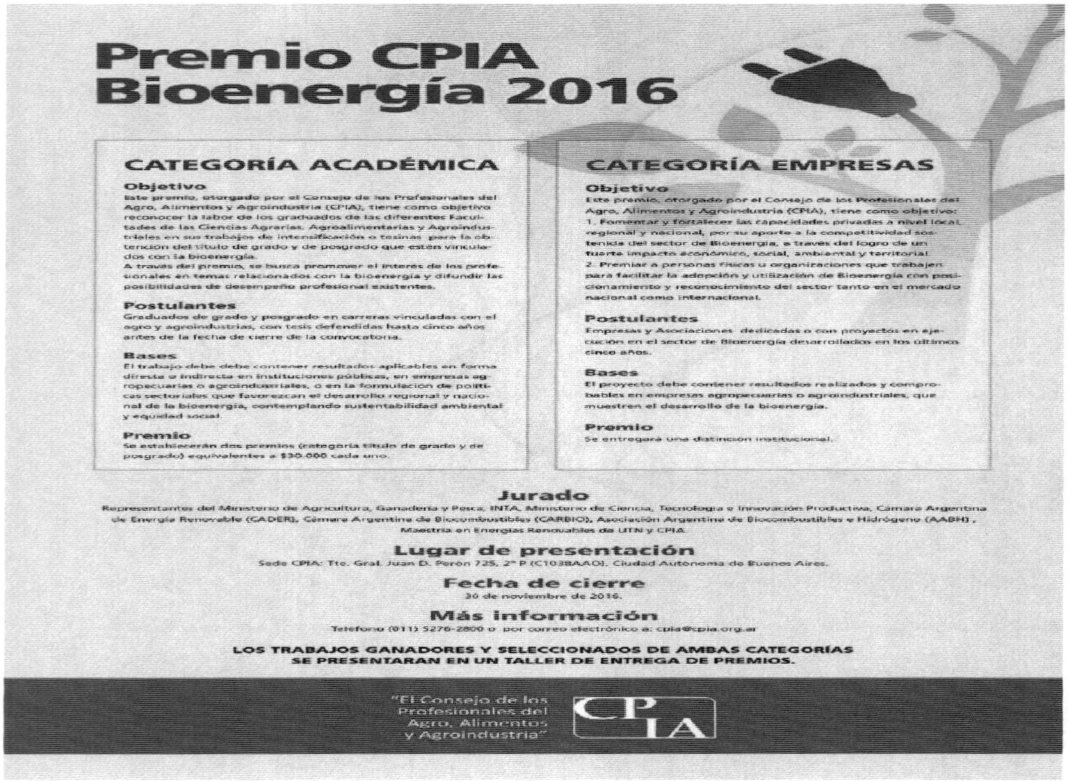
Premio CPIA Bioenergía 2016
CATEGORÍA ACADÉMICA
Objetivo
Este premio, otorgado por el Consejo de los Profesionales del Agro, Alimentos y Agroindustria (CPIA), tiene como objetivo reconocer la labor de los graduados de las diferentes Facultades de las Ciencias Agrarias, Agroalimentarias y Agroindustriales en sus trabajos de intensificación o tesinas para la obtención del título de grado y de posgrado que estén vinculados con la bioenergía.
A través del premio, se busca promover el interés de los profesionales en temas relacionados con la bioenergía y difundir las posibilidades de desempeño profesional existentes.
Postulantes
Graduados de grado y posgrado en carreras vinculadas con el agro y agroindustrias, con tesis defendidas hasta cinco años antes de la fecha de cierre de la convocatoria.
Bases
El trabajo debe debe contener resultados aplicables en forma directa o indirecta en instituciones públicas, en empresas agropecuarias o agroindustriales, o en la formulación de políticas sectoriales que favorezcan el desarrollo regional y nacional de la bioenergía, contemplando sustentabilidad ambiental y equidad social.
Premio
Se establecerán dos premios (categoría título de grado y de posgrado) equivalentes a $30.000 cada uno.
CATEGORÍA EMPRESAS
Objetivo
Este premio, otorgado por el Consejo de los Profesionales del Agro, Alimentos y Agroindustria (CPIA), tiene como objetivo:
1. Fomentar y fortalecer las capacidades privadas a nivel local, regional y nacional, por su aporte a la competitividad sostenida del sector de Bioenergía, a través del logro de un fuerte impacto económico, social, ambiental y territorial.
2. Premiar a personas físicas u organizaciones que trabajen para facilitar la adopción y utilización de Bioenergía con posicionamiento y reconocimiento del sector tanto en el mercado nacional como internacional.
Postulantes
Empresas y Asociaciones dedicadas o con proyectos en ejecución en el sector de Bioenergía desarrollados en los últimos cinco años.
Bases
El proyecto debe contener resultados realizados y comprobables en empresas agropecuarias o agroindustriales, que muestren el desarrollo de la bioenergía.
Premio
Se entregará una distinción institucional.
Jurado
Representantes del Ministerio de Agricultura, Ganadería y Pesca, INTA, Ministerio de Ciencia, Tecnología e Innovación Productiva, Cámara Argentina de Energía Renovable (CADER), Cámara Argentina de Biocombustibles (CARBIO), Asociación Argentina de Biocombustibles e Hidrógeno (AABH), Maestría en Energías Renovables de UTN y CPIA.
Lugar de presentación
Sede CPIA: Tte. Gral. Juan D. Perón 725, 2° P (C1038AAO), Ciudad Autónoma de Buenos Aires.
Fecha de cierre
30 de noviembre de 2016.
Más información
Teléfono (011) 5276-2800 o por correo electrónico a: cpia@cpia.org.ar
LOS TRABAJOS GANADORES Y SELECCIONADOS DE AMBAS CATEGORÍAS SE PRESENTARAN EN UN TALLER DE ENTREGA DE PREMIOS.
"El Consejo de los Profesionales del Agro, Alimentos y Agroindustria"
CPIA

PAN AMERICAN BIOFUELS AND BIOENERGY SUSTAINABILITY - AN NFS RESEARCH COORDINATION NETWORK

PAN-AMERICAN BIOFUELS & BIOENERGY SUSTAINABILITY

2ª RCN Conferencia Panamericana sobre
Sustentabilidad en Biocombustibles y Bioenergía
Buenos Aires, Argentina - Septiembre 13-16, 2016

¡Muchas Gracias!

Ing.Agr. M.Sc. Jorge A. Hilbert
Instituto de Ingeniería Rural CIA
Mail hilbert.jorge@inta.gob.ar
jorgeantoniohilbert@gmail.com

http://inta.gob.ar/bioenergia

Twitter https://twitter.com/INTABioenergia

UNIVERSIDAD TECNOLÓGICA NACIONAL

CPIA "El Consejo de los Profesionales del Agro, Agroalimentos y Agroindustria"

INTA

Interactions among bioenergy, food security and ecosystem services
Session 1: Biomass from crops and residues – effects on food and ecosystem services

Presenter: Keith L. Kline klinekl@ornl.gov
Center for Bioenergy Sustainability http://web.ornl.gov/sci/ees/cbes/

This presentation puts an expanding bio-based economy into the context of integrated food-energy-ecosystem servicing systems. The presentation reviews how bioenergy can and often does contribute to improving food security, reducing reliance on fossil fuels, and achieving multiple United Nations Sustainable Development Goals (Figure). Bioenergy and improved land management are fundamental components in every strategy that aims to reduce climate forcing from fossil fuels. This presentation shares conclusions from an international, multi-disciplinary team (Acknowledgments) about how to develop a bioeconomy in a manner that supports greater food security and resilience in times of crisis. The presentation includes recommendations to address the energy-food-climate-ecosystem challenges by: (1) engaging local stakeholders to design projects that effectively meet their needs, (2) identifying and encouraging flex-crops and strategies that add value and jobs while diversifying and stabilizing local markets, (3) applying good management practices and tools such as those provided by the Food and Agriculture Organization of the United Nations (FAO BEFS), (4) planning and implementing multiple-use landscapes that improve system efficiency and minimize waste, (5) communicating science-based analyses that address social priorities, and (6) strengthening collaborations.

Acknowledgements: Siwa Msangi, International Food and Policy Research Institute, Washington D.C.; Glaucia M. Souza, University of São Paulo, Brazil; Virginia H. Dale, Center for Bioenergy Sustainability, USA; Jeremy Woods, Imperial College London, UK; Patricia Osseweijer, Delft University of Technology and Joy S. Clancy, University of Twente, Netherlands; Jorge Antonio Hilbert, INTA, Buenos Aires, Argentina; Harriet K. Mugera, World Bank, Italy; Patrick C. McDonnell, Bee Energy, Mexico; Francis X. Johnson, World Agroforestry Centre (ICRAF) Nairobi Kenya.

Sustainable bioeconomy contributes to SDGs addressing #1 poverty, #2 food security and nutrition, #3 health, #5 gender, #6 water and sanitation, #7 **affordable and clean energy**, #8 **jobs**, #12 sustainable consumption/production, #13 **climate change**, #14 oceans, seas and marine resources, #15 terrestrial eco-systems, forests, land degradation and biodiversity, and #16 strengthened institutions.

Interactions among bioenergy, food security and ecosystem services

2nd Research Coordination Network Conference on Pan American Biofuels & Bioenergy Sustainability

Session 1: Biomass crops and residues

Keith L. Kline (klinekl@ornl.gov)
25 August 2016

Acknowledgements: National Science Foundation support via Michigan Technological University RCN and PIRE projects; DOE Bioenergy Technologies Office (BETO) Kristen Johnson and Jim Spaeth; Virginia Dale and team at Center for Bioenergy Sustainability and Climate Change Science Institute, ORNL; and Siwa Msangi (IFPRI) and coauthors (full list in final slide).

http://www.ornl.gov/sci/ees/cbes/ and http://climatechangescience.ornl.gov/

U.S. DEPARTMENT OF ENERGY

OAK RIDGE NATIONAL LABORATORY
MANAGED BY UT-BATTELLE FOR THE DEPARTMENT OF ENERGY

Outline

- **Key issues and concepts**
  - **Bioenergy sustainability and ecosystem services**
  - **Carbon cycle**
  - **Barriers**
    - **Price**
    - **Sustainability concerns**
    - **Policy/market uncertainties**
  - **Opportunities**
    - **Policies to boost bioenergy**
    - **Resource management needs**
    - **Linkages with SDGs**
- **Moving forward**
  - **Stakeholder engagement**
  - **Management practices**
  - **Integration, efficiency and flex crops**
  - **M&E for continual improvement**
  - **Better communications**

Photo by Kline: LUC near Atlanta, GA

OAK RIDGE National Laboratory

Environmental indicators for bioenergy sustainability & associated ecosystem services

Category	Ecosystem service: type	Indicator
Soil quality	**Supporting and regulating service:** soil quality	Total organic carbon (TOC)
		Total nitrogen (N)
		Extractable phosphorus (P)
		Bulk density
Water quality and quantity	**Regulating service:** drinking water; **Cultural service:** recreation	Nitrate concentration in streams
		Total phosphorus (P) concentration in streams
		Suspended sediment concentration in streams
		Herbicide concentration in streams
		Peak storm flow
		Minimum base flow
		Consumptive water use (incorporates base flow)
Green-house gases	**Regulating services:** carbon sequestration and climate regulation	CO_2 equivalent emissions (CO_2 and N_2O)
Bio-diversity	**Regulating services:** biodiversity, pollination, seed dispersal, pest mitigation; **Supporting service:** habitat	Presence of taxa of special concern
		Habitat area of taxa of special concern
Air quality	**Supporting and regulating service:** air quality	Tropospheric ozone
		Carbon monoxide
		Total particulate matter <2.5μm diameter ($PM_{2.5}$)
		Total particulate matter <10μm diameter (PM_{10})
Produc-tivity	**Production services:** food, feed, fiber and fuel	Yield

McBride et al. (2013) & Dale et al. (submitted 2016)

Socioeconomic indicators for bioenergy sustainability & associated ecosystem service

Category	Ecosystem service: type	Indicator
Social well-being	**Cultural services:** jobs and family income; **Provisioning service**: food	Employment
		Household income
		Work days lost due to injury
		Food security
Energy security	**Provisioning service:** energy	Energy security premium
		Fuel price volatility
External trade	**Provisioning services**: food, feed, fuel and fiber	Terms of trade
		Trade volume
Profitability	**Provisioning services**: food, feed, fuel and fiber	Return on investment (ROI)
		Net present value (NPV)
Resource conserva-tion	**Provisioning services:** fuel, chemicals, plastics	Depletion of non-renewable energy resources
		Fossil Energy Return on Investment (fossil EROI)
Social accepta-bility	**Provisioning services**: food, feed, fuel and fiber	Public opinion
		Transparency
		Effective stakeholder participation
		Risk of catastrophe

Dale et al. (2015 and submitted 2016)

GLOBAL CARBON PROJECT

Global carbon budget

http://www.globalcarbonproject.org/carbonbudget/15/files/GCP_budget_2015_v1.02.pdf

The carbon sources from fossil fuels, industry, and land use change emissions are balanced by the atmosphere and carbon sinks on land and in the ocean

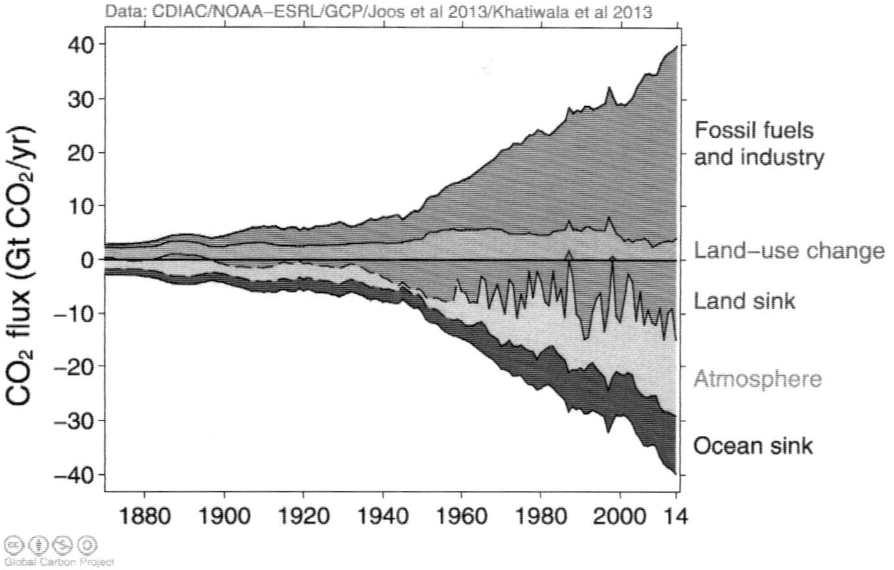

Source: CDIAC; NOAA-ESRL; Houghton et al 2012; Giglio et al 2013; Joos et al 2013; Khatiwala et al 2013; Le Quéré et al 2015; Global Carbon Budget 2015

Barriers are inter-related

1. **Price, economics**
   - Existing policies, infrastructure favor fossil
   - Social and environmental costs treated as externalities
2. **Bioenergy policy → market uncertainty**
   - Less investment
   - More difficult to compete with fossil
3. **Social acceptance and support**
   - Negative perceptions undermine efforts to improve markets and stabilize supportive policies
   - Lack of stable, supportive policies favors status quo (fossil dependence)

Land-use change (LUC) is at core of most concerns driving social opposition to biofuels

OAK RIDGE National Laboratory

Can barriers to acceptance of bioenergy crops be overcome?

IEA Economist (Laszlo Varro) "not optimistic" about supply of "genuinely sustainable biomass" *

- Science-based information to guide decisions
  - Indicators
  - Causal analysis
  - Standards
- Process of monitoring, analysis & continual improvement
- Discussion & examples
  - Natural climate mitigation
  - Food, fuel and other services
  - Climate-smart soil management

Photo by Kline: LUC near Atlanta, GA

* http://www.endseurope.com/article/46959/iea-not-optimistic-about-future-role-of-bioenergy

Policies to Boost Solid Biomass (IRENA)

How to get there?

- ***Accelerate improvement of crop yields***:
  - How can higher yields be promoted when prices to growers are low and 40-50% of food is wasted?
- Collect ***comprehensive data on land...***
  - Land is not global constraint [Souza et al. 2015 SCOPE report]
  - Need systems to improve land management, accelerate carbon storage *AND* productivity, and strengthen terrestrial sinks (e.g., Woodall et al. 2015)
- Research ***practices for cultivating rapidly growing trees and grasses*** on pastureland… – For what? What market specification? We throw away biomass.
  - Millions of hectares of pasture burn each year to eliminate encroaching vegetation
- Institute ***more secure land tenure*** and ***better governance*** to provide incentives for more intensive land management. – ***Yes***
- Provide Incentives to plant ***trees on degraded lands.***
  - ***What incentives work best?***
  - ***Who benefits?***
  - ***Can markets provide the right signals?***
  - ***Is certification the answer?***

Comments in black by Kline on the Policy suggestions in **blue** from Jeff Skeer, IRENA, 2016 presentation, "Boosting Biofuels" in http://www.eubce.com/parallel-events/workshops/the-world-needs-more-land-use-change.html

IEA Bioenergy Joint Task Meeting Question: "Can certification ensure sustainability?"

"No" because –

1. Nothing can ***ensure*** sustainability.
2. There are many opportunities for substitution in biomass markets
3. Transaction costs for certification, monitoring and verification are high relative to the value of the product (biomass)
4. Uncertainties about sustaining political will and market premiums
5. Even well-designed schemes can be "gamed," and a few well-publicized cases undermine credibility.
6. Certification is values-based with specific market intent

Photo VH Dale, 2016: Logging residues in East TN left to rot or burn because there is no market for biomass-bioenergy.

9 Managed by UT-Battelle for the U.S. Department of Energy

OAK RIDGE National Laboratory

Can standards-based rules and policies *facilitate* the transition toward sustainability?"

"Yes, *if*"

1. ***Developed with users*** as a cost-effective tool that meets their needs
2. Provide feedback about real conditions that can guide production toward ***continual improvement*** from users' perspectives
3. Designed to ***adapt*** to changing contexts and priorities
4. Are ***Inclusive***
5. Are broadly ***supported*** (civil society, government, producers, business, and financial markets)

10 Managed by UT-Battelle for the U.S. Department of Energy

OAK RIDGE National Laboratory

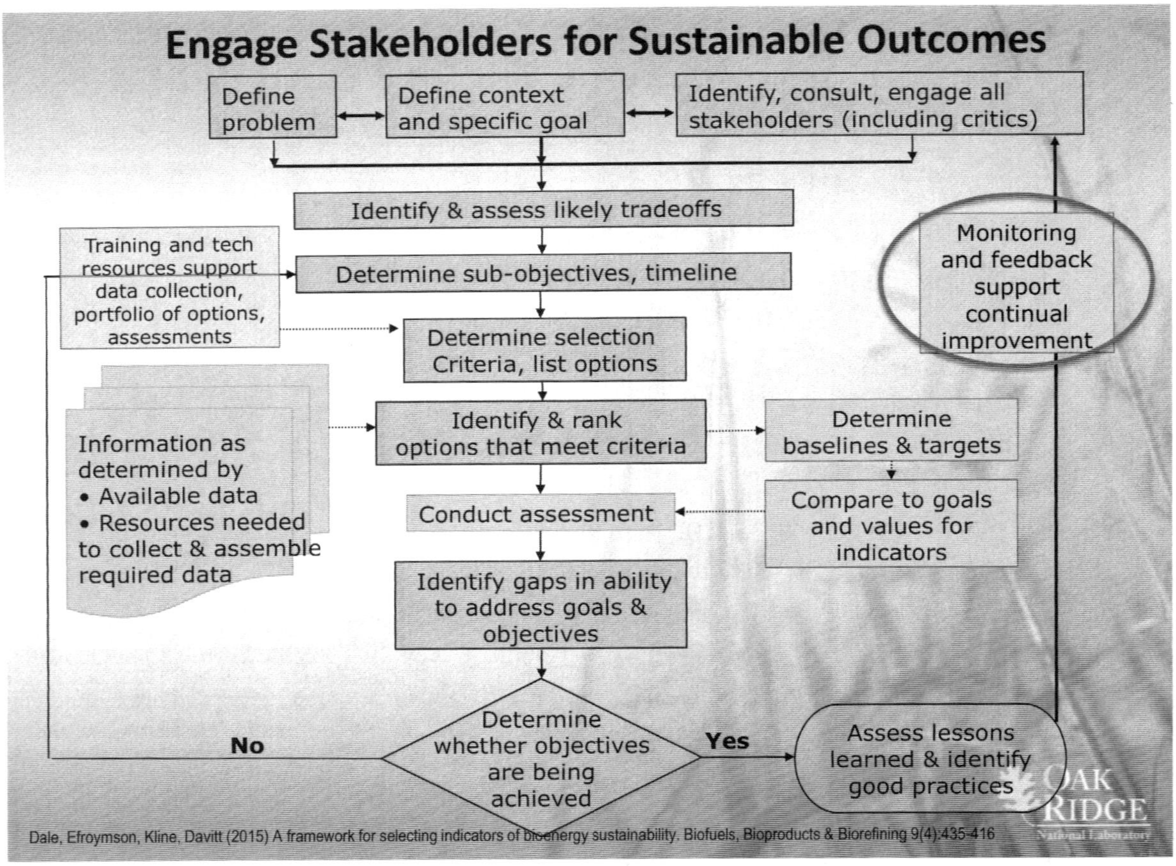
Engage Stakeholders for Sustainable Outcomes
Define problem
Define context and specific goal
Identify, consult, engage all stakeholders (including critics)
Identify & assess likely tradeoffs
Training and tech resources support data collection, portfolio of options, assessments
Determine sub-objectives, timeline
Monitoring and feedback support continual improvement
Determine selection Criteria, list options
Information as determined by
• Available data
• Resources needed to collect & assemble required data
Identify & rank options that meet criteria
Determine baselines & targets
Compare to goals and values for indicators
Conduct assessment
Identify gaps in ability to address goals & objectives
No
Determine whether objectives are being achieved
Yes
Assess lessons learned & identify good practices
OAK RIDGE
National Laboratory
Dale, Efroymson, Kline, Davitt (2015) A framework for selecting indicators of bioenergy sustainability. Biofuels, Bioproducts & Biorefining 9(4):435-416

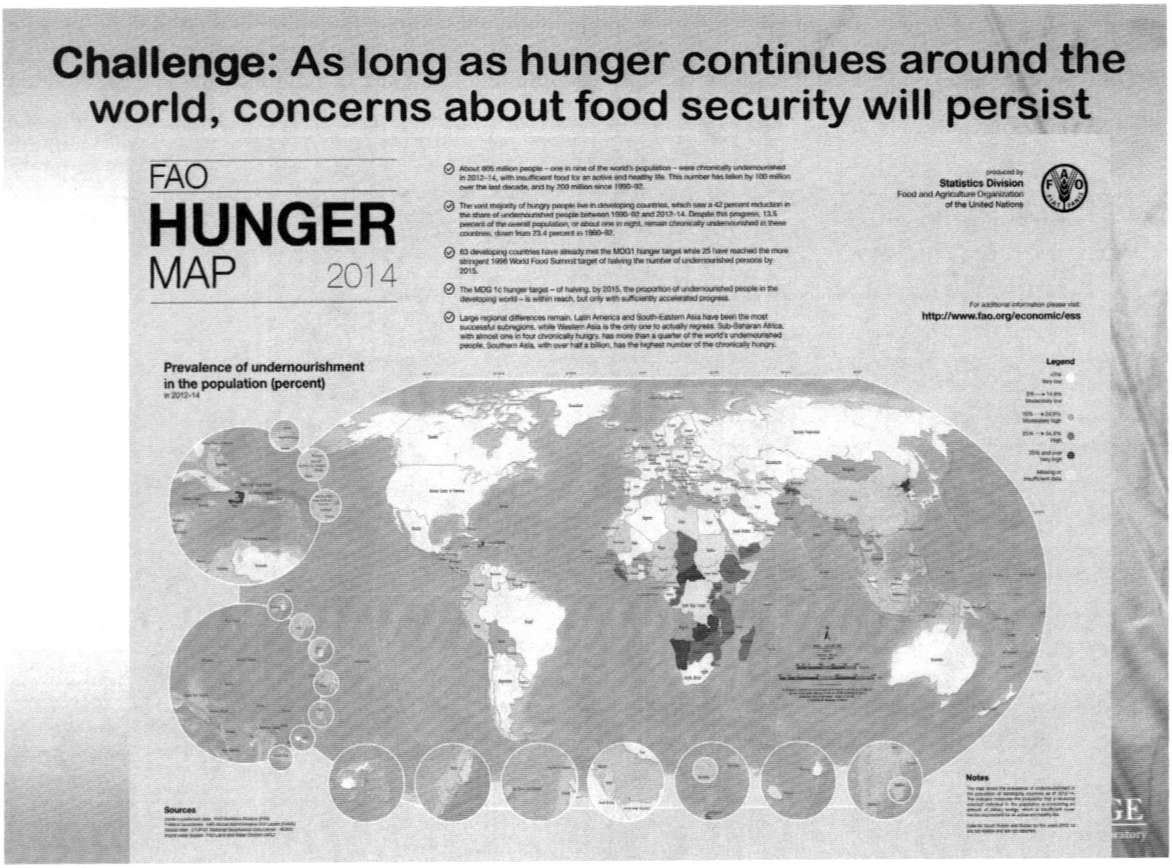
Challenge: As long as hunger continues around the world, concerns about food security will persist
FAO
HUNGER
MAP
2014
Statistics Division
Food and Agriculture Organization of the United Nations
http://www.fao.org/economic/ess
Prevalence of undernourishment in the population (percent)
Legend
Sources
Notes

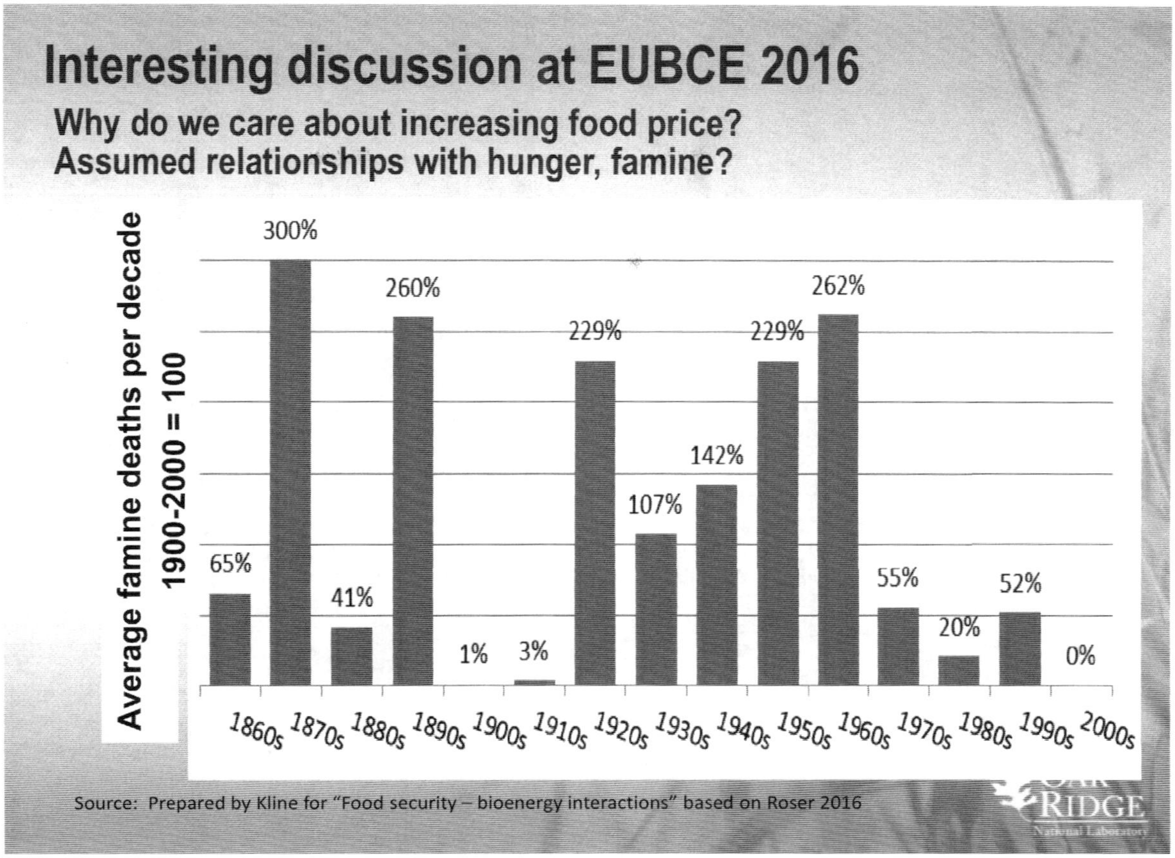

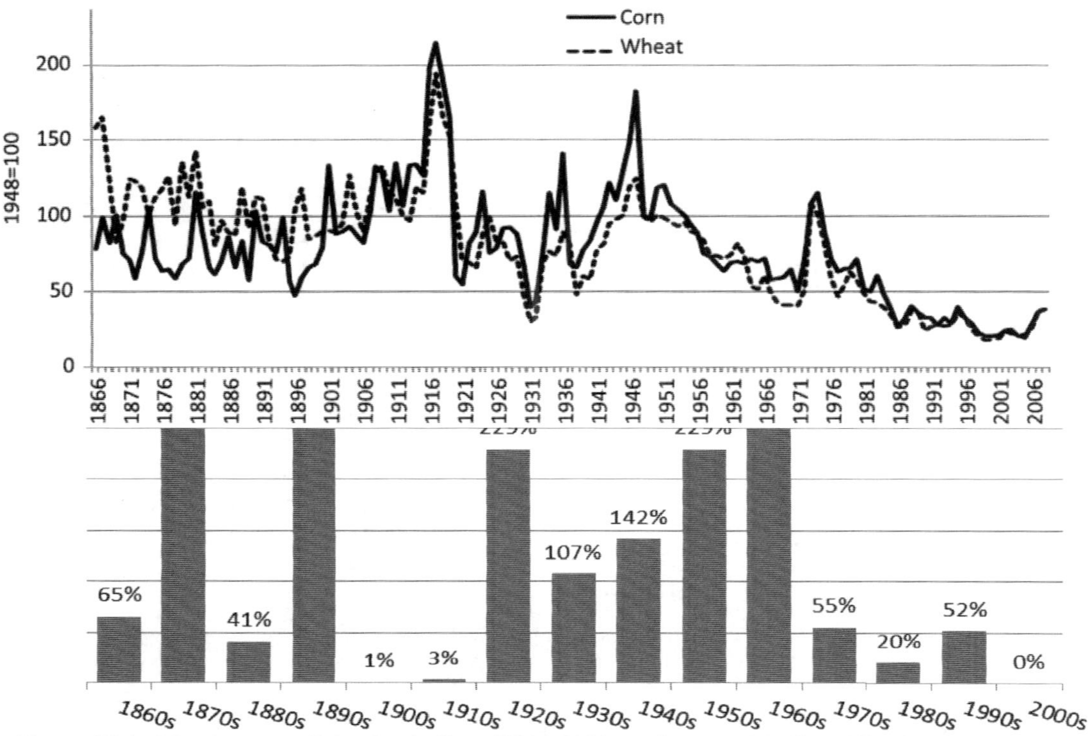

Figure Global food commodity price indices (1866-2008 per Sumner 2009) and Famine Index based on the average famine deaths reported by decade, (average for 1900-2010 = 100; index based on Roser 2015 data). Chart prepared by Kline for analysis in "Food Security-Bioenergy Interactions" (see refs)

Food security

International workshop set forth key issues

- **Identify synergies –** Flex crops can be used for food or fuel
  - Rural infrastructure supports food & fuel
  - Energy for food production, processing, transport and use
  - Sustainable resource management
- **Frame the problem: Ask the questions that matter**
- **Use clear terminology**
  - Workshop report and publication in GCB-Bioenergy (Food Security and Bioenergy, Kline et al., 2016)

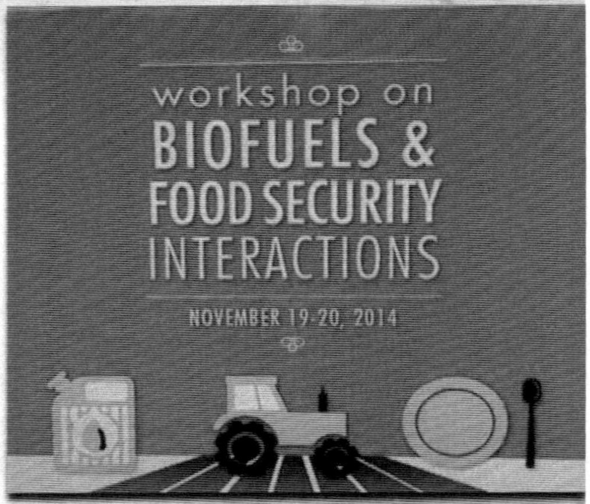

http://onlinelibrary.wiley.com/doi/10.1111/gcbb.12366/full
http://www.ifpri.org/event/workshop-biofuels-and-food-security-interactions

Attributes of the nexus

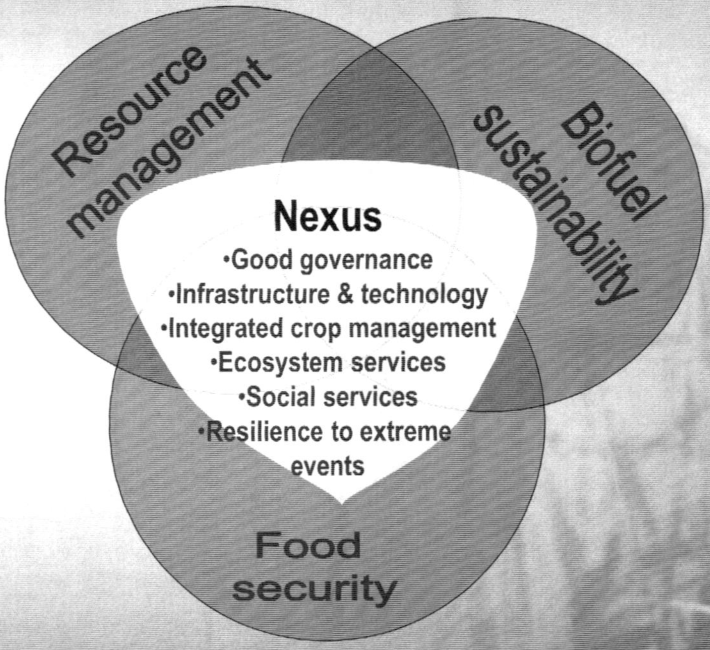

(Kline et al., 2016 GCB-Bioenergy)

At the nexus of food security and sustainable bioenergy in much of the world:

- ➢Address rural poverty
- ➢Diversify crops and sources of income
- ➢Improve local productivity
- ➢Cooperation among sectors

USAID photo - Feed the Future annual report 2015

OAK RIDGE National Laboratory

Challenge: As long as deforestation continues, concerns about LUC will persist.

Maya Biosphere Reserve, Guatemala: Habitat loss, contamination of water and soil, and new settlements are legacies of oil, not agriculture.

Map 4. Frequency of wildfires for 2003, 2005 and 2007 fire seasons in the MBR.

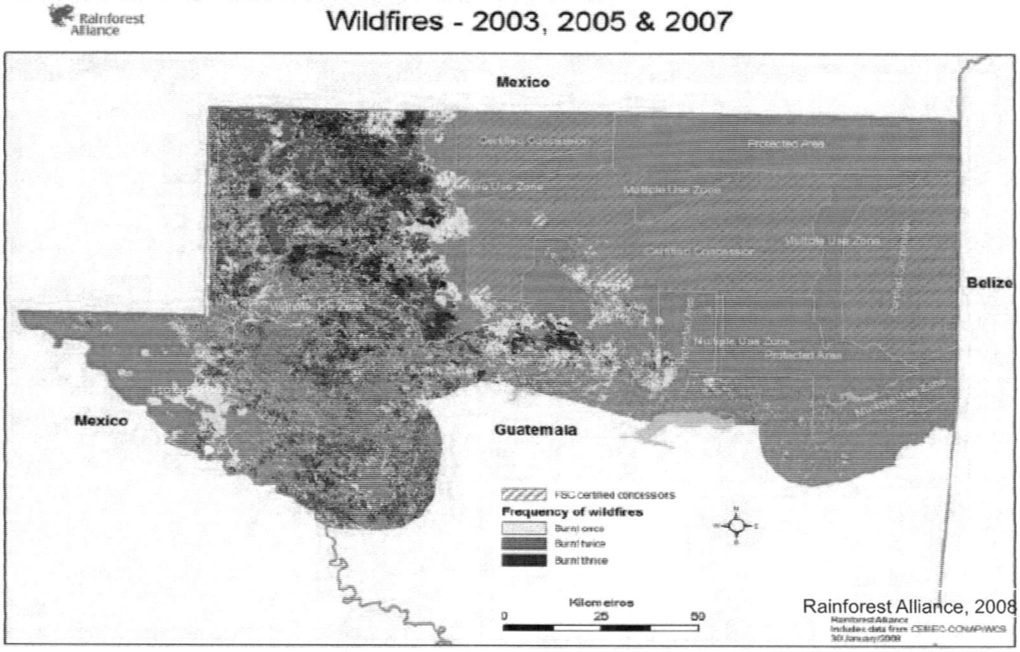

Ecosystem services – climate-smart soils

Markets for biomass support green mitigation practices (right column)

- Apply practices that increase
  - Productivity
  - Soil carbon
- Research needed on inter-cropping and long-term rotation of trees with annuals to store more carbon, deeper, in more stable forms, than what would occur without 'LUC' (e.g., Woodall et al. 2015)
- Clarify that it's not "LUC" that matters, but how matter changes that matters (e.g., science-based approaches)

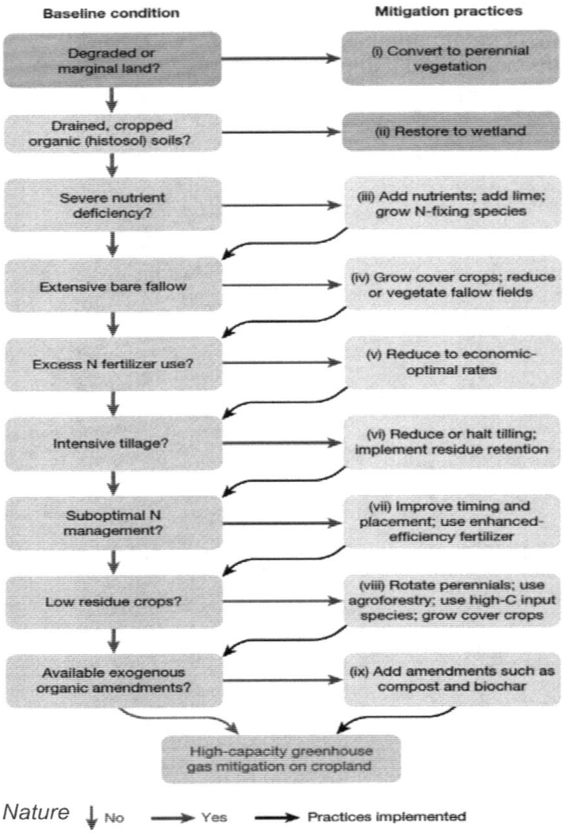

Chart from Paustian et al. (2016) *Nature*

Science-based analysis to guide decisions

Science: systematic methodology based on evidence and observation

- ✓ Start with clear definition of problem
- ✓ Ask right questions
- ✓ Test hypotheses
- ✓ Conduct critical analysis
- ✓ Determine cause and effect
- ✓ Document verifiable, replicable results
- ✓ Build on experience and learn from others

Challenges

- Science evolves as new data and understanding become available
- Targeted data collection
- Confounding data and terminology
  - ✓ Land cover versus land uses (multiple) and management
  - ✓ Crop price and trade versus total production and actual uses, losses
  - ✓ Correlation versus causation

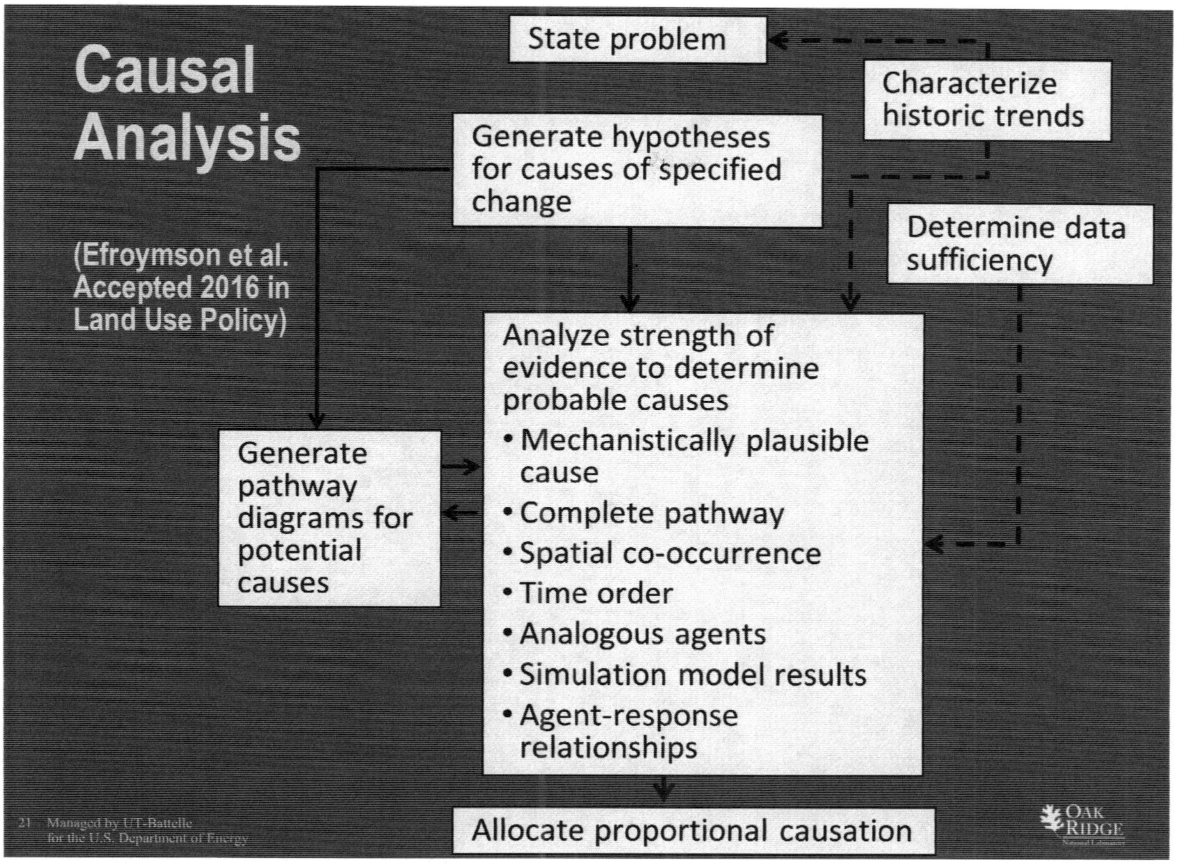

Steps to encourage beneficial LUC

- **Give biomass value! Reduce losses**
  - Fires and other disturbances
  - Wastes
- **Legal and financial motivations to invest in and adopt better management practices**
- **Pathways to "natural climate mitigation"**
  - Better management of occupied lands
  - Conserve remaining forests (no new roads in wilderness areas)
  - Restore and replant former forest lands
  - Create value-added jobs and services that reduce pressure on isolated forest frontiers (reduced deforestation) in LDCs
- **Accelerate shifts to ever higher performing urban and integrated agro-silvo-pastoral systems**
- **Increase scrutiny, awareness and enforcement to end illicit land-management activities**
- **Apply same performance criteria to all sources of energy and all land management (food, feed, fiber, energy...)**

Conclusion: Plenty of biomass, inadequate rewards for good management

We need to

- Learn from experiences
- Build local partnerships
- Develop and apply a suite of metrics that reflect local stakeholder priorities for "sustainability"

Source: Kline training seminar for Advanced School on Present and Future of BioEnergy; ESPCA – FAPESP – University of Campinas, 10-17 October, 2014. Campinas, SP Brazil.

➢ **And link to the SDGs:**

SUSTAINABLE DEVELOPMENT GOALS

17 GOALS TO TRANSFORM OUR WORLD

>160 indicators

http://sd.iisd.org/news/iaeg-sdgs-sets-workplan-for-finalizing-indicators/

Sustainable bioeconomy contributes to SDGs addressing #1 poverty, #2 food security and nutrition, #3 health, #5 gender, #6 water and sanitation, #7 **affordable and clean energy**, #8 **jobs**, #12 sustainable consumption/production, #13 **climate change**, #14 oceans, seas and marine resources, #15 terrestrial eco-systems, forests, land degradation and biodiversity, and #16 strengthened institutions.

Souza et al. 2015

References

- Dale B et al. 2014. Take a closer look: biofuels can support environmental, economic and social goals. ES&T48(13):7200-7203
- Dale VH et al. 2013. Indicators for assessing socioeconomic sustainability of bioenergy systems: A short list of practical measures. Ecological Indicators 26: 87-102. http://dx.doi.org/10.1016/j.ecolind.2012.10.014
- Dale VH et al. 2016. Incorporating bioenergy into sustainable landscape designs. *Renewable & Sustainable Energy Reviews* 56:1158-1171. http://authors.elsevier.com/sd/article/S1364032115014215
- Dale VH et al.(Submitted 2016 for Biomass & Bioenergy Special Issue on Biofuels and Ecosystem Services) Selecting indicators of changes in ecosystem services due to cellulosic-based biofuels in the midwestern US.
- Efroymson RA et al. 2013. Environmental indicators of biofuel sustainability: What about context? Environmental Management 51(2): 291-306. http://web.ornl.gov/sci/ees/cbes/Publications/Efroymsonetal2012biofuelindicatorcontextEMfinal10%201007_s00267-012-9907-5.pdf
- FAO (2015a) Hunger Map 2015. FAO Statistics Division, Rome. Available: http://www.fao.org/hunger/en/
- FAO (2015b) Forty-second Session Report, Committee on World Food Security, Rome, Italy, 12-15 October 2015. Global Strategic Framework for Food Security & Nutrition (GSF) Available at: http://www.fao.org/3/a-mo187e.pdf Last Accessed 10/10/2015.
- FAO, IFAD, WFP (2013) The State of Food Insecurity (SOFI) in the World 2013 -The multiple dimensions of food security. And FAO, IFAD, WFP (2014) SOFI Strengthening the enabling environment for food security and nutrition. FAO Rome. And FAO, IFAD, WFP (2015) SOFI Meeting the 2015 international hunger targets: taking stock of uneven progress. FAO, Rome, Italy
- Kline KL, Dale VH (2008) Biofuels, causes of land-use change, and the role of fire in greenhouse gas emissions. Science, 321, 199.
- Kline KL, Dale VH, Lee R, Leiby P (2009) In Defense of Biofuels, Done Right. Issues in Science and Technology, 25(3), 75-84
- Kline KL, Oladosu GA, Dale VH, McBride AC (2011) Scientific analysis is essential to assess biofuel policy effects. Biomass and Bioenergy, 35, 4488-4491
- Kline KL (2014) Advanced School on Present and Future of BioEnergy, ESPCA–FAPESP–University of Campinas, 10-17 October, 2014. Campinas, SP Brazil.
- Kline KL et al. (2016) Reconciling biofuels and food security: priorities for action. GCB-Bioenergy. http://onlinelibrary.wiley.com/doi/10.1111/gcbb.12366/full
- McBride A et al. (2011) Indicators to support environmental sustainability of bioenergy systems. *Ecological Indicators* 11(5) 1277-1289.
- Parish ES et al. (2012) Multimetric spatial optimization of switchgrass plantings across a watershed. *BioFPR*. 6(1):58-72
- Parish ES, Kline KL, Dale VH, Efroymson RA, et al., (2013) Comparing Scales of Environmental Effects from Gasoline and Ethanol Production. Environmental Management 51(2):307-338
- REN 21 Renewables (2016) and (2014) Global Status Report Paris, REN21 Secretariat. http://www.ren21.net/status-of-renewables/global-status-report/
- IRENA (Jeff Skeer) (2016) Boosting Biofuels: Sustainable paths to greater energy security. www.irena.org
- Rainforest Alliance (2008) Impact of FSC Certification on Deforestation and the Incidence of Wildfires in the Maya Biosphere Reserve. http://www.rainforest-alliance.org/forestry/documents/peten_study.pdf
- Roser M (2015) Our World in Data. www.OurWorldinData.org
- Souza et al. (eds.) 2015. Scientific Committee on Problems of the Environment (SCOPE), Bioenergy & Sustainability: bridging the gaps. SCOPE 72. Paris, France and Sao Paulo, Brazil. ISBN: 978-2-9545557-0-6.
- Sumner DA (2009) Recent commodity price movements in historical perspective. American Journal of Agricultural Economics, 91(5) 1250-1256
- UNEP (2016) Unlocking the Sustainable Potential of Land Resources: Evaluation Systems, Strategies and Tools. Working Group on Land and Soils, International Resource Panel (IRP UNEP). Herrick, JE, O Arnalds, B Bestelmeyer, S Bringezu, G Han, MV Johnson et al. , ISBN: 978-92-807-3578-9
- USDA Economic Research Service (2015) Definitions of Food Security: Ranges of Food Security and Food Insecurity. U.S. Department of Agriculture
- Woodall et al. 2015. Monitoring Network Confirms Land-Use Change is a Substantial Component of the Forest Carbon Sink in eastern United States

OAK RIDGE National Laboratory

Also References

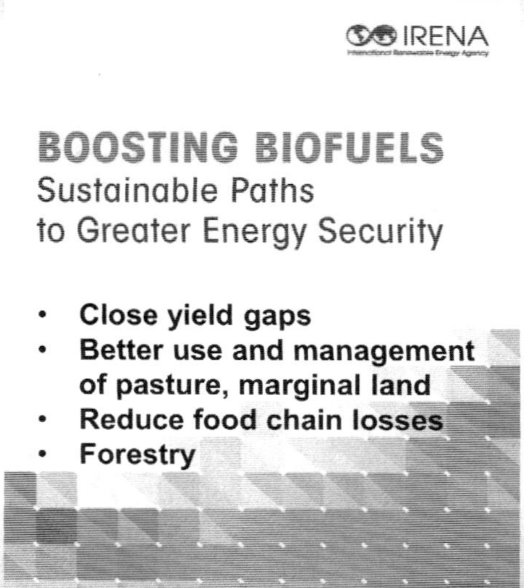

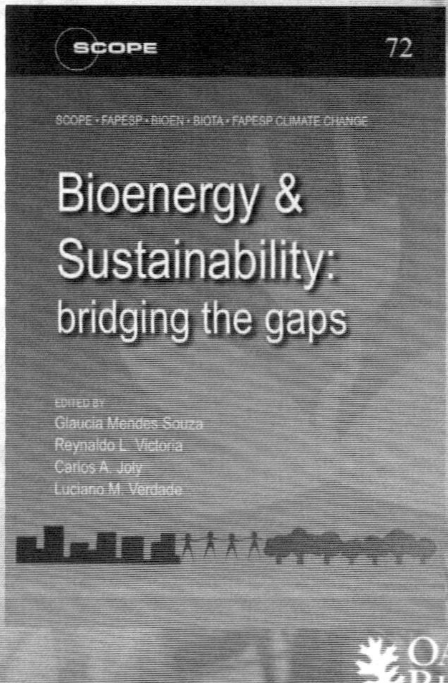

26

OAK RIDGE National Laboratory

Thank you

Center for Bioenergy Sustainability
http://www.ornl.gov/sci/ees/cbes/

See CBES website for
- Reports
- Forums on current topics
- Recent publications

See the new publication, ***Reconciling food security and biofuels***, now available:
http://onlinelibrary.wiley.com/doi/10.1111/gcbb.12366/full

This research is supported by the U.S. Department of Energy (DOE) Bio-Energy Technologies Office and performed at Oak Ridge National Laboratory (ORNL). Oak Ridge National Laboratory is managed by the UT-Battelle, LLC, for DOE under contract DE-AC05-00OR22725.

The views in this presentation are those of the author/presenter who is responsible for any errors or omissions.

CBES
Center for BioEnergy Sustainability

Acknowledgements: Co-authors: Siwa Msangi, International Food and Policy Research Institute, Washington D.C.; Glaucia M. Souza, University of São Paulo, Brazil; Virginia H. Dale, Center for Bioenergy Sustainability, USA; Jeremy Woods, Imperial College London, UK; Patricia Osseweijer, Delft University of Technology and Joy S. Clancy, University of Twente, Netherlands; Jorge Antonio Hilbert, INTA, Buenos Aires, Argentina; Harriet K. Mugera, World Bank, Italy; Patrick C. McDonnell, Bee Energy, Mexico; Francis X. Johnson, World Agroforestry Centre (ICRAF) Nairobi Kenya. Papers supported by: US Department of Energy Bioenergy Technologies Office (BETO), the US National Science Foundation (NSF IIA #1243444) PIRE for Environmental and Social Sustainability Assessment of Bioenergy in Pan America and ORNL.

OAK RIDGE
National Laboratory

Ministerio de Ciencia,
Tecnología e Innovación Productiva
Presidencia de la Nación

Ministerio de Ciencia,
Tecnología e Innovación Productiva
Presidencia de la Nación

What is CONAE?

- CONAE is the only Federal Government agency competent to propose policies to promote and execute activities in the space area with peaceful purposes throughout Argentina territory (Creation Decree: 995/91).

- CONAE is assigned to propose and implement a National Space Plan as state policy with a national priority.

(http://www.conae.gov.ar/index.php/espanol/)

2

Main Objectives of CONAE

- ✓ To develop knowledge and technology in the space field contributing to social, economic and productive sectors
- ✓ To promote the development of the national industry
- ✓ To satisfy demands and needs from economic sectors and society, with respect to spatial information, providing the necessary training

National Space Plan →

3

Ministerio de Ciencia,
Tecnología e Innovación Productiva
Presidencia de la Nación

National Space Plan

Strategic Areas

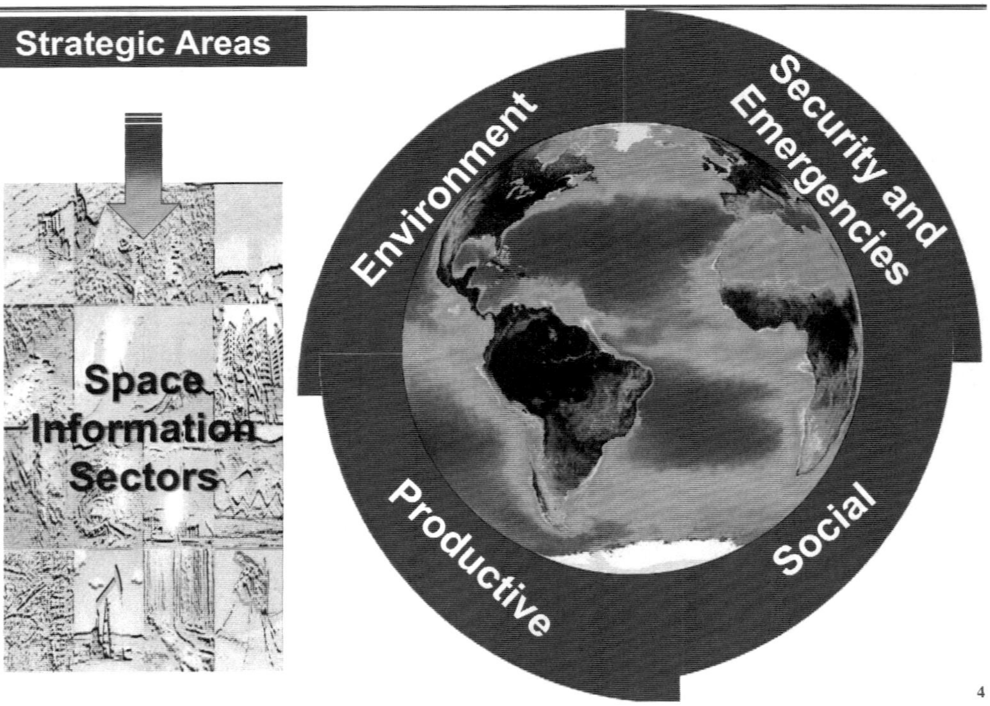

4

Ministerio de Ciencia, Tecnología e Innovación Productiva
Presidencia de la Nación

Facilities: Infrastructure

CONAE

- ✓ SAC-D/Aquarius
- ✓ Landsat 7 y 8
- ✓ Spot 5
- ✓ Eros 1B
- ✓ Terra
- ✓ Aqua
- ✓ NOAA
- ✓ NPP
- ✓ METOP
- ✓ Feng-Yun
- ✓ GOES 13
- ✓ COSMO Skymed 1-2-3 & 4
- ✓ ERS
- ✓ Radarsat-1
- ✓ ALOS-1
- ✓ UAVSAR
- ✓ SARAT

d a t a

Ministerio de Ciencia, Tecnología e Innovación Productiva
Presidencia de la Nación

Facilities: Training

CONAE

Gulich-Higher Level Courses: Master

2MP

6

Ministerio de Ciencia,
Tecnología e Innovación Productiva
Presidencia de la Nación

Facilities: Training

Gulich-Higher Level Courses: Master 2MP

- Master on Applications derived from Spatial Information (UNC/CONAE/ASI)
- Satellite Instruments(UTNM, Universidad Tecnológica Nacional Facultad Regional Mendoza)
- Software development for Spatial Applications (UNLAM, Universidad Nacional de La Matanza)
- Satellite Technology (UTNC, Universidad Tecnológica Nacional Facultad Regional Córdoba)

7

Ministerio de Ciencia,
Tecnología e Innovación Productiva
Presidencia de la Nación

Satellite Missions

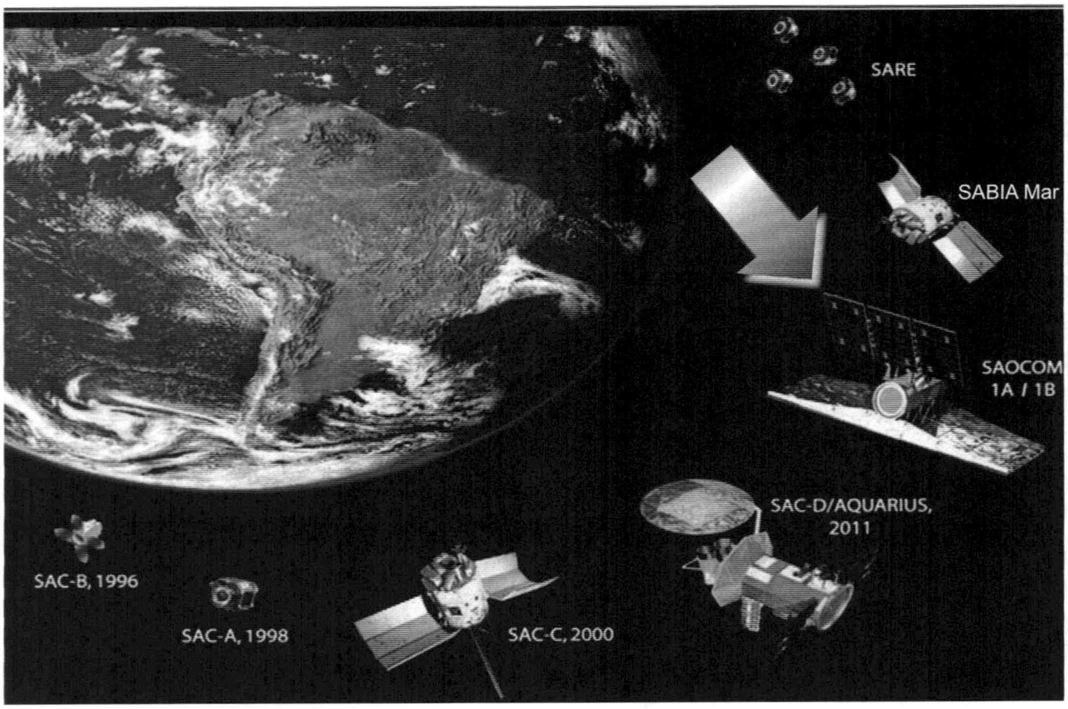

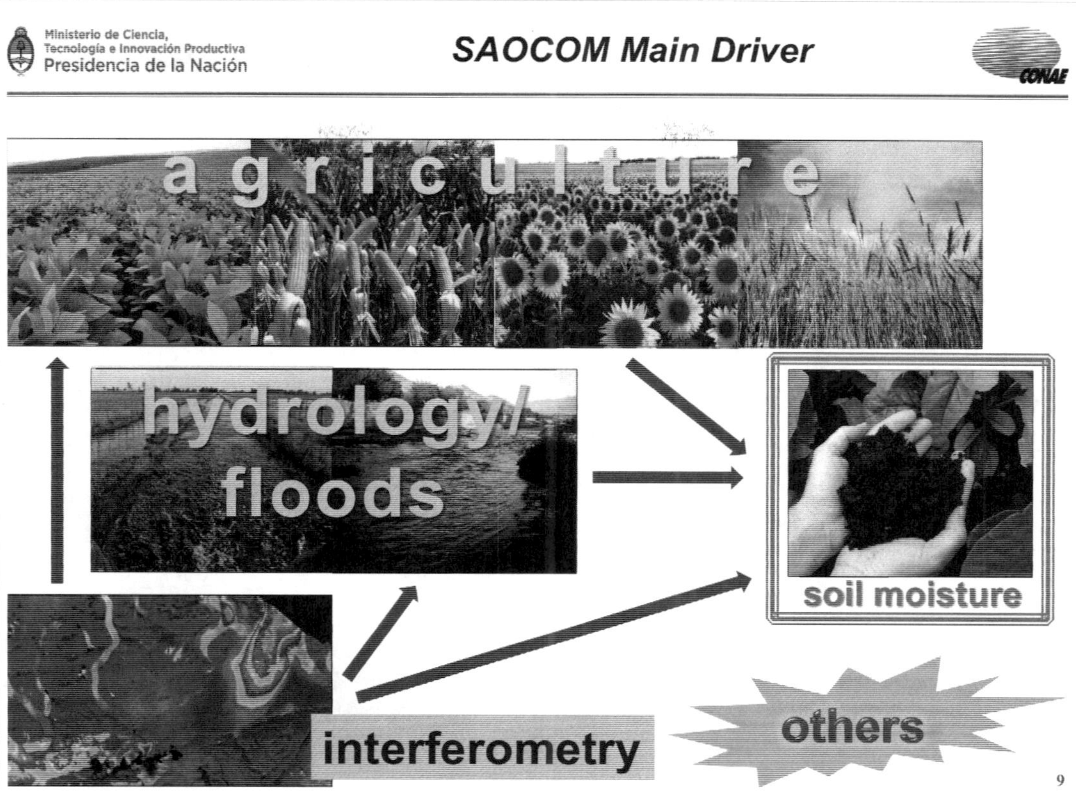
Ministerio de Ciencia,
Tecnología e Innovación Productiva
Presidencia de la Nación
SAOCOM Main Driver
CONAE
agriculture
hydrology/
floods
soil moisture
interferometry
others
9

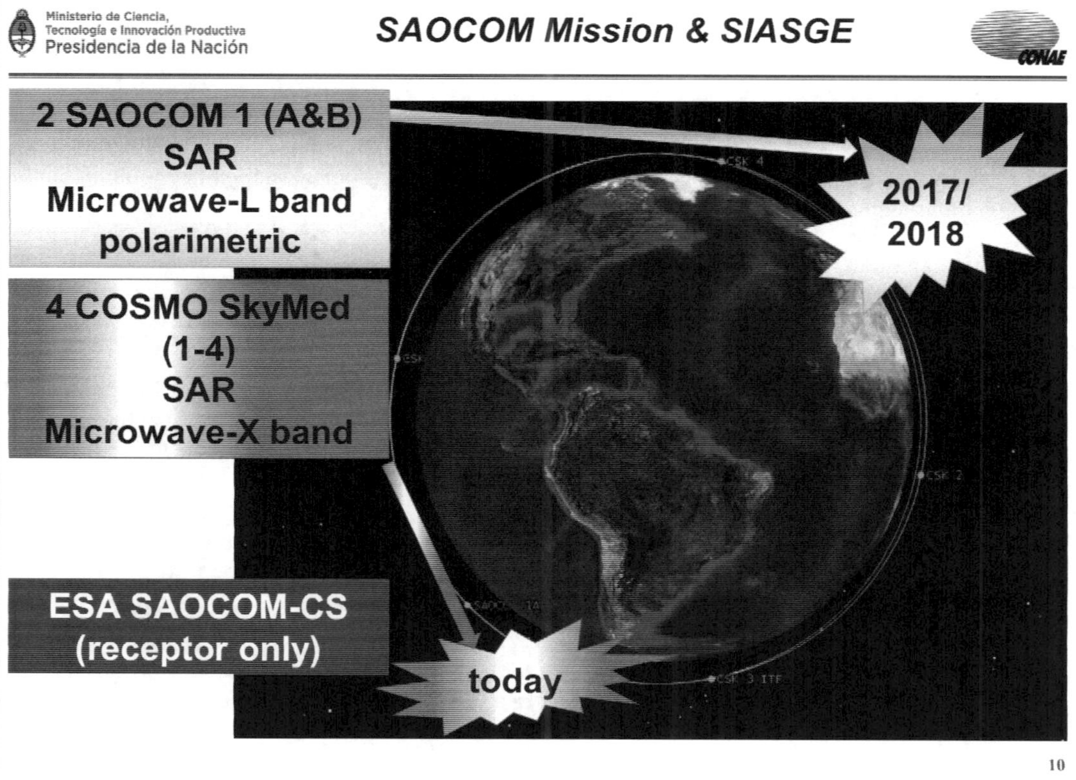
Ministerio de Ciencia,
Tecnología e Innovación Productiva
Presidencia de la Nación
SAOCOM Mission & SIASGE
CONAE
2 SAOCOM 1 (A&B)
SAR
Microwave-L band
polarimetric
4 COSMO SkyMed
(1-4)
SAR
Microwave-X band
ESA SAOCOM-CS
(receptor only)
2017/
2018
today
10

Benefits of the Final Configuration

- Images fron the same place acquired:
  - two bands X and L (polarimetric)
  - cuasi-simultaneous
  - with the same observation geometry
  - different spatial resolutions
  - revisit of 12 hours

SIASGE

- Different positions close to SAOCOM 1B linked with different phases during the life time:
  - Tomographic phase
  - Bistatic phase
  - Specular phase

SAOCOM/
SAOCOM-CS

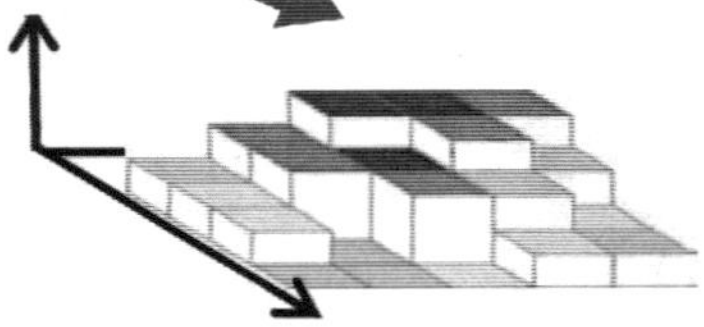

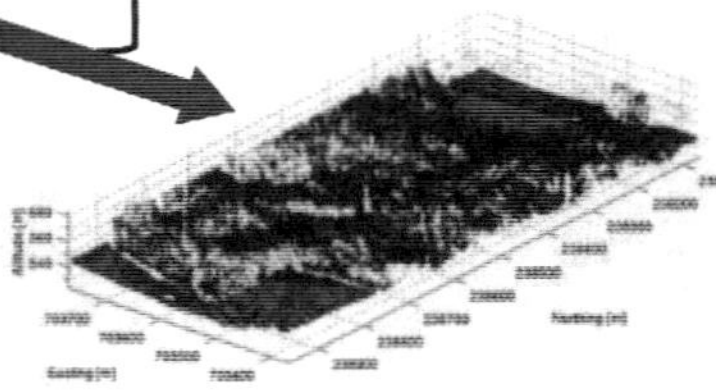

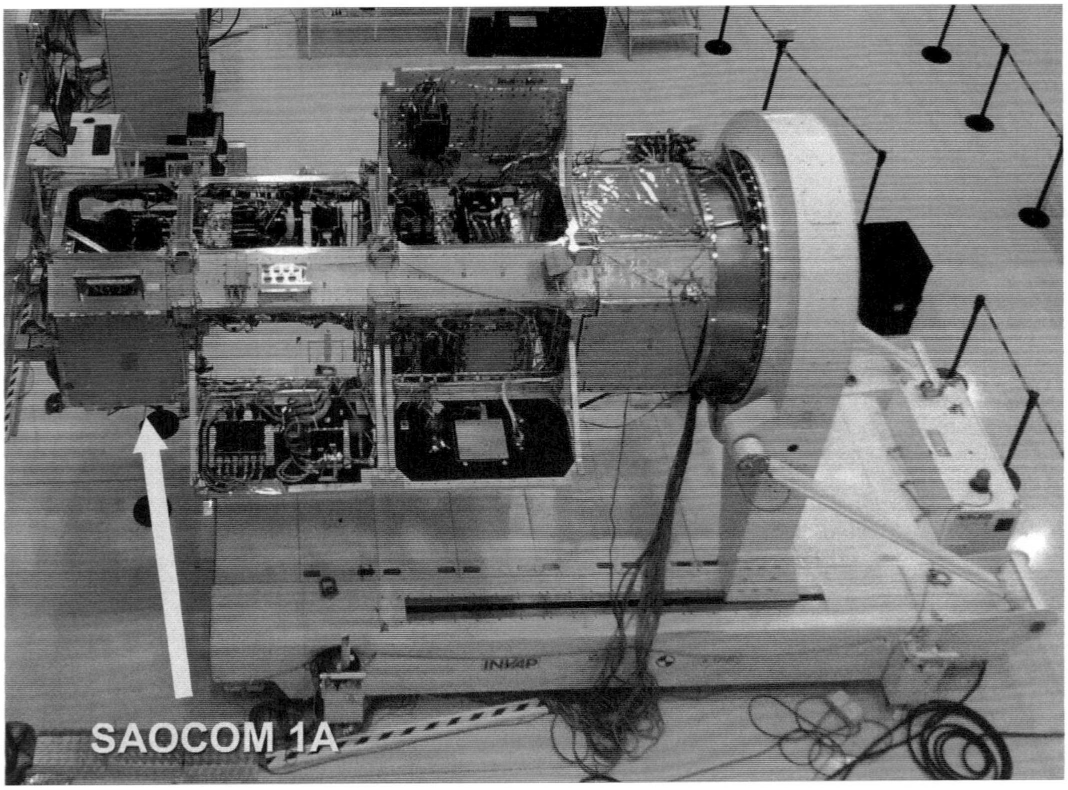

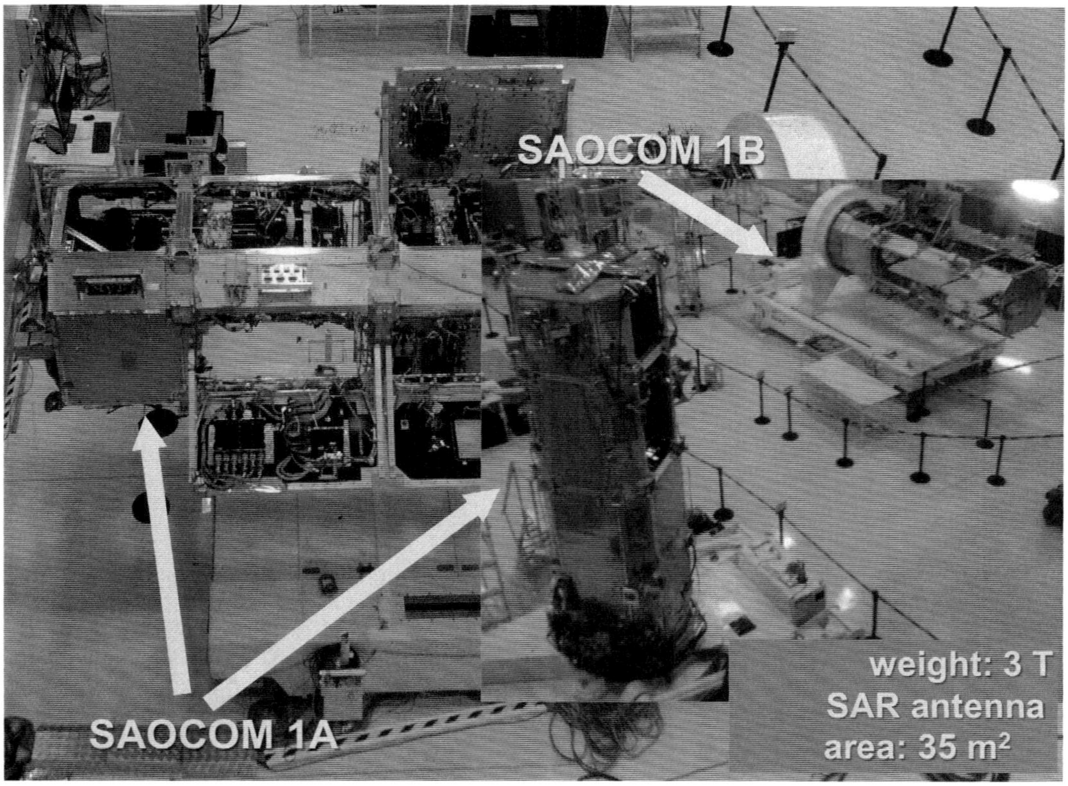

Ministerio de Ciencia, Tecnología e Innovación Productiva
Presidencia de la Nación

SAOCOM General Features

orbit type	sun synchronous, polar
acquisition capability	day/night/any weather conditions
altitude	620 km
local time of asc. node	06:12 am
repeat cycle	16 days (1 satellite) 8 days (constellation)
looking direction	right (nominal) / left (capability)
acquisition modes	real time/stored
coverage	world wide
field work support	soil moisture measurements over Pampa´s Region

Ministerio de Ciencia, Tecnología e Innovación Productiva
Presidencia de la Nación

SAOCOM Nominal Modes Features

CONAE

acquisition mode	polarization mode	swath width	spatial resolution	minimum incidence angle range
StripMap	SP: HH or HV or VH or VV	> 40 km	< 10 m	21° - 50°
	DP: HH/HV or VV/VH	> 40 km	< 10 m	21° - 50°
	QP: HH/HV/VH/VV	> 20 km	< 10 m	20° - 35°
TOPSAR Narrow	SP: HH or HV or VH or VV	> 150 km	< 30 m	25° - 45°
	DP: HH/HV or VV/VH	> 150 km	< 30 m	25° - 45°
	QP: HH/HV/VH/VV	> 100 km	< 50 m	20° - 35°
TOPSAR Wide	SP: HH or HV or VH or VV	> 350 km	< 50 m	25° - 45°
	DP: HH/HV or VV/VH	> 350 km	< 50 m	25° - 45°
	QP[(1)]: HH/HV/VH/VV	> 220 km	< 100 m	20° - 35°
	CL-POL: RH/RV or LH/LV	> 350 km	< 50 m	25° - 45°

(1) TOPSAR Wide QP assigned for Strategic Applications

Ministerio de Ciencia, Tecnología e Innovación Productiva
Presidencia de la Nación

Techniques

CONAE

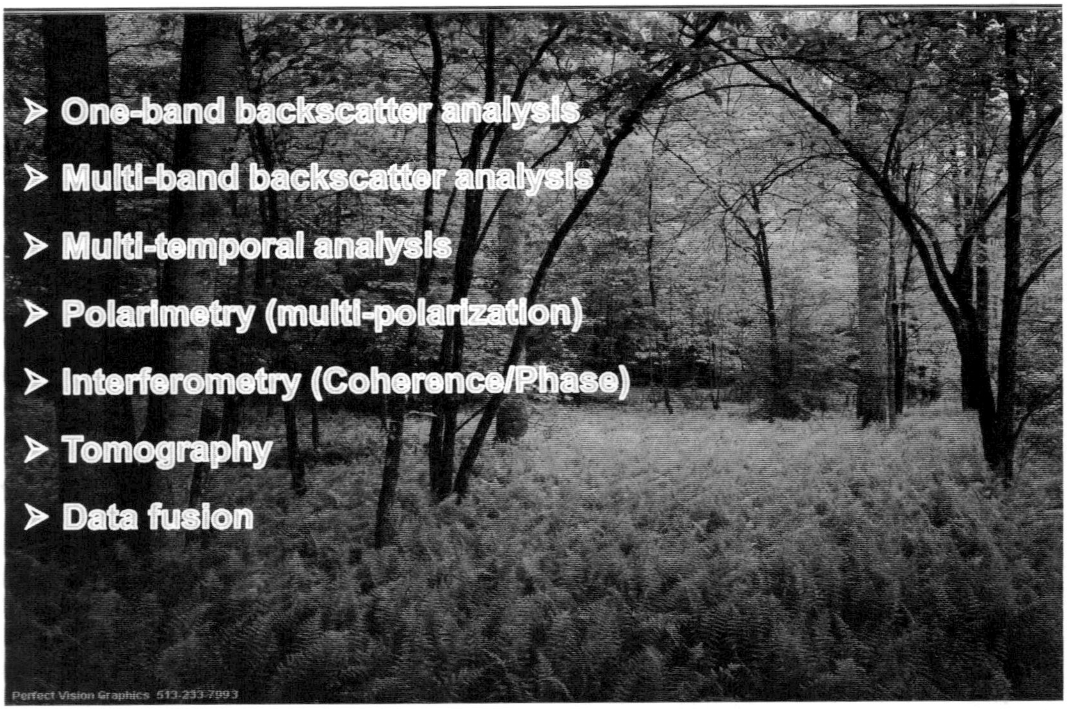

Ministerio de Ciencia, Tecnología e Innovación Productiva
Presidencia de la Nación

One-Band/Multi-band Backscatter Analysis(1/2)

One-Band: interaction with surface/roughness

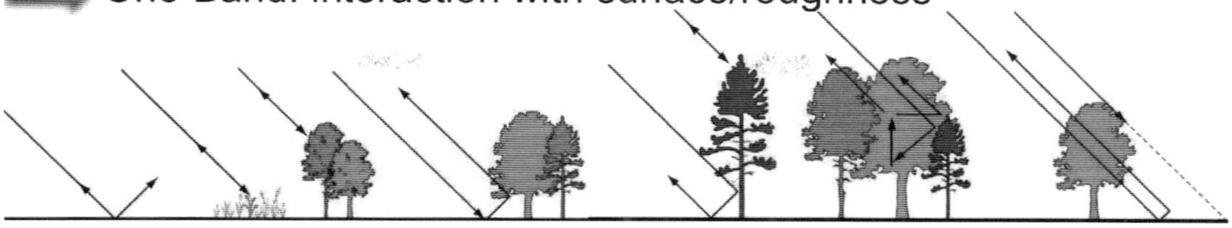

Multi-Band: interaction with surface/penetration

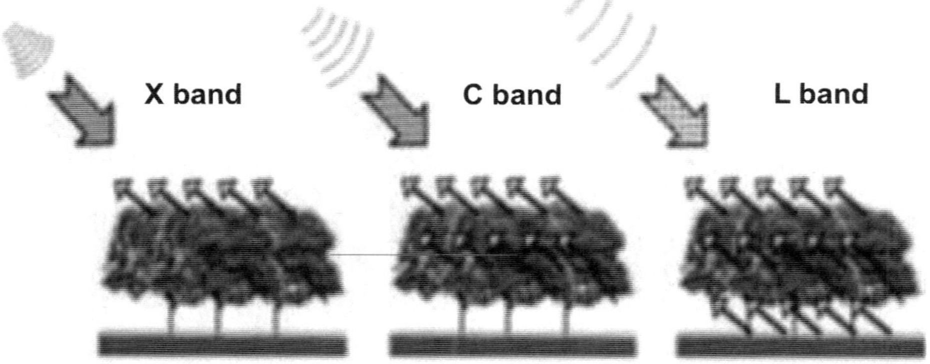

17

Ministerio de Ciencia, Tecnología e Innovación Productiva
Presidencia de la Nación

One-Band/Multi-band Backscatter Analysis(1/2)

Penetration capabilities

band	X	C	L	P
main scatterers	leaves, twigs	leaves, small branches	branches	branches, trunks

18

Ministerio de Ciencia, Tecnología e Innovación Productiva
Presidencia de la Nación

One-Band/Multi-band Backscatter Analysis(1/2)

Penetration capabilities

- species
- growth stages

main scatterers	leaves, twigs	leaves, small branches	branches	branches, t

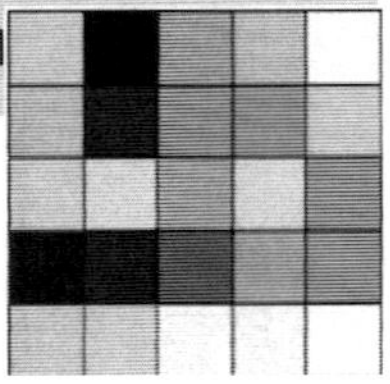

Ministerio de Ciencia, Tecnología e Innovación Productiva
Presidencia de la Nación

Multi-temporal backscatter analysis

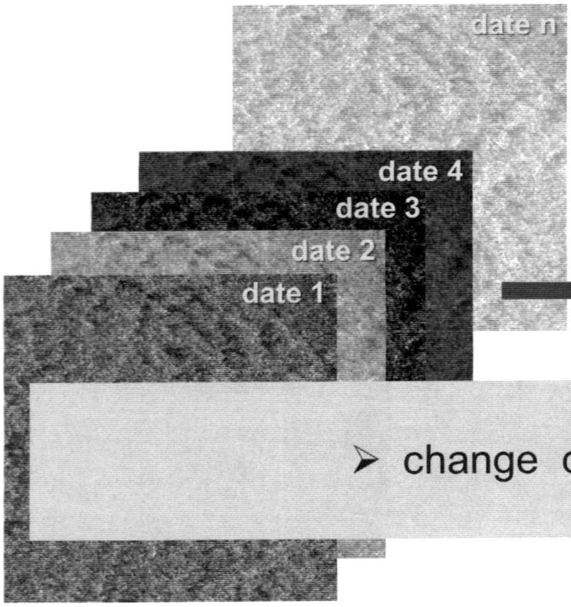

- change detection

20

Ministerio de Ciencia, Tecnología e Innovación Productiva
Presidencia de la Nación

Polarimetry (Multi-polarization)

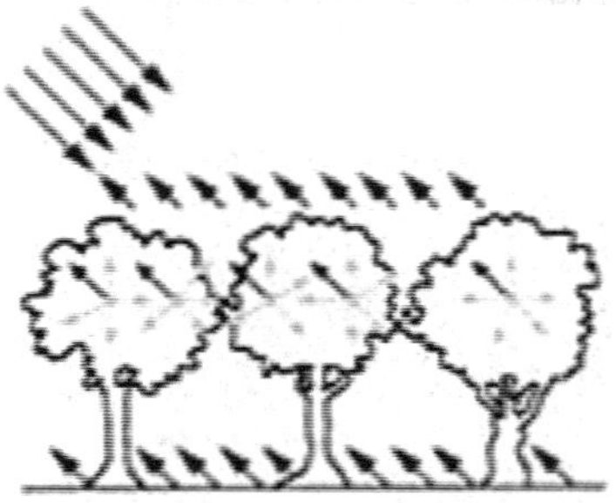

— single polarization
— cross-polarization

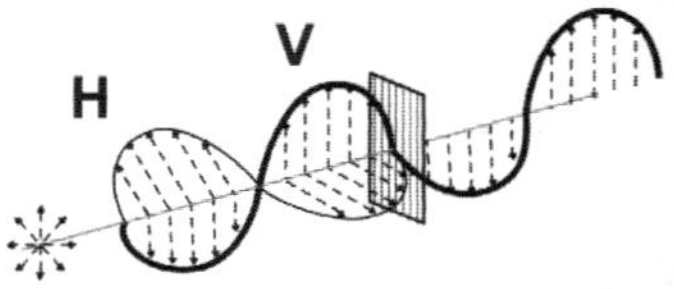

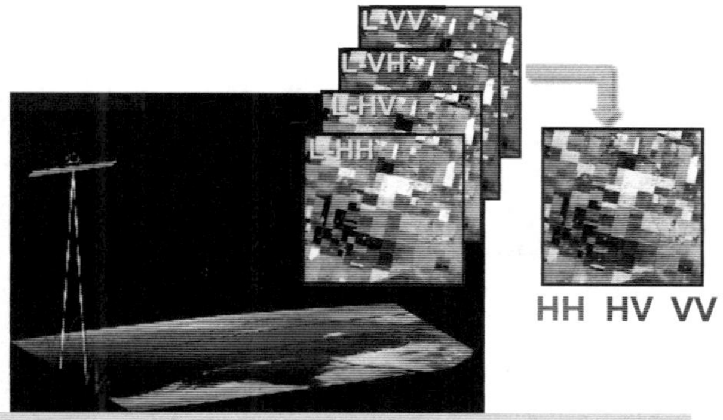

- ➢ species identification
- ➢ changes at surface conditions

21

Ministerio de Ciencia, Tecnología e Innovación Productiva
Presidencia de la Nación

Interferometry (Coherence/Phase)

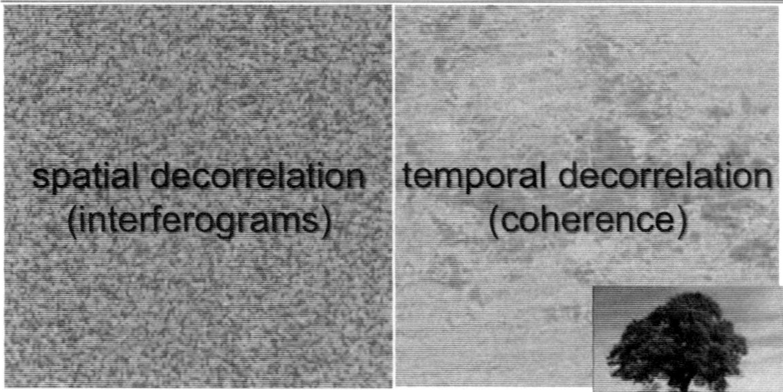

- ➢ change detection

22

Ministerio de Ciencia,
Tecnología e Innovación Productiva
Presidencia de la Nación

Tomography

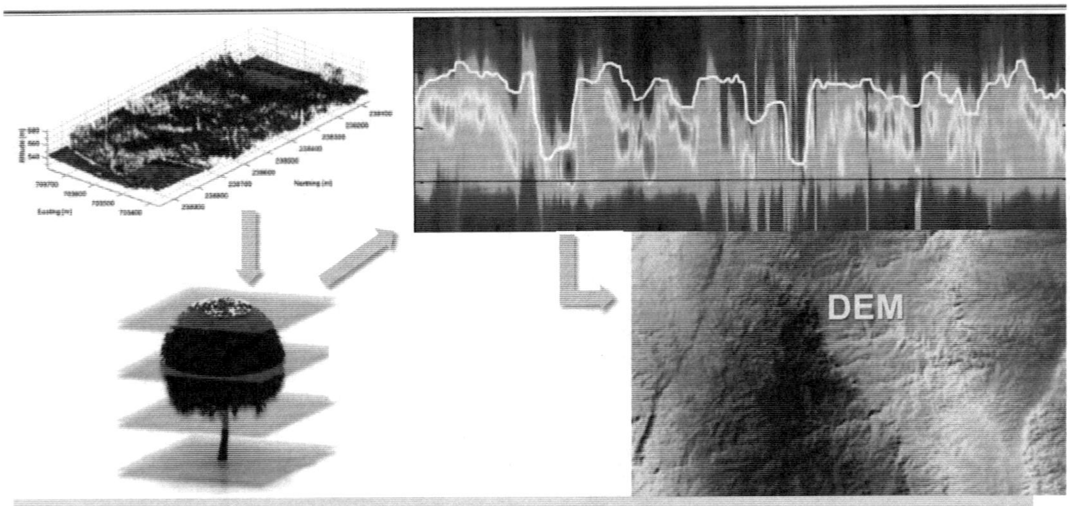

- height profiles
  - volumes

23

Ministerio de Ciencia,
Tecnología e Innovación Productiva
Presidencia de la Nación

Data Fusion (Optical / SAR Multi-polarization / SAR Multi-band)

Iguazú falls (Misiones)

- species identification
- changes at surface conditions

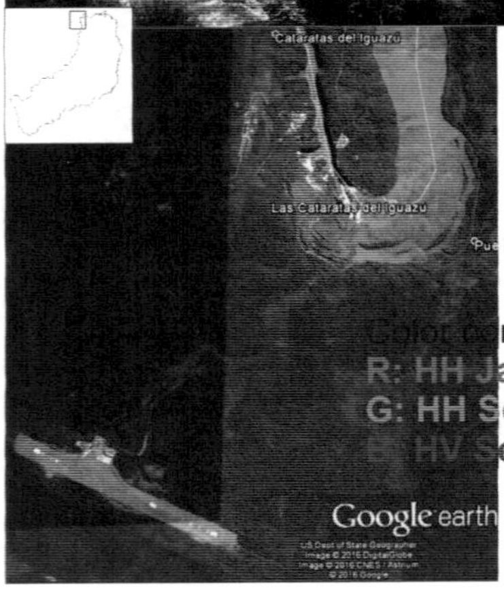

mposition
anuary
eptember
eptember

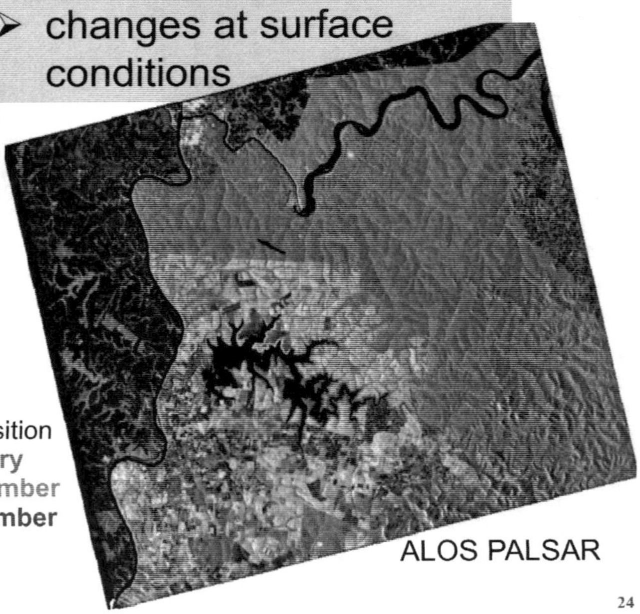

ALOS PALSAR

24

CONAE COMISION NACIONAL DE ACTIVIDADES ESPACIALES

SIASGE PRODUCTS (SAOCOM&COSMO)

Developed & Under Development

Ministerio de Ciencia, Tecnología e Innovación Productiva
Presidencia de la Nación

Processors for SAOCOM/SIASGE Products

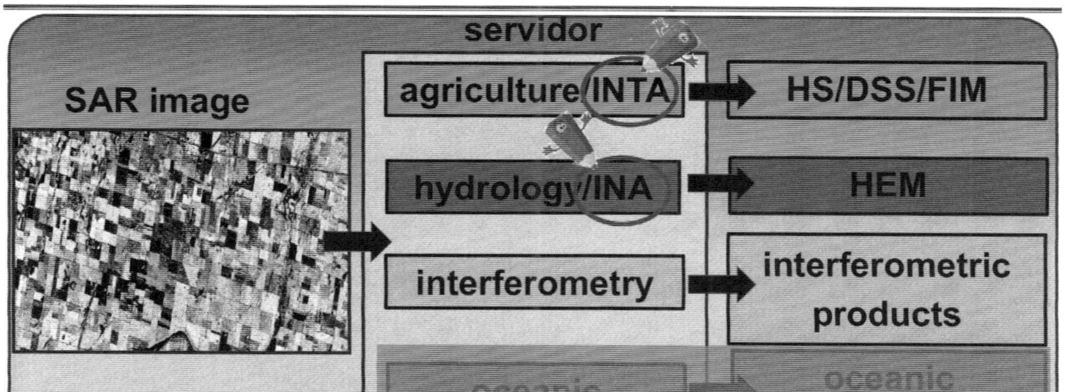

HS: soil moisture map
DSS: support system for decision-making in agriculture
FIM: system for *fusariosis* forecast in wheat
HEM: hydrological risk management

26

Ministerio de Ciencia, Tecnología e Innovación Productiva
Presidencia de la Nación

Soil Moisture Maps(1/2)

CONAE

Ministerio de Ciencia, Tecnología e Innovación Productiva
Presidencia de la Nación

Soil Moisture Maps(2/2)

CONAE

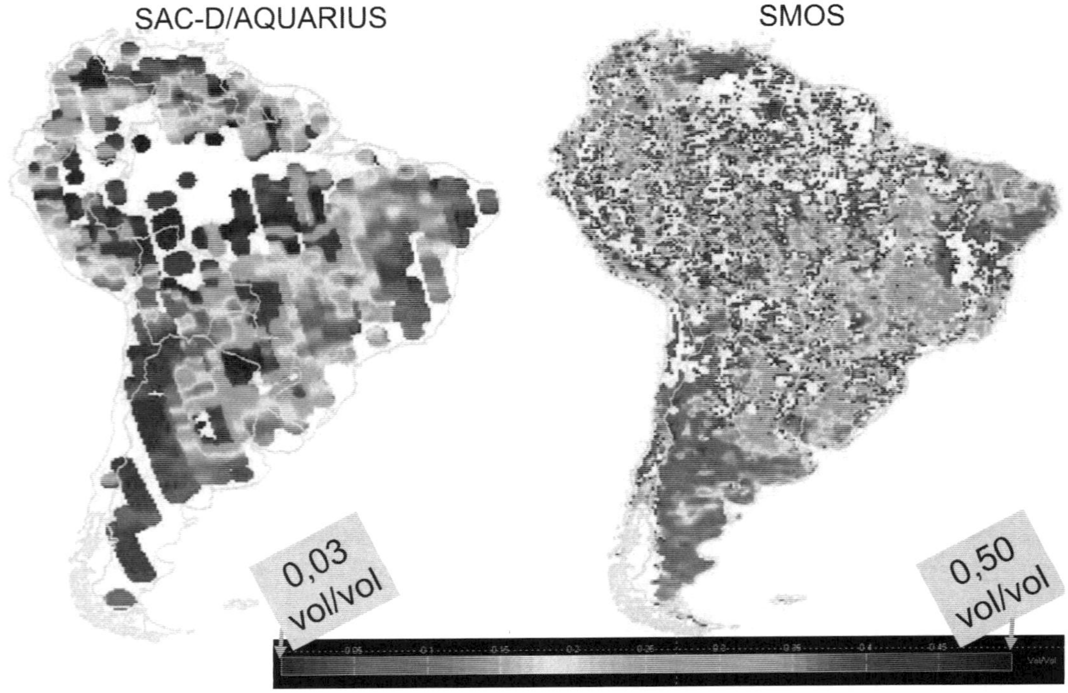

Ministerio de Ciencia, Tecnología e Innovación Productiva
Presidencia de la Nación

Soil Roughness

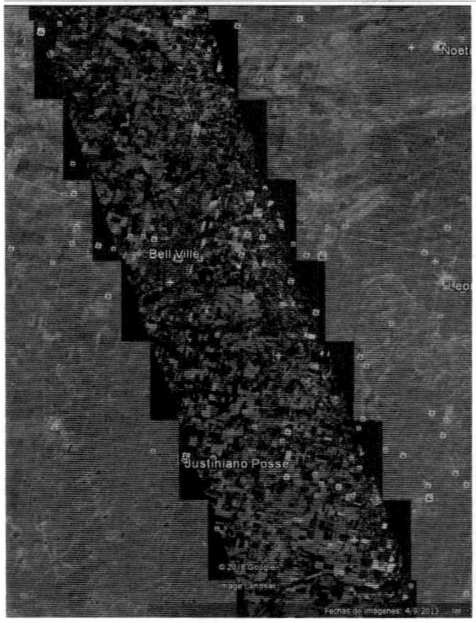

UAVSAR-JPL
March 2015

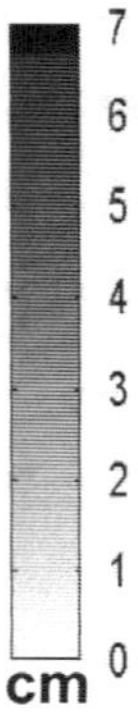

ALOS PALSAR
November 2010

29

Ministerio de Ciencia, Tecnología e Innovación Productiva
Presidencia de la Nación

Radar Vegetation Indices

COSMO SkyMed

30

Ministerio de Ciencia, Tecnología e Innovación Productiva
Presidencia de la Nación

Decision Support to Fertilize at Planting

CONAE

- ✓ different scenarios at yield
- ✓ geographic coordinates
- ✓ species (corn, wheat, sunflower, …)
- ✓ seed type
- ✓ time window

INPUT

RESULTS

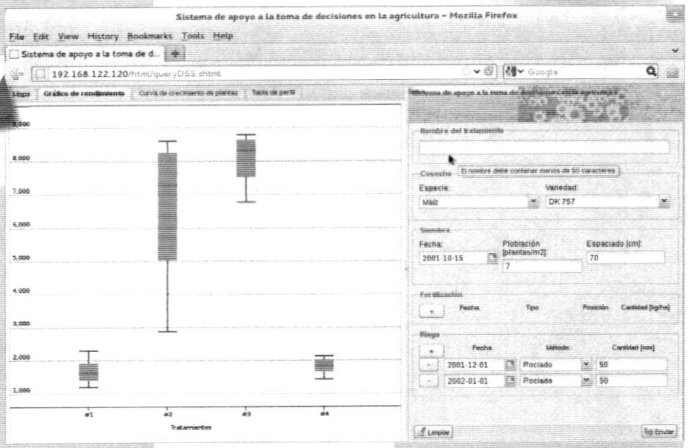

Decision support to:
- ➢ planting
- ➢ fertilization
- ➢ watering
- ➢ optimization in the use of fertilizers
- ➢ yield estimation
- ➢ soil moisture profile estimation → up to 2 m (in depth)

Ministerio de Ciencia, Tecnología e Innovación Productiva
Presidencia de la Nación

Fusariosis Estimation

CONAE

- ✓ geographic coordinates
- ✓ planting data
- ✓ time window

INPUT

RESULTS

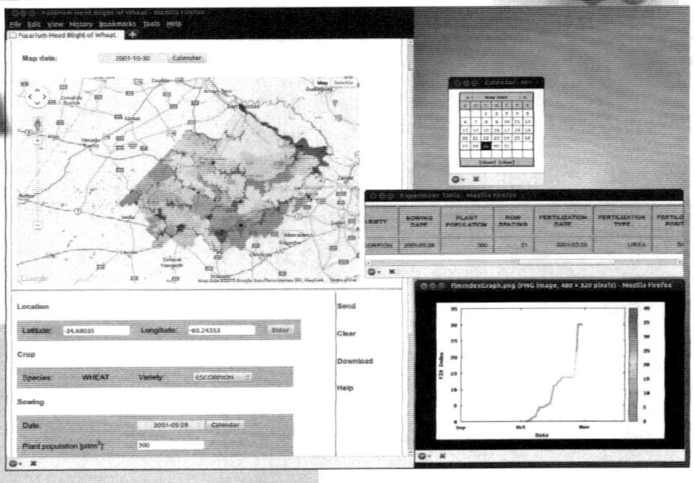

Analysis:
- ➢ illness degree
- ➢ *fusariosis* status
- ➢ agrochemicals use
- ➢ illness progress estimation

32

Identification, Mapping and Crop Monitoring

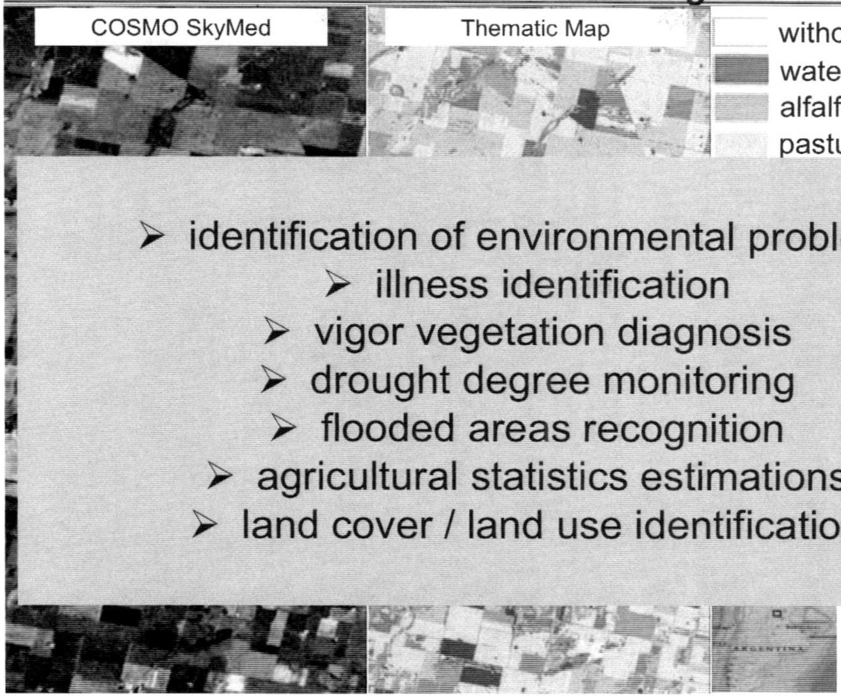

- identification of environmental problems
- illness identification
- vigor vegetation diagnosis
- drought degree monitoring
- flooded areas recognition
- agricultural statistics estimations
- land cover / land use identification

33

Multi-frequency Benefits

C-HH

L-HH

Multi-frequency colour composition Courtesy: JPL

Flevoland Holand

P-HH

34

Ministerio de Ciencia,
Tecnología e Innovación Productiva
Presidencia de la Nación

Vegetation Stress: Orange Crop (Baradero-Bs. As.)

Data fusion:
COSMO SkyMed(VV),
imagen SPOT 7

COSMO SkyMed, modo Spotlight (VV)

Early warning for crop recovery system

Ministerio de Ciencia,
Tecnología e Innovación Productiva
Presidencia de la Nación

Fired Areas, Dry Soils and/or Soils Under Desertification Process

➡ Monitoring and affected area maps

R: January 16, 1996
(after fire)
G: Februay 8, 1995
(after fire)
B: January 19, 1993
(before fire)

Courtesy Dr. Héctor del Valle - CENPAT

36

Ministerio de Ciencia, Tecnología e Innovación Productiva
Presidencia de la Nación

Field Plantations Monitoring

Landsat 5
August 8, 2004
TM 226/75

SIR-C, April 11, 1994
L HH
X VV
L VV

N

Ministerio de Ciencia, Tecnología e Innovación Productiva
Presidencia de la Nación

Identification and Maps Generation of Forested/Deforested Areas

April 11, 1994

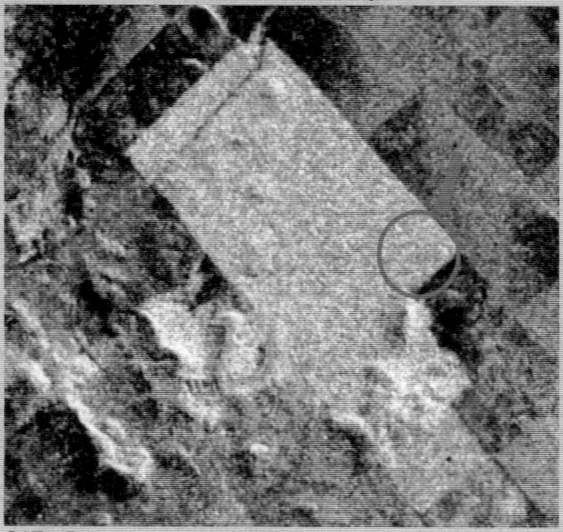

SIR-C L HH, X VV, L VV

N

Landsat 5 TM 226/75

August 8, 2004

38

Ministerio de Ciencia,
Tecnología e Innovación Productiva
Presidencia de la Nación

Forested Areas Monitoring

SAR
COSMO SkyMed
HIMAGE, 2014
2.5 m

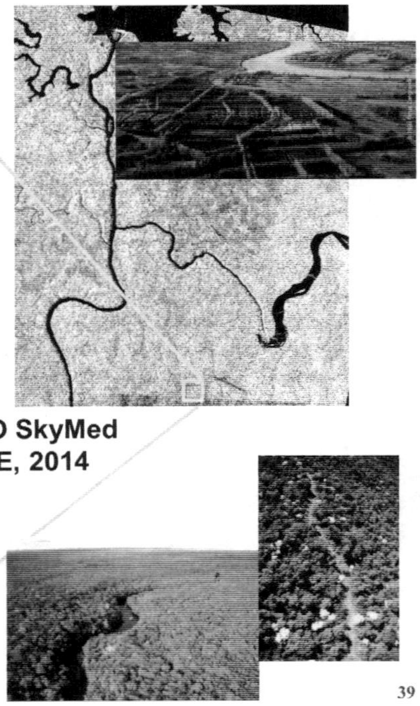

39

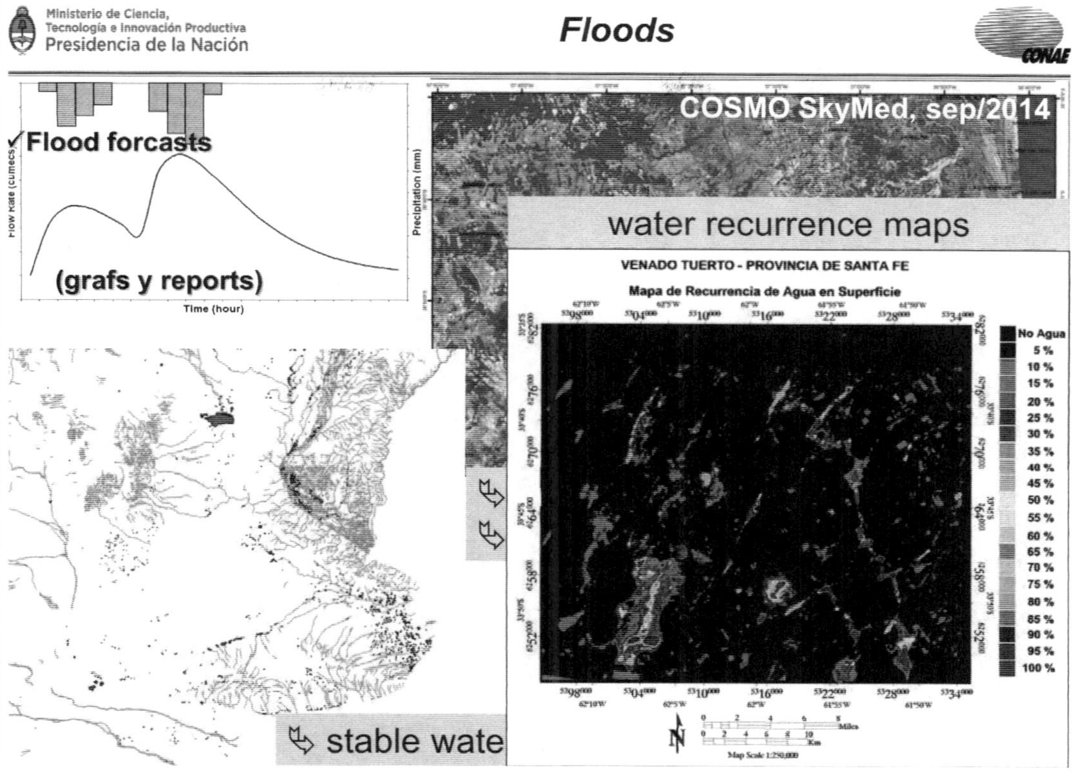
Ministerio de Ciencia, Tecnología e Innovación Productiva
Presidencia de la Nación
Floods
CONAE
✓Flood forcasts
(grafs y reports)
Flow Rate (cumecs)
Precipitation (mm)
Time (hour)
COSMO SkyMed, sep/2014
water recurrence maps
VENADO TUERTO - PROVINCIA DE SANTA FE
Mapa de Recurrencia de Agua en Superficie
No Agua
5 %
10 %
15 %
20 %
25 %
30 %
35 %
40 %
45 %
50 %
55 %
60 %
65 %
70 %
75 %
80 %
85 %
90 %
95 %
100 %
stable wate

Ministerio de Ciencia, Tecnología e Innovación Productiva
Presidencia de la Nación
Wetlands Monitoring
CONAE
ALOS PALSAR ScanSAR
(SAR-L band)

42

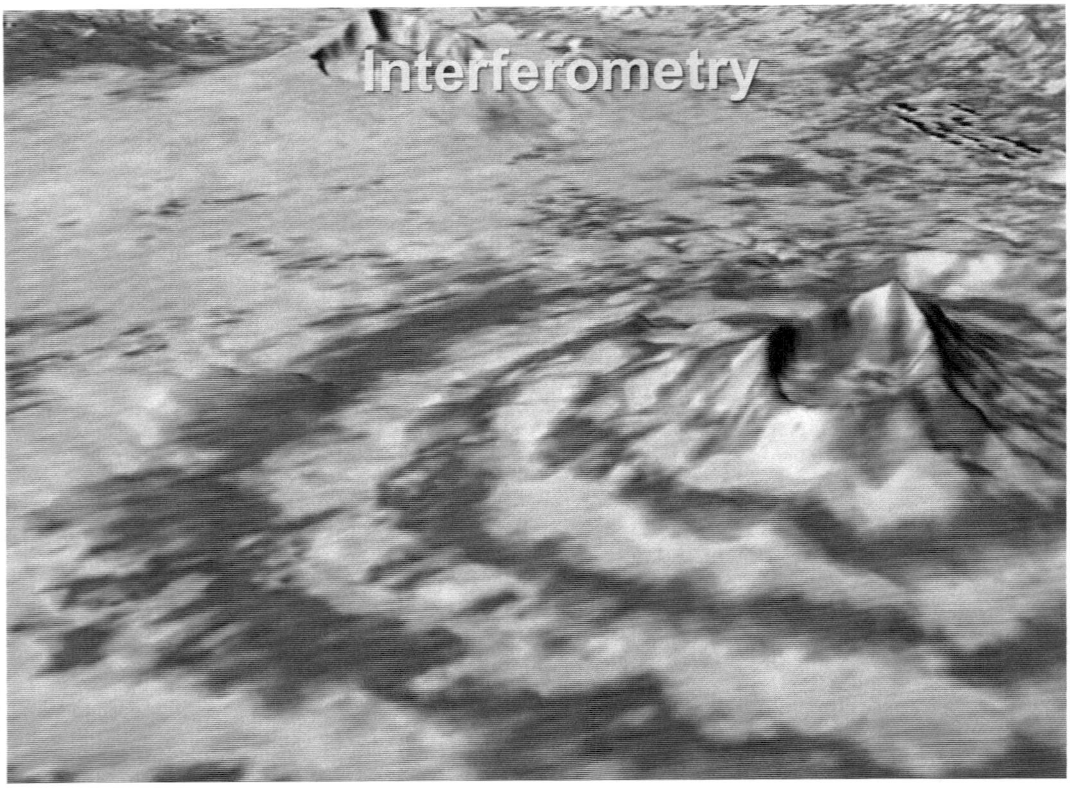

Ministerio de Ciencia,
Tecnología e Innovación Productiva
Presidencia de la Nación

Interferometric Products

- ✓ pair of corregistered images
- ✓ corregistered inrterferometric pair
- ✓ Interferometric phase image
- ✓ interferometric coherence image
- ✓ subsidences maps
- ✓ DEMs

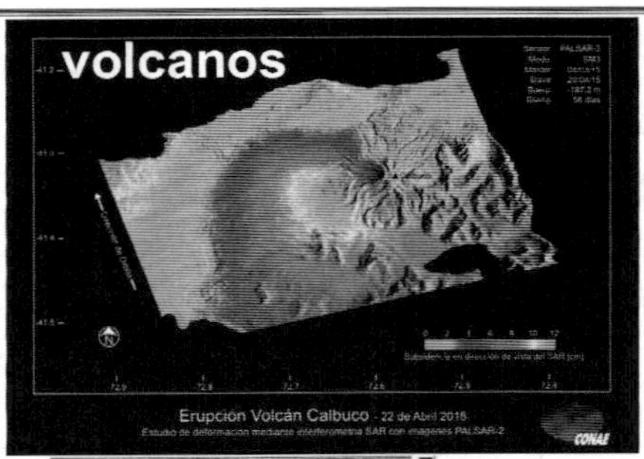

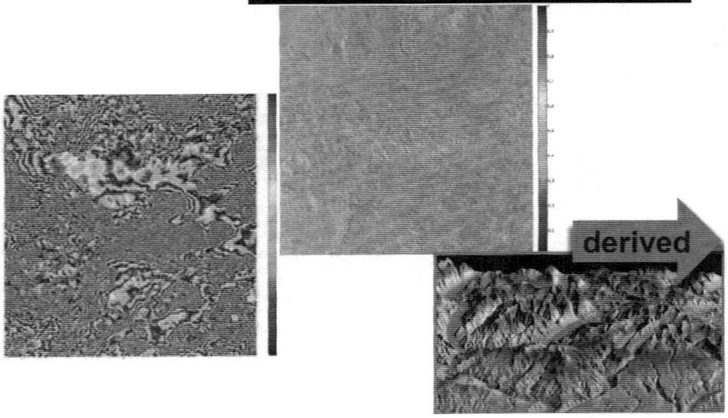

Ministerio de Ciencia, Tecnología e Innovación Productiva
Presidencia de la Nación

Slope Maps

USGS DEM SRTM 30 m)

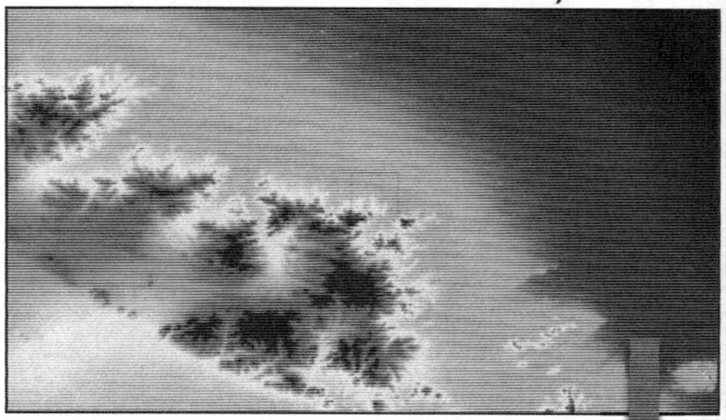

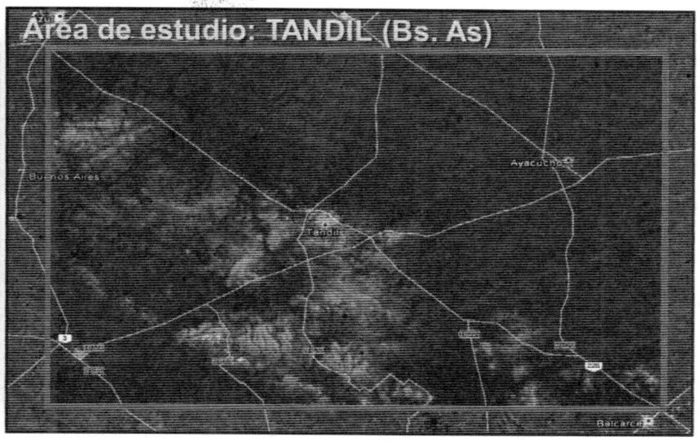

Slope map

Slopes Statistics [%]			
Min	Max	Mean	Stdev
0.00	141.023	5.10	4.34

Slopes Statistics (angles [°])			
Min	Max	Mean	Stdev
0.00	54.70	2.91	2.42

45

Ministerio de Ciencia, Tecnología e Innovación Productiva
Presidencia de la Nación

Thank you

lfrulla@conae.gov.ar

46

PAN AMERICAN BIOFUELS AND BIOENERGY SUSTAINABILITY

AN NSF RESEARCH COORDINATION NETWORK

The Impact of Future Climate and Socioeconomic Changes on Irrigated Sugarcane Production and on the Willingness-to-pay (WTP) for water in Brazil

INNOVATE

Márcia M.G.Alcoforado de Moraes[1], Ana Cristina G. Carneiro[1], Gerald N. S. da Silva[1], Anne Biewald[2], Hermann Lotze-Campen[2] , Alexander Popp[2]

[1]Federal University of Pernambuco, Brazil

[2]Potsdam Institute for Climate Impact Research – PIK

2nd RCN Conference

Buenos Aires, September 13-16th, 2016

CNPq

Conselho Nacional de Desenvolvimento Científico e Tecnológico

Universidade Federal de Pernambuco

PIK

Outline

- Study Area
- Rationale
- Methods
- Results
- Discussion

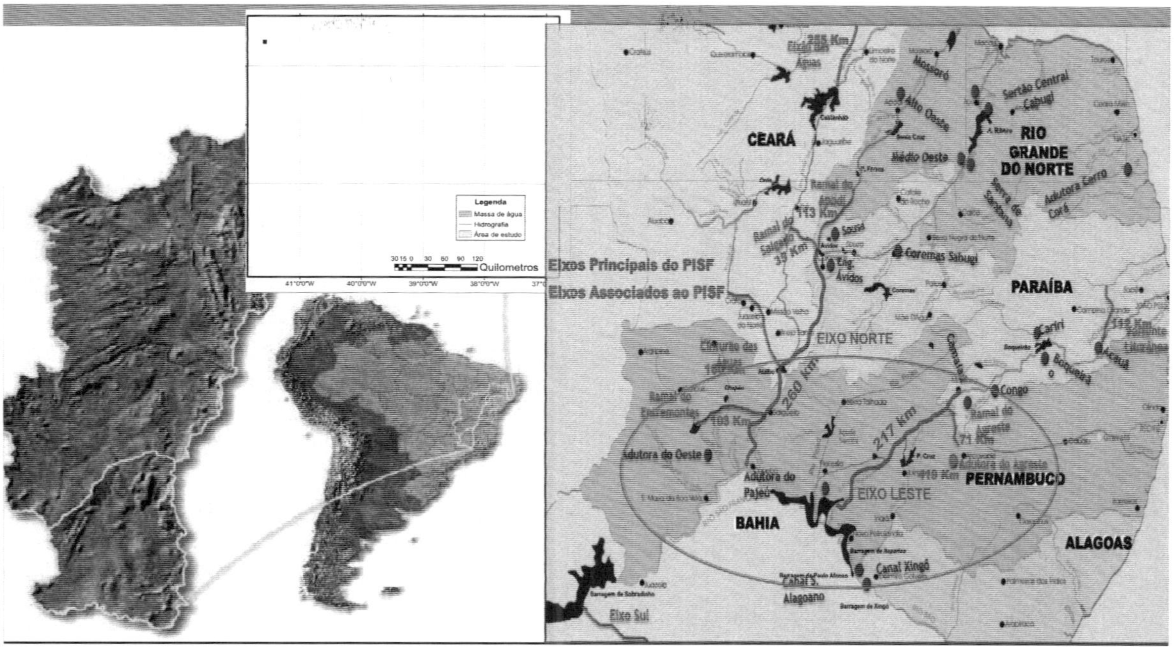

Rationale

A global agro-economic land- and water use model (MagPIE) was regionalized for SFRB

→ Results of the MagPIE regionalized version show a significant increase in the production of irrigated sugarcane in the basin in the coming decades (under A1 and A2 SRES scenarios)

→ The increase in production of irrigated sugarcane in the basin and expansion of the sugar and alcohol sector have been associated with positive socio-economic impacts(Herrera-Martinez et al, 2013)

→ It also brings concerns about water scarcity and economic value of it , and the possible intensification of conflicts between uses in the basin

Sugarcane Irrigated Production in NE Brazil

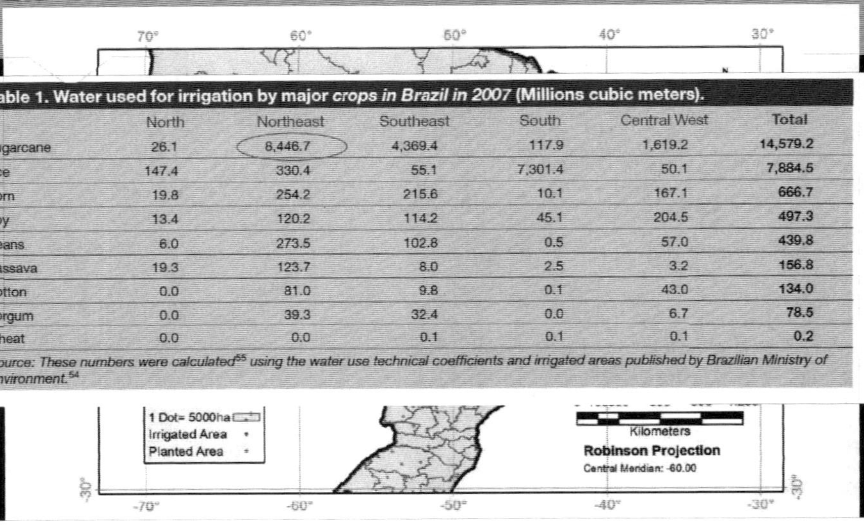

Table 1. Water used for irrigation by major crops in Brazil in 2007 (Millions cubic meters).

	North	Northeast	Southeast	South	Central West	Total
sugarcane	26.1	8,446.7	4,369.4	117.9	1,619.2	**14,579.2**
rice	147.4	330.4	55.1	7,301.4	50.1	**7,884.5**
corn	19.8	254.2	215.6	10.1	167.1	**666.7**
soy	13.4	120.2	114.2	45.1	204.5	**497.3**
beans	6.0	273.5	102.8	0.5	57.0	**439.8**
cassava	19.3	123.7	8.0	2.5	3.2	**156.8**
cotton	0.0	81.0	9.8	0.1	43.0	**134.0**
sorgum	0.0	39.3	32.4	0.0	6.7	**78.5**
wheat	0.0	0.0	0.1	0.1	0.1	**0.2**

Source: These numbers were calculated[55] using the water use technical coefficients and irrigated areas published by Brazilian Ministry of Environment.[54]

Methods: Modelling land use in the SFRB with the Model of Agricultural Production and its Impact on the Environment (MagPIE)

MagPIE was used in order to project crop-specific agricultural land use patterns for the SFRB into the future;

It was taken into account the impact of global socio-economic changes such as population growth, trade liberalization and changes in overall dietary patterns;

On the biophysical side, the model is linked to the grid-based dynamic vegetation model LPJmL which simulates crop yields depending on climatic conditions on a 0.5 degree resolution;

The objective function of MAgPIE minimizes global costs, which involves among others production costs for the agricultural commodities.

Methods:
Modelling land use in the SFRB with the Model of Agricultural Production and its Impact on the Environment (MagPIE)

Parameters	With CC	Without CC	Data source
Population (in mill. people)	299 in Brazil 900 in Latin America fast increasing		GLUES (data)
GDP per capita in US$/a	9680 Latin America (poor)		GLUES (data)
Livestock share in the overall diet	0.22		
Climate	with CC	w/o CC constant values 1995	
Kcal per capita and day	3129		Own calculations based on Bodirsky et al. 2015
Livestock share in the overall diet	0.22		
Water availability in km^3	134	167	GLUES (LPJmL, average of 3 GCMS)

Methods:
Positive Mathematical Programming (PMP) for estimating Willingness-to-pay for water of the main agricultural Users at Sub-Middle SFRB

This valuation technique proposed by Howitt (1995) uses Mathematical Programming as a modeling method for the water demand for irrigation.

The method is based on the behaviour of the agricultural producers that are always trying to maximize their profits using limited resources.

Simulating less availability of water becomes possible identify their willingness-to-pay for that resource.

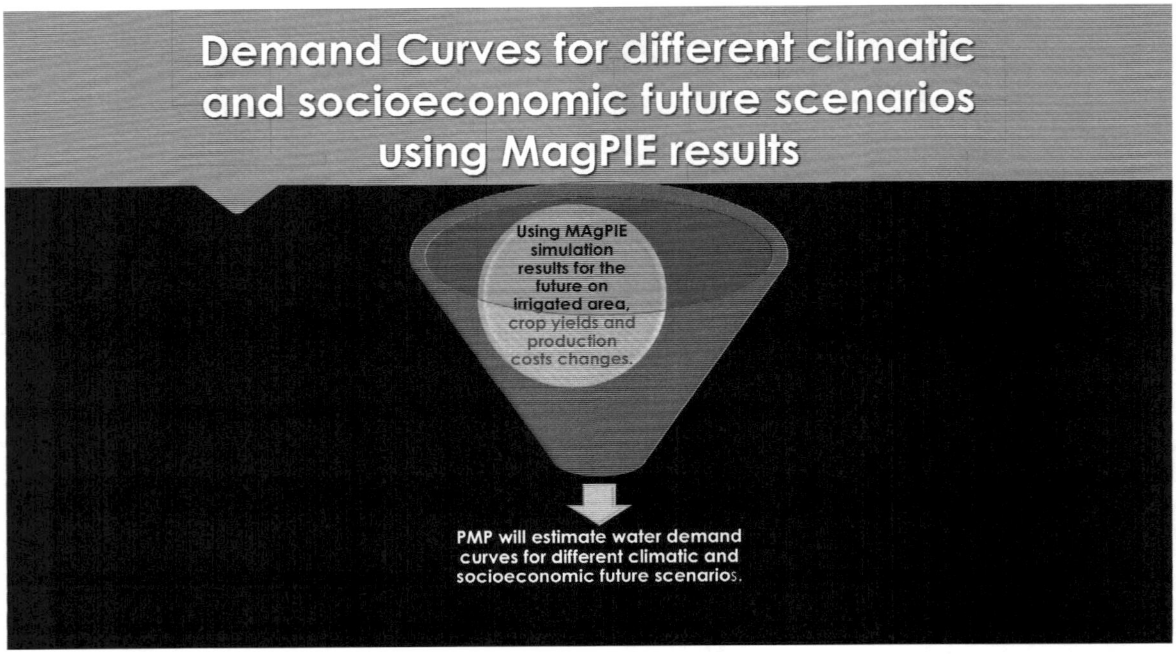
Demand Curves for different climatic and socioeconomic future scenarios using MagPIE results
Using MAgPIE simulation results for the future on irrigated area, crop yields and production costs changes.
PMP will estimate water demand curves for different climatic and socioeconomic future scenarios.

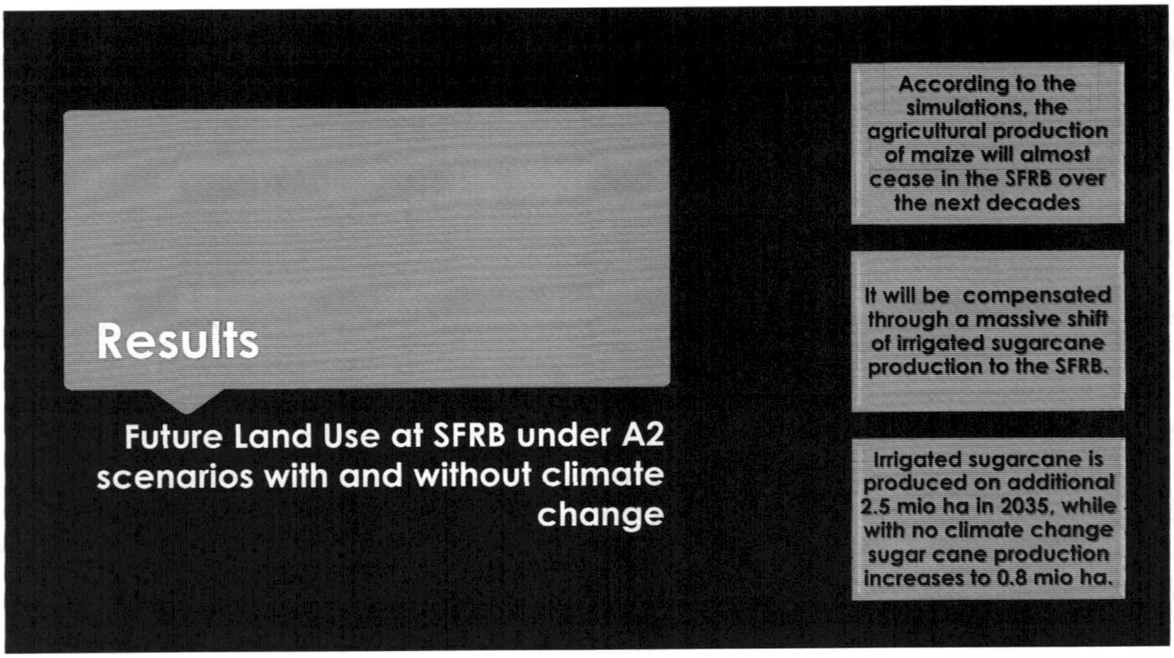
Results
Future Land Use at SFRB under A2 scenarios with and without climate change
According to the simulations, the agricultural production of maize will almost cease in the SFRB over the next decades
It will be compensated through a massive shift of irrigated sugarcane production to the SFRB.
Irrigated sugarcane is produced on additional 2.5 mio ha in 2035, while with no climate change sugar cane production increases to 0.8 mio ha.

Results

Distribution of sugar cane areas simulated by MagPie under A2 scenarios at SM-SFRB and its PIS

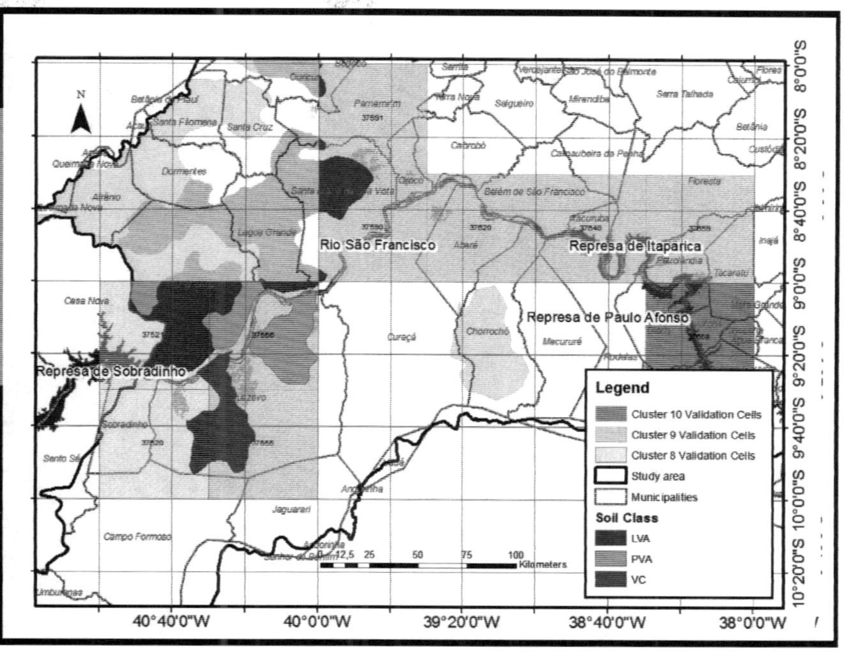

Distribution of sugar cane areas simulated by MagPIE under A2 scenarios among Municipalities at SM-SFRB

Municipality	A2_cc (hectares)	A2_nocc(hectares)
Casa Nova	2,058	299
Juazeiro	40,219	26,297
Pedro Alexandre	12,049	9,590
Santa Brígida	2,872	2,286
Afrânio	1,042	498
Dormentes	520	438
Lagoa Grande	1,508	1,270
Ouricuri	28	24
Petrolina	6,123	3,989
Santa Maria da Boa Vista	942	793
Canindé de São Francisco	2,818	2,243
Poço Redondo	3,679	2,928

Distribution of sugar cane areas simulated by MagPIE under A2 scenarios among Public Irrigation Schemes (PIS) at [illegible]-SFRB

Municipality	Public Irrigation Scheme(PIS)	Total Area of PIS (ha)	MagPie 2035 sugarcane results by PIS (ha)		Other crops (ha)	
			A2_cc	A2_nocc	A2_cc	A2_nocc
Casa Nova	Cruz das Almas	28000	335	49	27665	27951
Casa Nova	Sertão Pernambucano	37070	443	64	36627	37006
Petrolina	Terra Nova	30000	689	449	29311	29551
Petrolina	Pontal Sobradinho	27930	642	418	27288	27512
Petrolina	Pontal	7885	1820	1186	6065	6699
Petrolina	NiloCoelho	20053	461	300	19592	19753
Petrolina	Bebedouro	2418	558	364	1860	2054
Juazeiro	Serra da Batateira	3700	109	71	3591	3629
Juazeiro	Salitre	31305	9349	6113	21956	25192
Juazeiro	Mandacaru	419	219	143	200	276
Juazeiro	Maniçoba	4293	2240	1465	2053	2828
Juazeiro	Tourão	14695	14695	14695	0	0
Juazeiro	Curaçá	4350	2270	1484	2080	2866
Santa Maria da Boa Vist	Brejo de Santa Maria	3800	176	149	3624	3651
Santa Maria da Boa Vist	Caraibas	5223	243	204	4980	5019
Orocó	Brigida	1435	0	0	1435	1435
Curaçá	Pedra Branca	2442	0	0	2442	2442
Belém de São Francisco	Manga de baixo	93	0	0	93	93
Rodelas	Rodelas	1664	0	0	1664	1664
Petrolândia	Apolonio Sales	808	0	0	808	808
Petrolândia	Barreiras I e II	1154.5	0	0	1155	1155
Petrolândia	Ico-Mandantes	2280	0	0	2280	2280
Glória	Gloria	384	0	0	384	384
Glória	Dois Irmãos	55000	0	0	55000	55000
Glória	Paulo Afonso	7000	0	0	7000	7000

Results

Demand Curves under A2 scenarios

With these estimates of irrigated sugarcane and fruit areas for each of the PIS , there is need for updating the other production factors (water, labor and supplies production factors)

For that we use regional data in sugarcane and fruits from Agricultural Census (2006) and update them with IPCC and MagPIE previewed changes under A2 scenarios;

As MagPie not includes market in it, we didn´t have prices for the agricultural products in A2 scenarios.

Regional data (2006) was updated for 2035 using MagPIE and IPCC results by crop and region

MagPIE and IPCC results

- Sugarcane Irrigated area
  - MagPIE 2035 A2_cc and A2_nocc
- Supplies and labor requirements
  - MagPIE Crop yield changes under A2 scenarios
- Water use
  - IPCC temperatures and precipitation rate changes under A2 scenarios
- Variable Costs of Production factors and Prices
  - MagPie production factor cost changes under A2 scenarios

Preliminary Results

Obtaining production factors amounts associated with new land use configuration for future scenarios estimated by MagPIE to 2035

CULTURA	PERIMETROS	LAND (HA)	WATER (MIL M³)	LABOR (TRABALHADOR)	SUPPLY (MIL R$)
can	P1	331	5641	67	545
can	P2	438	7469	89	722
can	P3	95	2459	10	80
can	P4	88	2289	9	75
can	P5	1670	43363	175	1413
can	P6	63	1644	7	54
can	P7	512	13298	54	433
can	P8	100	2601	20	165
can	P9	8573	222906	1750	14132
can	P10	200	5213	41	330
can	P11	2054	53408	419	3386
can	P12	23314	606150	4758	38429
can	P13	2081	54117	425	3431
can	P14	157	4082	16	133
can	P15	216	5611	23	183

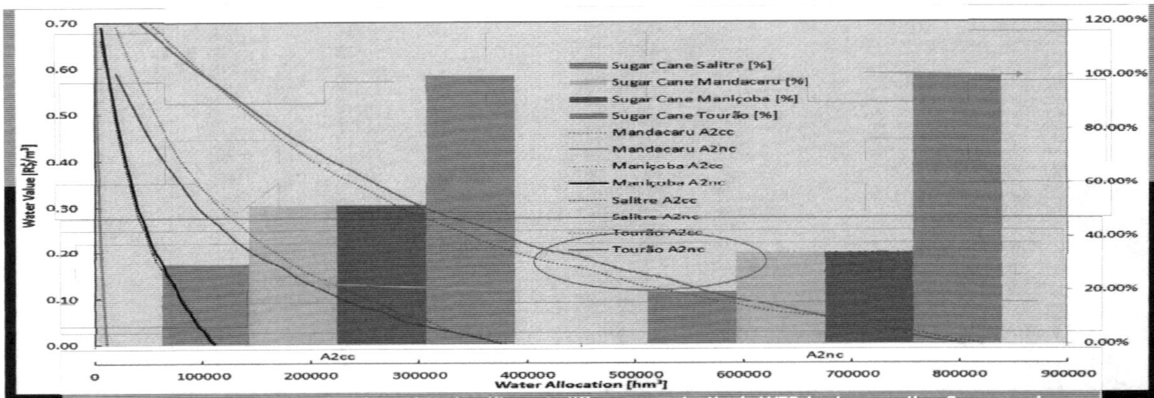

There were two PIS in Juazeiro with significant differences in their WTP between the 2 scenarios.

Tourão with just sugarcane crops under the two scenarios had higher WTP under climate change scenario . Salitre with fruits and sugarcane (even with an increase in sugarcane areas w cc) has lower WTP under climate change.

Preliminary Results:
Comparing demand curves under A2 scenario with climate change and the baseline

For all PIS , the demand curves indicate higher economic value for all quantities of water in the future given climate change related to the baseline.

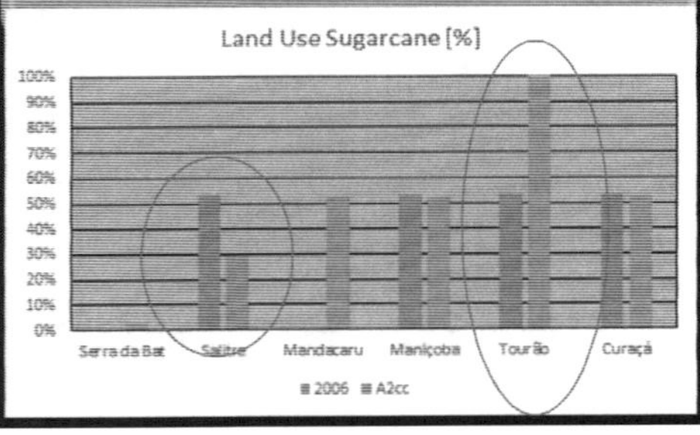

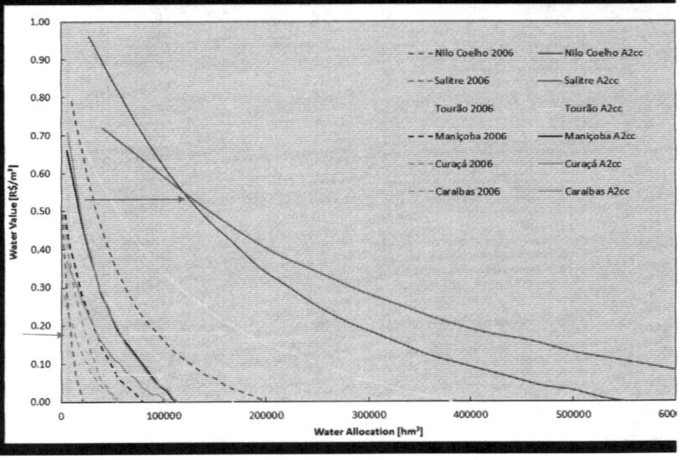

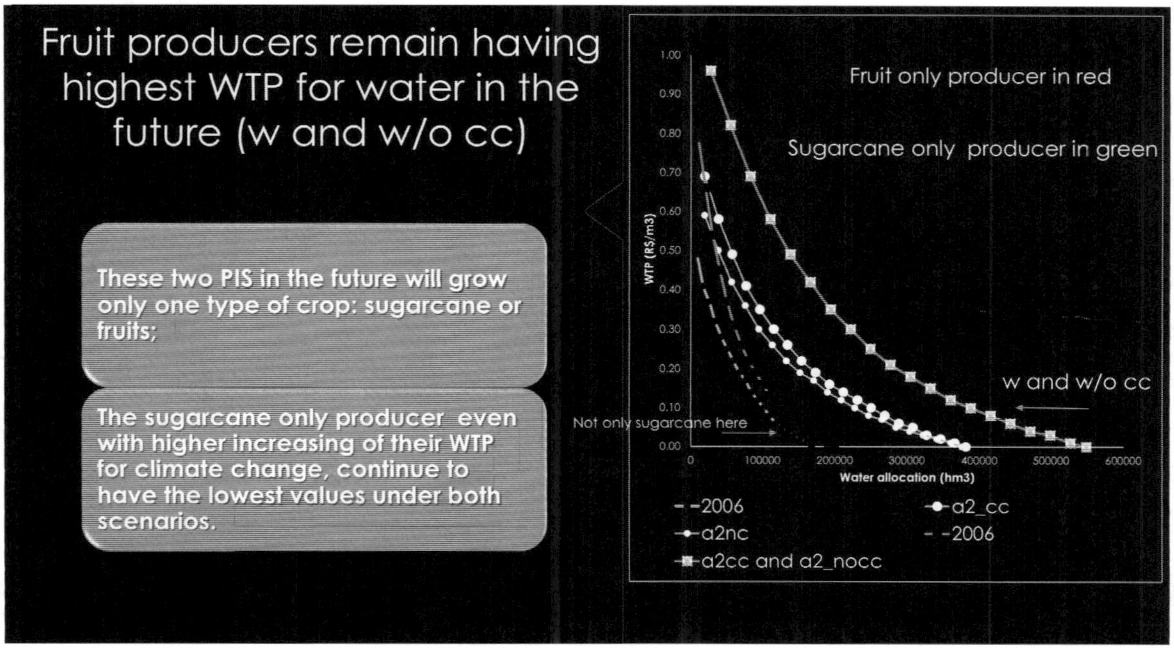
Fruit producers remain having highest WTP for water in the future (w and w/o cc)
These two PIS in the future will grow only one type of crop: sugarcane or fruits;
The sugarcane only producer even with higher increasing of their WTP for climate change, continue to have the lowest values under both scenarios.
Fruit only producer in red
Sugarcane only producer in green
w and w/o cc
Not only sugarcane here
WTP (R$/m3)
Water allocation (hm3)
- -2006
a2nc
a2cc and a2_nocc
a2_cc
- -2006

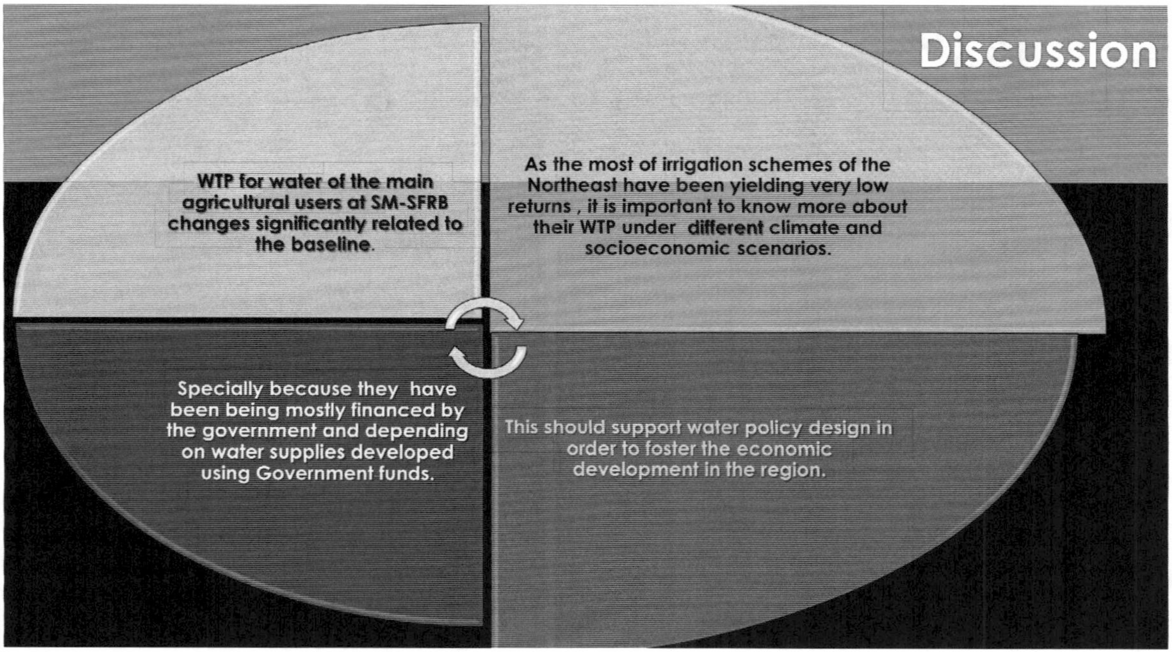
Discussion
WTP for water of the main agricultural users at SM-SFRB changes significantly related to the baseline.
As the most of irrigation schemes of the Northeast have been yielding very low returns , it is important to know more about their WTP under different climate and socioeconomic scenarios.
Specially because they have been being mostly financed by the government and depending on water supplies developed using Government funds.
This should support water policy design in order to foster the economic development in the region.

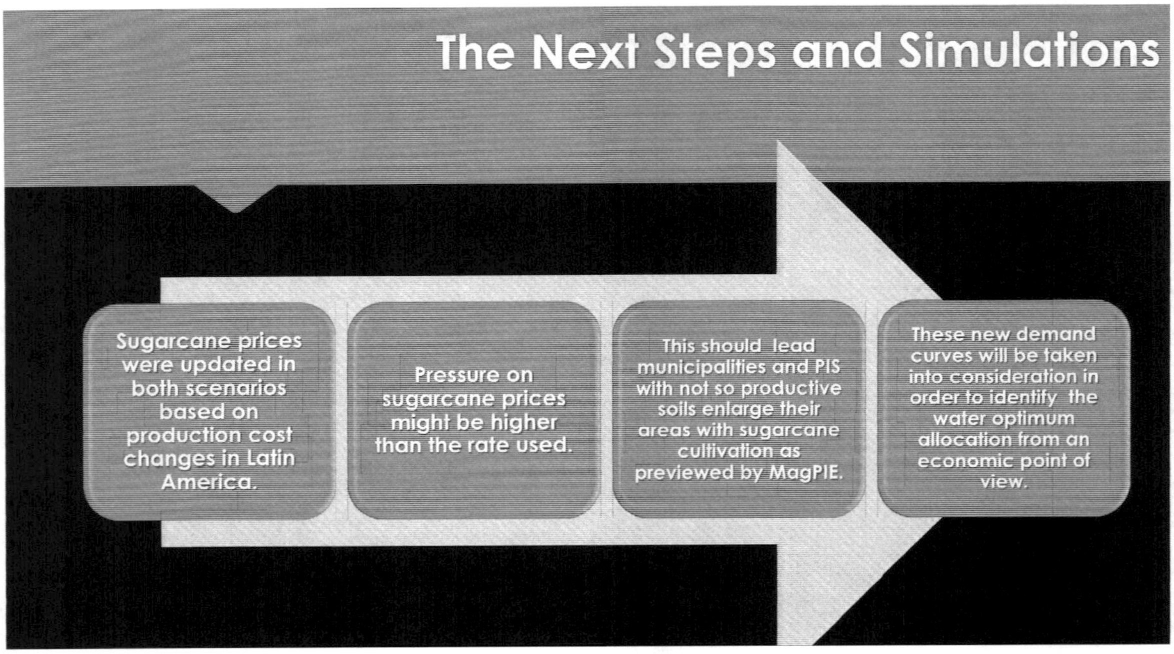

Thanks!

Márcia Ma. G. Alcoforado de Moraes

Departamento de Economia/ PIMES/ PPGEC

Universidade Federal de Pernambuco (UFPE)

marcia.alcoforado@gmail.com

PAN AMERICAN BIOFUELS AND BIOENERGY SUSTAINABILITY
AN NSF RESEARCH COORDINATION NETWORK

Decision Support for the management of water resources in the Sub-middle São Francisco river basin using integrated hydro-economic modeling and scenarios for land use change

Souza da Silva, G.N & Alcoforado de Moraes, M.M.G.

Content

1. Introduction
2. Framework
3. Methodology
4. Results

Introduction

Hydro-economic models can measure the economic effects of different operating rules, environmental restrictions, ecosystems services, technical constraints and institutional constraints.

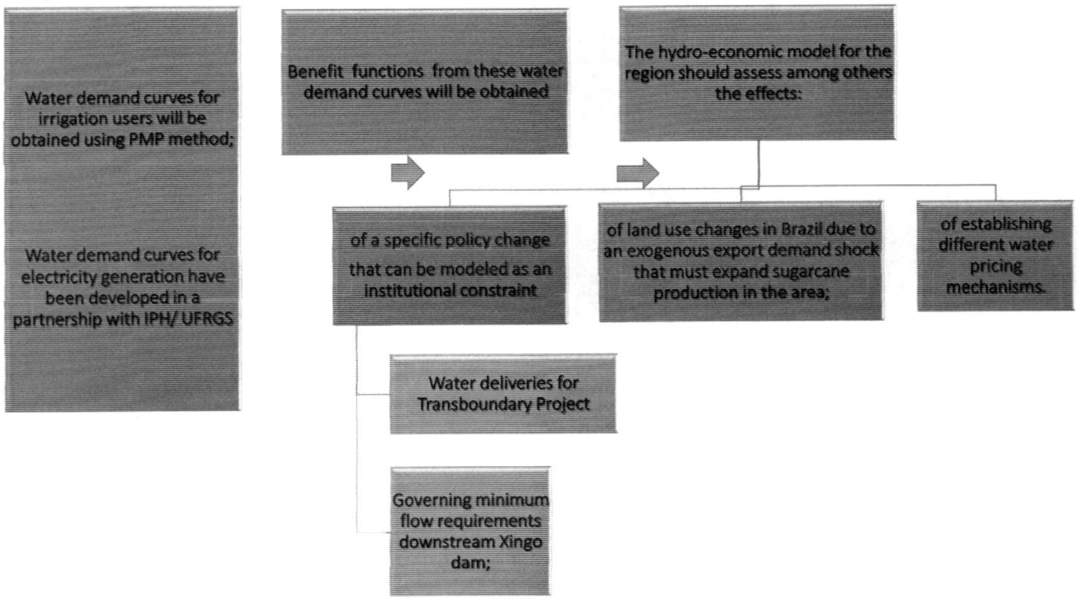

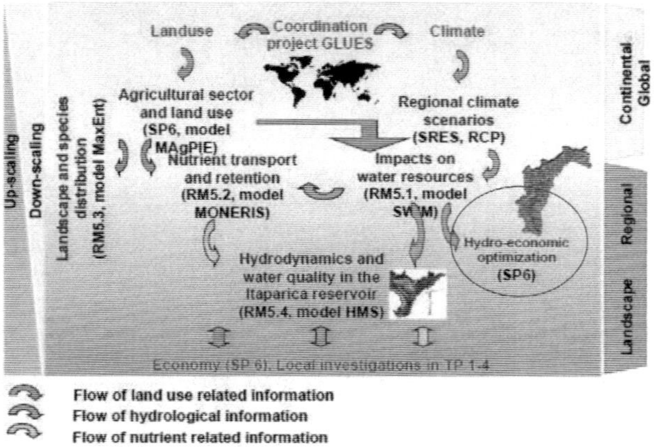

Framework INNOVATE

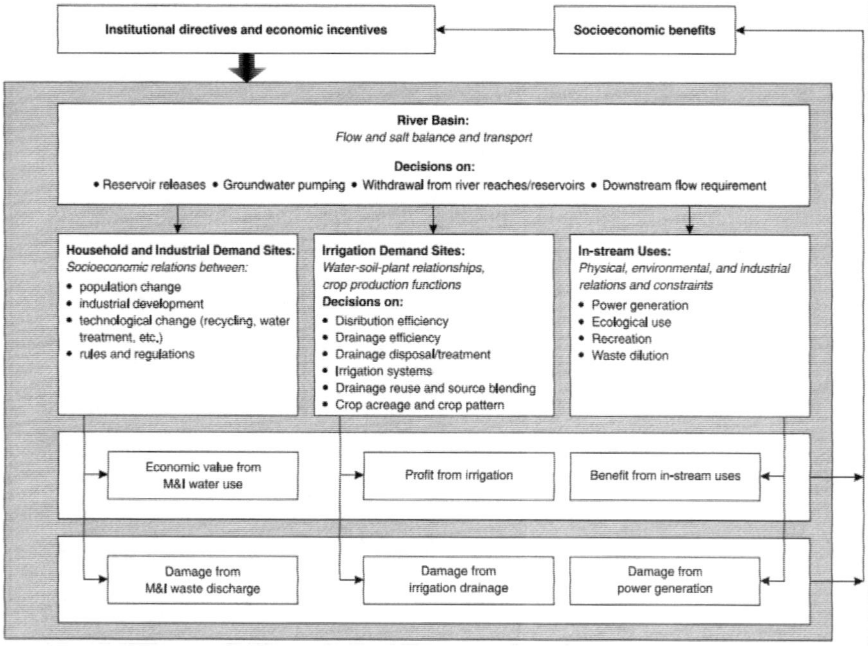

Methodology

Conceptual model

IWROM

Integrated water resources optimization models

McKinney et al (1999)

Methodology: Demand curves

The model uses demand curves for the irrigation projects, small farmers and human supply, rather than fixed requirements for water resources.

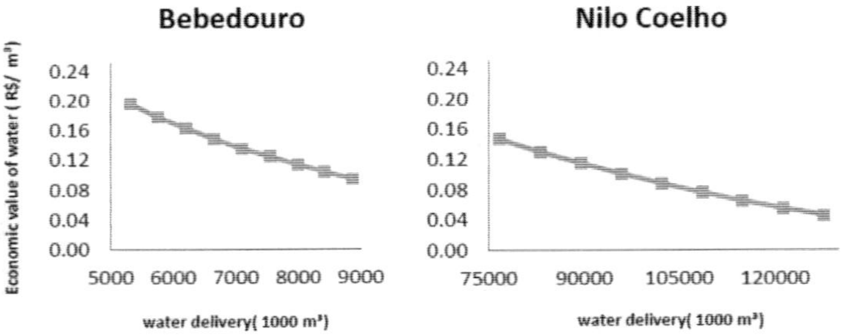

*Figura obtida a partir dos resultados do modelo aplicado no GAMS™.

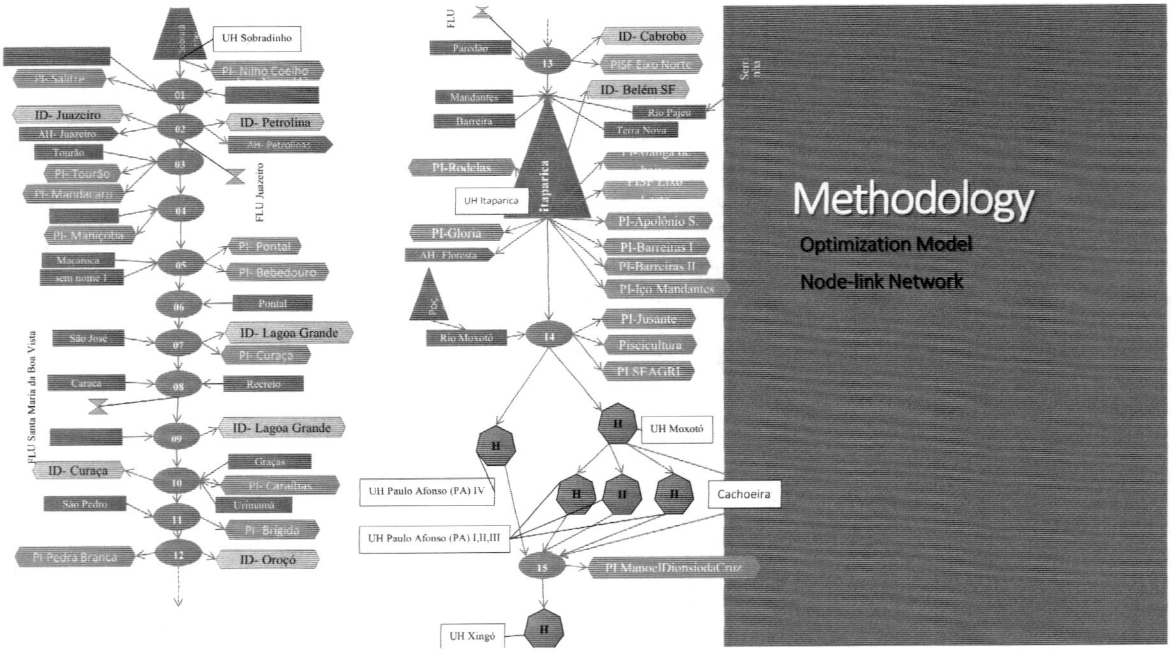

Methodology: Optimization Model

Water allocation optimization for various water uses, OF:

$$Minimizar\begin{pmatrix} \sum_{n\in pi}\sum_{a}\sum_{m} C_{\ n,a,m} + \sum_{n\in id}\sum_{a}\sum_{m} C_{\ n,a,m} + \sum_{n\in ah}\sum_{a}\sum_{m} C_{\ n,a,m} \\ -\sum_{n\in ener}\sum_{a}\sum_{m} B_{\ n,a,m} + \sum_{n\in ener}\sum_{a}\sum_{m}(E_{\ assegurada(n)} - E_{n,a,m})^2 \end{pmatrix}$$

Where: *n User: pi (irrigated area), id (diffuse irrigation), ah (human supply), ener (energy production)*
C shortage cost [R $]
B Benefit user [R $]
E Energy [MW]
a, m Year, Month

Hydro Economic modelling
Sub-Middle SFRB in NE BRAZIL

The objective function minimizes the water's scarcity cost or maximizes the net benefits to main users, subject to physical and legal constraints in the system

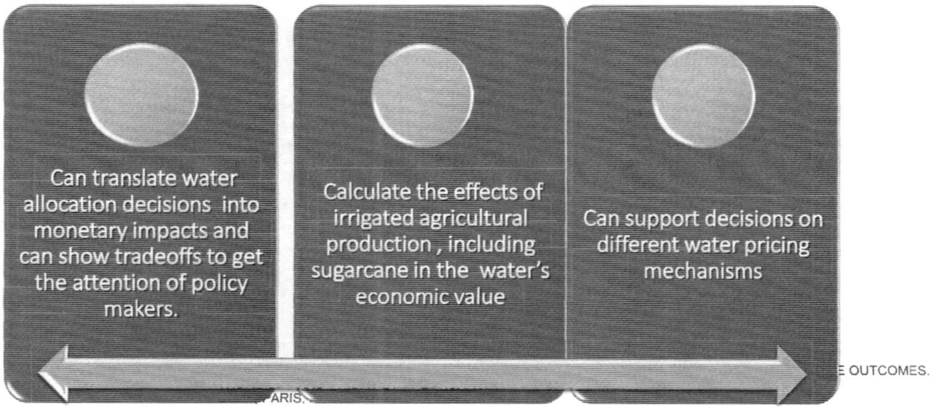

Results

Demand curves for the current and future scenarios

Positive Mathematical Programming (PMP)

- ✓Public irrigation schemes
- ✓diffuse irrigation

Point Expansion Method

- ✓diffuse irrigation
- ✓human consumption

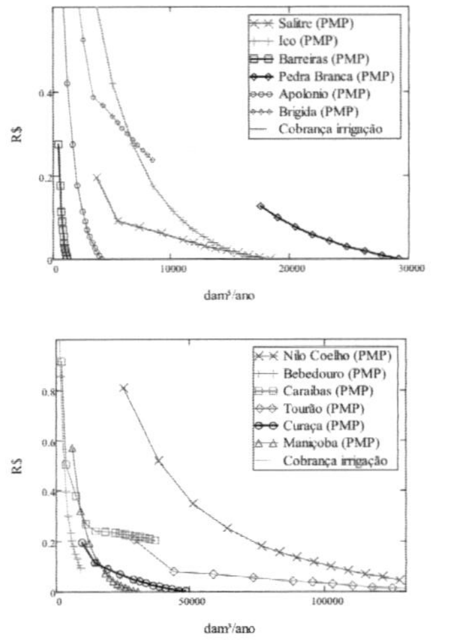

Current Results

Scenario:

Reference model (baseline) - present, years 2000-2006

Constraints

Technical restrictions (dam, powerhouse, spillway etc.)

Flood protection

Release Sobradinho > 1300m³/s

Release Xingo > 1300m³/s

Objective Function

Minimizing: scarcity costs from Irrigation Sites, Small Farmers and Human Supply, difference between e energy production and firm energy

Maximizing: benefits energy production

Simulation 1: Baseline w/o priority for domestic supply

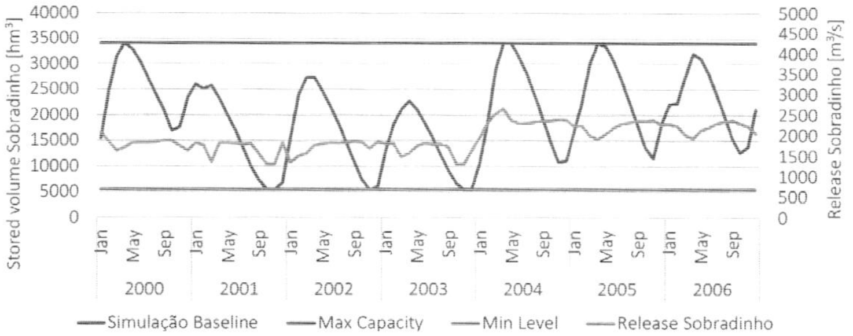

Results for Sobradinho Reservoir Storage - 2000-2006

Simulation 1: Baseline w/o priority for domestic supply

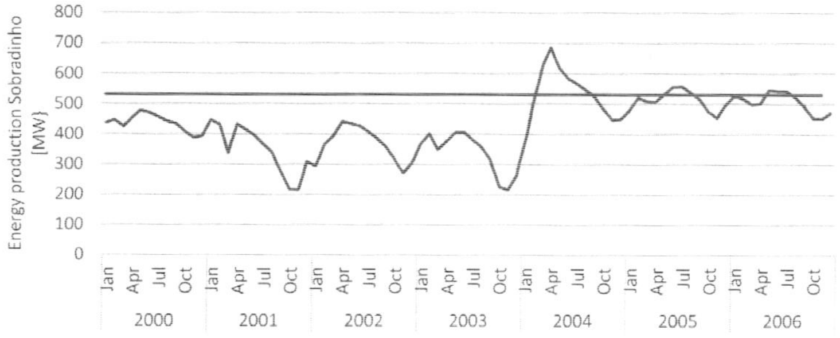

Energy Production and Firm Energy Sobradinho

Simulation 1: Baseline w/o priority for domestic supply

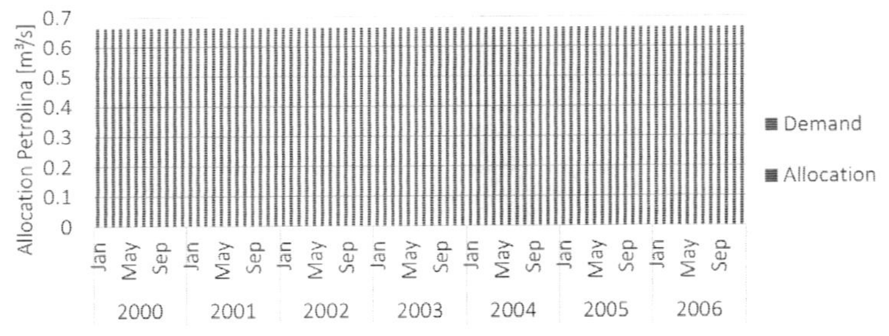

Allocation for the supply of the municipality Petrolina

Simulation domestic use w/o priority and w/o losses

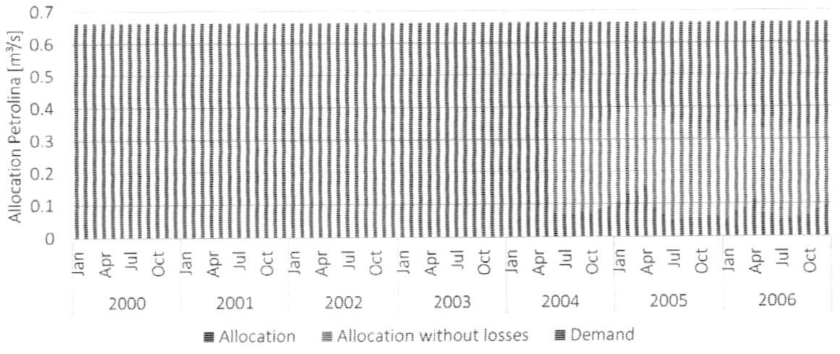

Water allocation human supply with and without distribution losses

Simulation 1: Baseline w/o priority for domestic supply

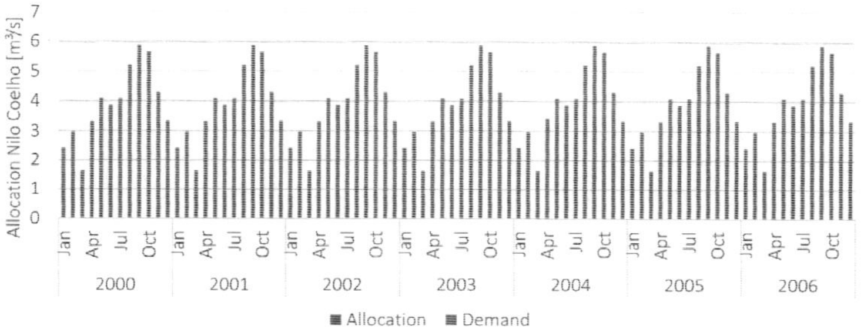

Water Allocation for Public Irrigation Scheme (Nilo Coelho) 2000-2006

Simulation 1: Baseline w/o priority for domestic supply

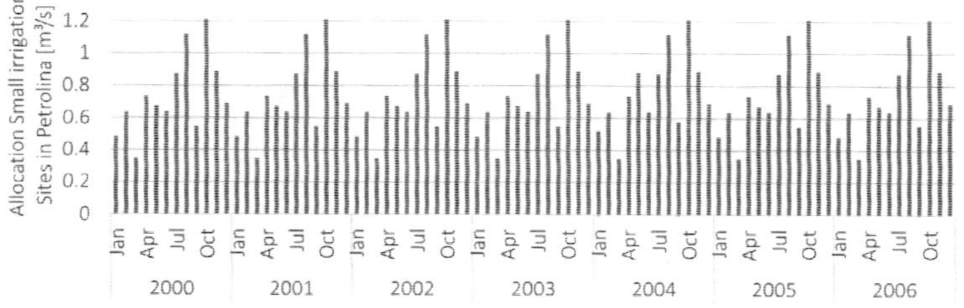

Groups of Small farmers along the river in Petrolina

Current Results

Simulation 2: *Baseline with environmental flow constraint*

Reference model (baseline) - present, years 2000-2006, Human supply without losses

Constraints

Technical restrictions (dam, powerhouse, spillway etc.)

Flood protection

Release Sobradinho > 1300m³/s

Release Xingo > 1300m³/s

Hidrograma Ambiental

month [m³/s	Jan	Feb	Mar	Apr	May	Jun	Jul	Aug	Sep	Oct	Nov	Dec
normal year	2754.00	3150.00	3097.00	2685.00	1727.00	1588.00	1448.00	1300.00	1300.00	1300.00	1647.00	2234.00
dry year	2020.00	2300.00	2100.00	1837.00	1271.00	1218.00	1100.00	1100.00	1100.00	1100.00	1331.00	1740.00

Objective Function

Minimizing: scarcity costs from Irrigation Sites, Small Farmers and Human Supply, difference between e energy production and firm energy

Maximizing: benefits energy production

Simulation: "Environmental Hidrograph"

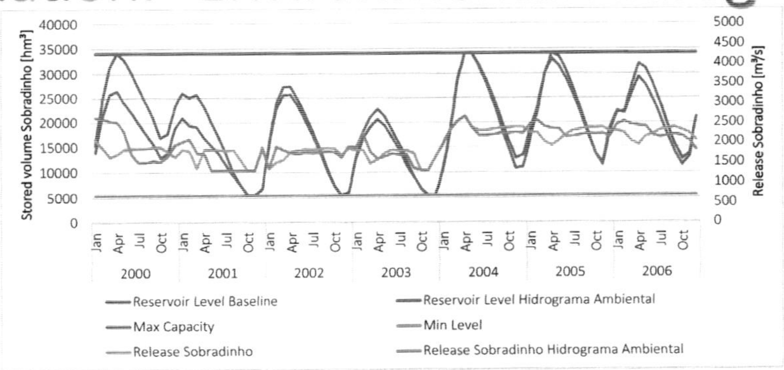

Storage and Release with "Hidrograma Ambiental" Sobradinho

Simulation: "Environmental Hidrograph"

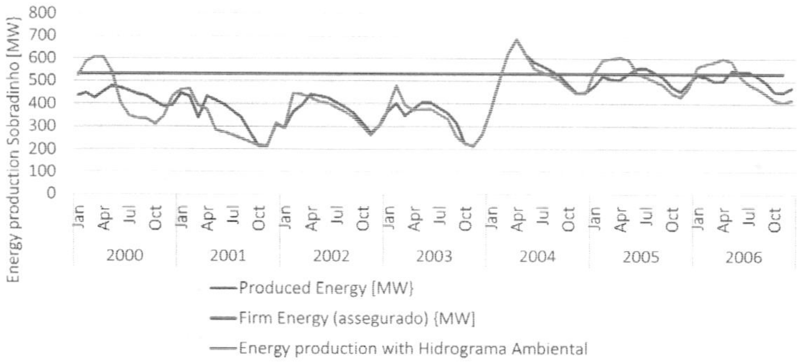

Storage and Release with "Hidrograma Ambiental" Sobradinho

Main Current Results

Costs/Benefits	Prioridade abast.	Transposição SF	Human Supply without losses (allocation)	Human Supply without losses 100% allocation	Hidrograma Ambiental	Level control	Reduced capacity*	Sobradino 1100 m³/s
Scarcity costs *human supply*			-11,71%					
Scarcity costs *small farmes*	2,86%	6,91%	1,94%	2,74%	39,89%	46,91%	58,78%	2,60%
Scarcity costs *Irrigation projetcts*	0,43%	8,64%	1,40%	2,73%	40,02%	41,91%	63,52%	1,96%
Benefits *Sobr./Itaparica*	-0,04%	-0,85%	0,01%	0,00%	-1,34%	-6,73%	-7,79%	-0,01%
Benefits *Moxoto, PA*	-0,05%	-0,98%	0,01%	-0,01%	-1,30%	-6,07%	-8,98%	-0,02%
Shadow Price *Defl. Sobradinho*	-34,73%	7,37%	-35,45%	-34,84%	765,14%	1432,76%	2499,27%	-87,56%

The Next Steps and Simulations

Irrigated Sugarcane Production and the Willingness-to-pay (WTP) for water of the agricultural users in SM-SFRB might change under scenarios of Climate and Socioeconomic Changes

These new demand curves will be taken into consideration in order to identify the water optimum allocation from an economic point of view in future scenarios.

This should support water policy design that foster the economic development in the region, considering the influence of global forces on the use of local land and water.

PAN AMERICAN BIOFUELS AND BIOENERGY SUSTAINABILITY
AN NSF RESEARCH COORDINATION NETWORK

UNIVERSIDADE FEDERAL DE PERNAMBUCO

CNPq
Conselho Nacional de Desenvolvimento Científico e Tecnológico

INNOVATE

Thanks!

Gerald Souza da Silva
Federal University of Pernambuco (UFPE)
geraldsouzadasilva@gmail.com
Márcia Maria Guedes Alcoforado de Moraes
Federal University of Pernambuco (UFPE) - Department of Economics
marcia.alcoforado.ma@gmail.com

24

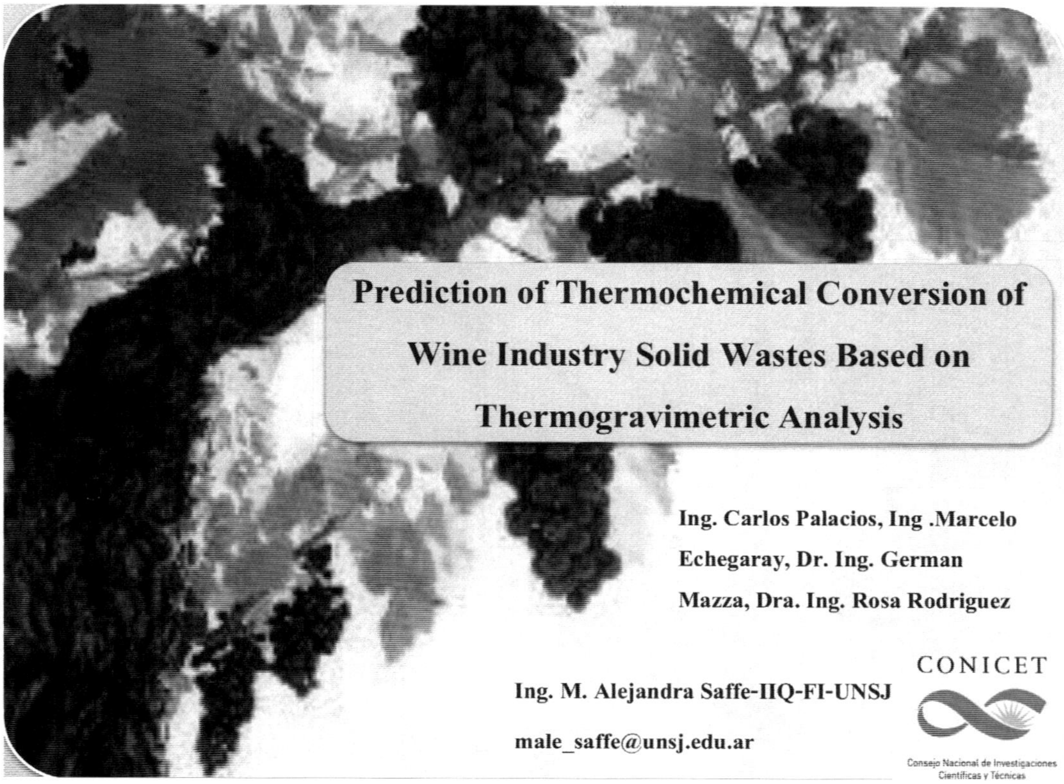

Fundaments

- Scarcity of fossil fuels
- Lower emissions of greenhouse gases
- Exploitation of renewable energy sources

  " Waste biomass "
- The agro-industrial produces a significant environmental impact due to waste produced

Agro-industrial wastes

Solid state material in disuse for the process that generated it

- **Biofuels are clean energy sources**
- **Low nitrogen and sulfur contents**
- **High oxygen content and volatile matter content**
- **HHV**

Marc  + ***Stalk***  = **37800tons/ year only in Region of Cuyo**

2nd RCN Conference on Pan American Biofuels & Bioenergy Sustainability

Thermal process

PYROLYSIS
GASIFICATION
COMBUSTION

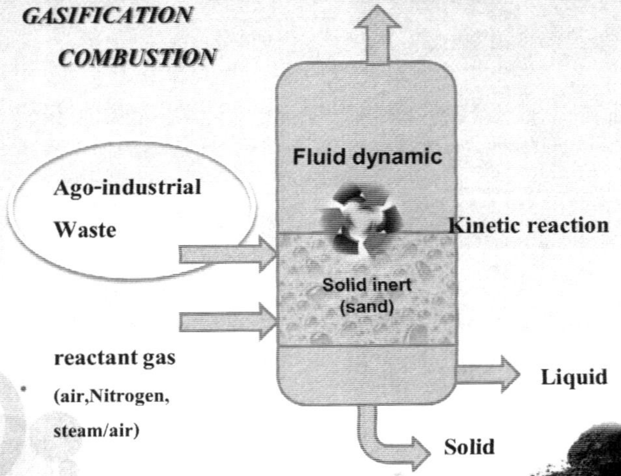

2nd RCN Conference on Pan American Biofuels & Bioenergy Sustainability

Characterize the agro-industrial wastes

Characterization by Standard Methods

	Marc	Stalk
Volatile Matter (tw%)	68.6	55.84
Ash (d.b %)	8.81	10.16
Fixed Carbon (d.b %)	21.98	23.07
Moisture (tw %)	8.38	7.7
C (d.b %)	52.91	46.14
H (d.b %)	5.93	5.74
O (d.b %)	30.41	37.54
Hemicellulose (%)	4.96	5.78
Cellulose (%)	15.31	16.02
Lignin (%)	37.97	30.79

2nd RCN Conference on Pan American Biofuels & Bioenergy Sustainability

Thermogravimetric analysis

This technique can be the simplest and most effective way to observe the profile of the thermal decomposition of a solid fuel. TGA is based on measuring the change in the sample mass, in the time, when subjected to a temperature program in a controlled atmosphere

2nd RCN Conference on Pan American Biofuels & Bioenergy Sustainability

Characterize the agro-industrial wastes

Characterization by thermogravimetric analysis (TGA)

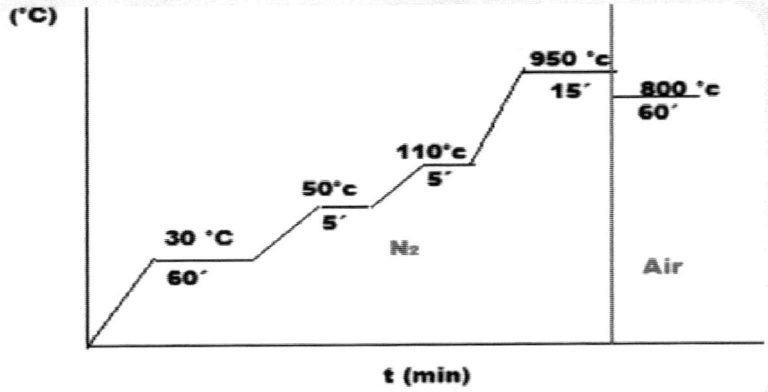

2nd RCN Conference on Pan American Biofuels & Bioenergy Sustainability

Proximate analysis

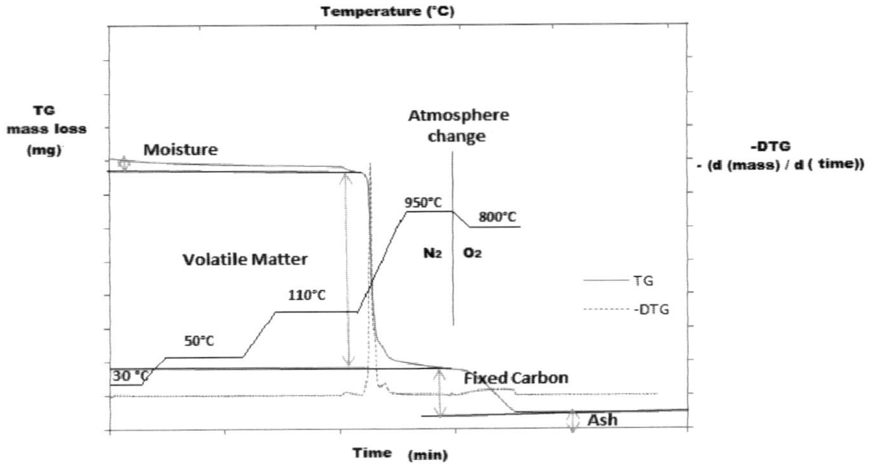

2nd RCN Conference on Pan American Biofuels & Bioenergy Sustainability

Ultimate analysis and HHV

$$C = 0{,}635\ FC + 0{,}460\ VM - 0{,}095\ ASH$$
$$H = 0{,}059\ FC + 0{,}060\ VM + 0{,}010\ ASH$$
$$O = 0{,}340\ FC + 0{,}469\ VM - 0{,}023\ ASH$$
$$HHV = 0{,}3536\ FC + 0{,}1559\ VM - 0{,}0078\ ASH$$

Parikh J, Channiwala SA, Ghosal GK. A correlation for calculating HHV from proximate analysis of solid fuels. Fuel 2005;84:487–94.

Parikh J, Channiwala SA, Ghosal GK. A correlation for calculating elemental composition from proximate analysis of biomass materials. Fuel 2007;86: 1710–9.

Lignin, cellulose and hemicellulose contents

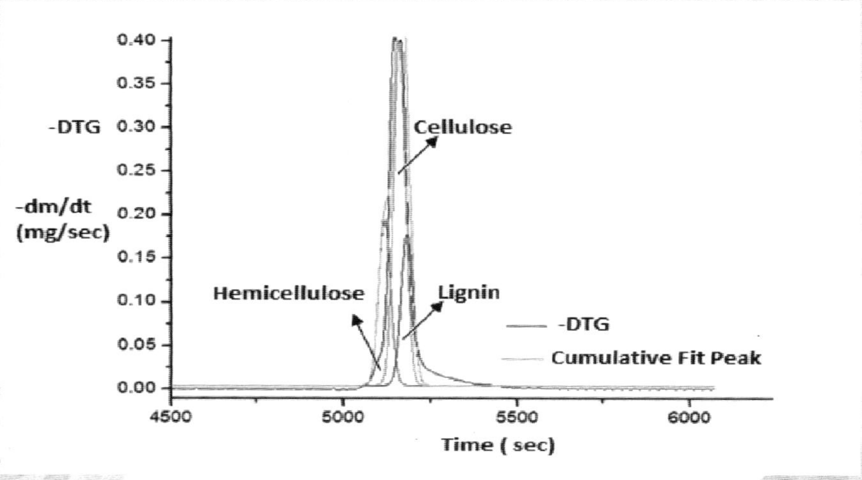

Lignocellulosic contents

- The DTG curves deconvolution were carried out using the Origin 8.6 software

$$\frac{-dm}{dt} = \frac{-dm_o}{dt} + \frac{A}{w\sqrt{\pi/4Ln(2)}} e^{\left[\frac{-4\ln(2)\,(t-tc)^2)}{W^2}\right]}$$

- This integration was carried out using Mathcad 1.5 software.

- Where dm/dt ,dm_0/dt, t, tc, A and W are the mass variation with the time at t=t and t=0, time, time of maximum dm/dt values and two constant parameters of model

2nd RCN Conference on Pan American Biofuels & Bioenergy Sustainability

Characterize the agro-industrial wastes

Characterization by TGA

	Marc	Stalk
Volatile Matter (tw %)	72.86	65.50
Ash (d.b %)	6.40	9.72
Fixed Carbon (d.b %)	16.85	20.83
Moisture (tw %)	5.07	5.70
C (d.b %)	55.88	52.76
H (d.b %)	5.60	5.44
O (d.b %)	33.85	33.16
HHV (MJ/Kg)	29.72	27.73
Hemicellulose (%)	4.92	6.18
Cellulose (%)	15.20	16.14
Lignin (%)	38.25	28.11

2nd RCN Conference on Pan American Biofuels & Bioenergy Sustainability

RESULTS

- High water content increases the energy requirements to carry out the thermal treatment, rises the residence time for drying and reduces the temperature, resulting in incomplete conversion, decrease the process efficiency.
- Considering the obtained results, *these biomass wastes can be used for energy production from different thermal treatments, without previous drying process.*
- The analyzed wastes showed a high volatile matter content, generally more the 65% .
- High volatile levels make the ignition easier at low temperatures, implying a high reactivity and hence improving the thermal process.

2nd RCN Conference on Pan American Biofuels & Bioenergy Sustainability

RESULTS

- The higher volatile matter/fixed carbon ratio improve the contact of reactants and reduce the residence time for the combustion process, indicating a higher reactivity degree of the biomass. This calculated ratio is 3.12 and 2.42 for marc and stalk t respectively.
- The ash affects the process costs and the environment, by the increase quantity of dust emission and reduced efficiency of thermal process because increasing its content causes the HHV decreases.
- The ash disposal and reuse may also be affected by the composition, and in some cases these wastes can be used as fertilizer or in the concrete industry for the construction.

2nd RCN Conference on Pan American Biofuels & Bioenergy Sustainability

RESULTS

- **The ultimate analysis indicates that studied wastes are environmental friendly, with only trace amounts of nitrogen and sulfur because their values are depreciable.**
- **The HHV are 29.72, 27.73 MJ/kg for marc and stalk respectively.**
- **Therefore, a higher oxygen concentration has a negative impact on the HHV.**
- **The operation variables of determined thermal process should adapt to the variation of the biomas composition for a better performance.**

2nd RCN Conference on Pan American
Biofuels & Bioenergy Sustainability

Conclusions

- ***The TGA is a fast and feasible alternative in order to determine the results of proximate and ultimate analysis and to predict the HHV. This is relevant because the elementary composition is one of the most important properties for the use of biomass as fuel.***
- ***The experimental determination of the ultimate analysis data requires special instrumentation, while the proximate analysis data can be easily obtained using common equipment.***
- ***The standard techniques require long operative time, so thermogravimetric analysis presents a great advantage over it.***
- ***This method of analysis is much simpler, faster and cheaper than the standard procedures for the characterization of lignocellulosic solid wastes used for the bioenergy production.***

2nd RCN Conference on Pan American
Biofuels & Bioenergy Sustainability

Thanks for your attention....!
2nd RCN Conference on Pan American
Biofuels & Bioenergy Sustainability

PAN AMERICAN BIOFUELS AND BIOENERGY SUSTAINABILITY
AN NSF RESEARCH COORDINATION NETWORK

Enzimatic processes development for oil degumming residues utilization

Ma. Laura Pisarello, Carlos Querini, Valentina Rossi, Maira Maquirriain, Melissa Rubel

INCAPE

UNL CONICET

Instituto de Investigaciones en Catálisis y Petroquímica
Santa Fe, Argentina

Problemática

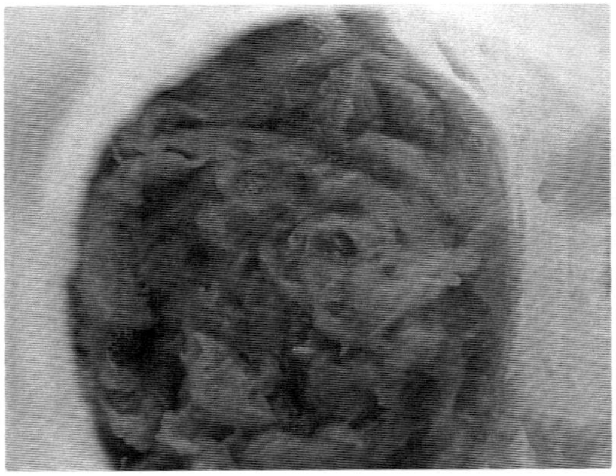

Residuos del desgomado de aceite

Demandado por empresas…

Una planta de gran capacidad genera 100 toneladas diarias

30% Aceite
40% Fosfolípidos
30 % Agua

Usos:

- ✓ Agregado en harinas
  NO ES POSIBLE
- ✓ Producción de lecitina
  SOBREOFERTA
- ✓ Oleínas (ataque ácido)
  problemas ambientales

Se propone...

Residuos del desgomado de aceite

30% Aceite ocluido
40% Fosfolípido } aprovechar la fracción lipídica

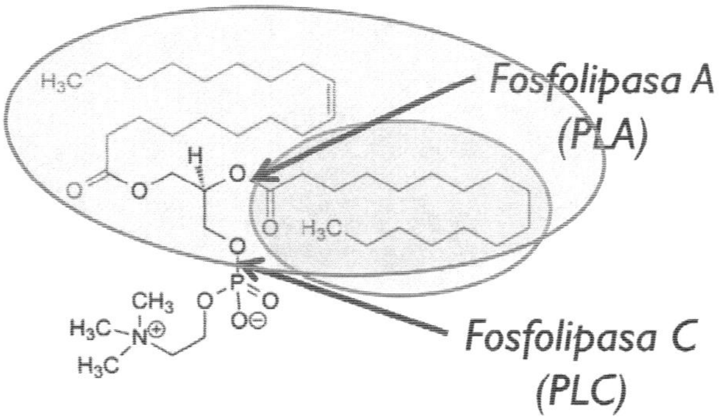

Aceite

Diglicérido

Ácido graso

Se propone...

Residuos del desgomado de aceite

Enzimas PLC/PLA

→ Enzimas comerciales

→ Producción recombinante de enzimas específicas

Aceite / Diglicérido / Ácido graso

Se propone…

Residuos del desgomado de aceite

Enzimas
PLC/PLA

Aceite / Diglicérido / Ácido graso

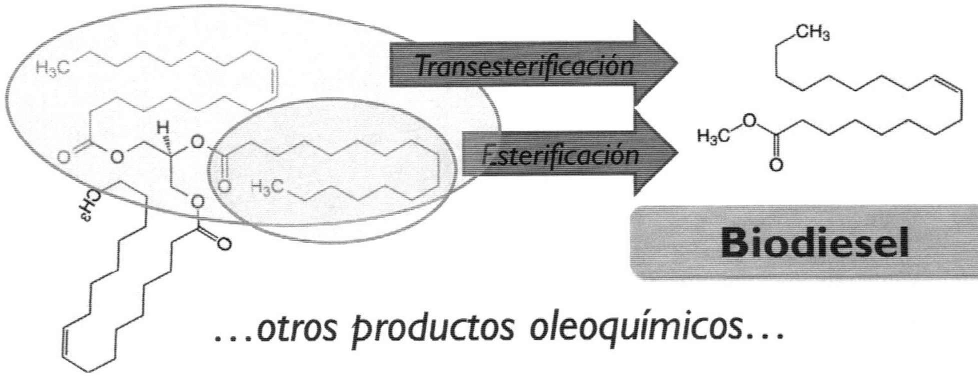

Biodiesel

…otros productos oleoquímicos…

Experiencias con residuo de desgomado

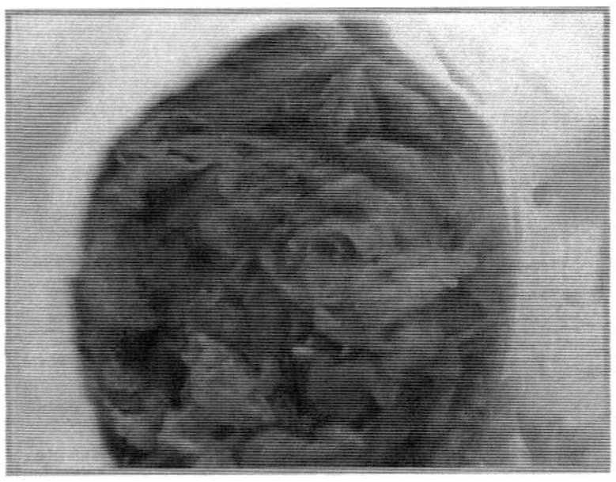

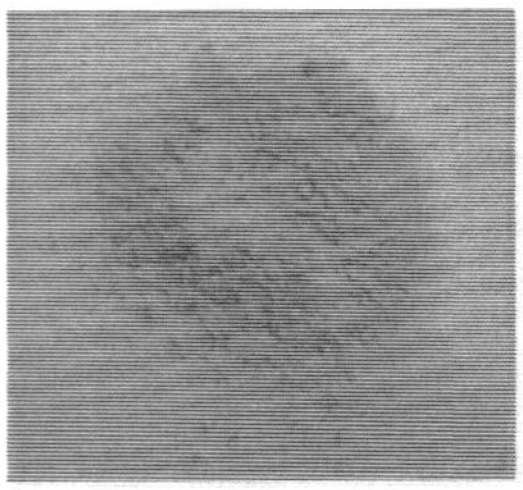

Caracterización

- ✓ Implementación de nuevas técnicas analíticas
- ✓ Verificación con diferentes técnicas analíticas

Fosfolípidos puros

- ✓ Aislación del fosfolípido del residuo de desgomado
- ✓ Materia prima de experiencias patrones

Experiencias con residuo de desgomado

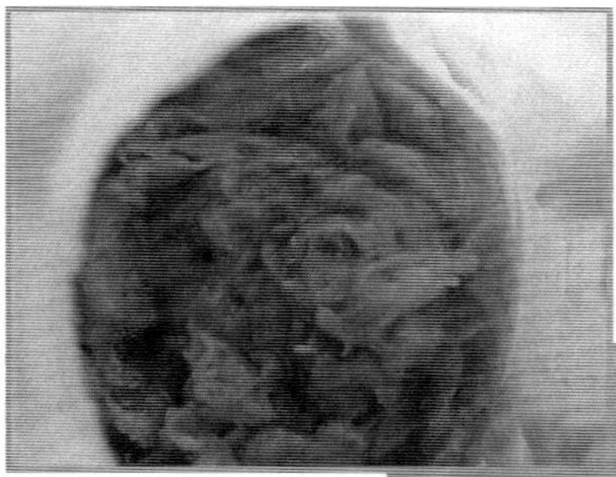

Problemas de mezclado

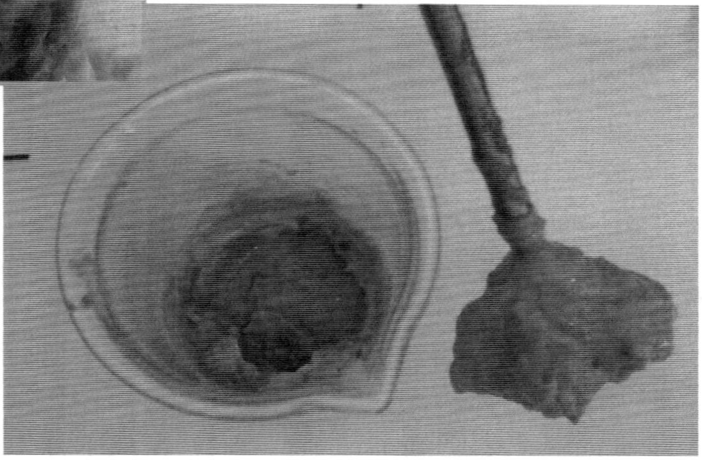

Uso de solventes

Experiencias con catálisis enzimática

- Parámetros de reacción evaluados
  - ✓ Temperaturas: 45, 50 y 55°C — *Insensible*
  - ✓ Tipo de agitación: magnética y mecánica de altas revoluciones — *Buenos resultados*
  - ✓ Materias primas: diferentes muestras de gomas / mezclas con solventes / fosfolípidos aislados — *En función de: Fosfolípidos y agua*
  - ✓ Concentración de la mezcla de enzimas: entre el 0,5% y 4,5% referida a la masa de gomas. — *Concentración óptima: 1,5%*

Experiencias con catálisis enzimática

- Análisis de productos de reacción:
  - ✓ Ácidos grasos: análisis químico
  - ✓ Diglicéridos: Cromatografía gaseosa (GC) y líquida (HPLC)
  - ✓ Fosfolípidos / Fósforo: análisis químicos, espectroscopia infrarroja (FTIR) y fluorescencia de rayos X (FRX)

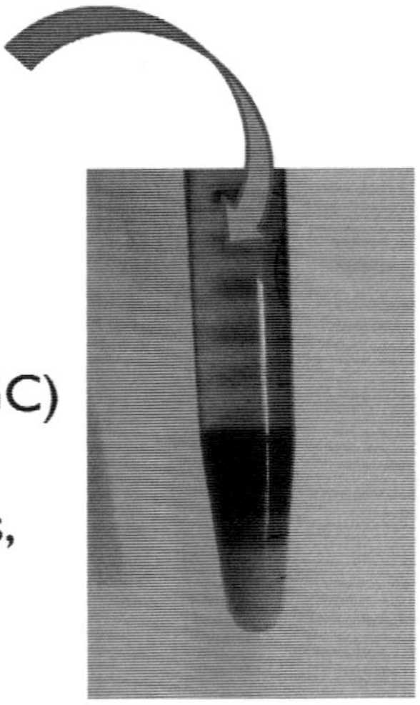

Experiencias con catálisis enzimática

Cromatografía gaseosa

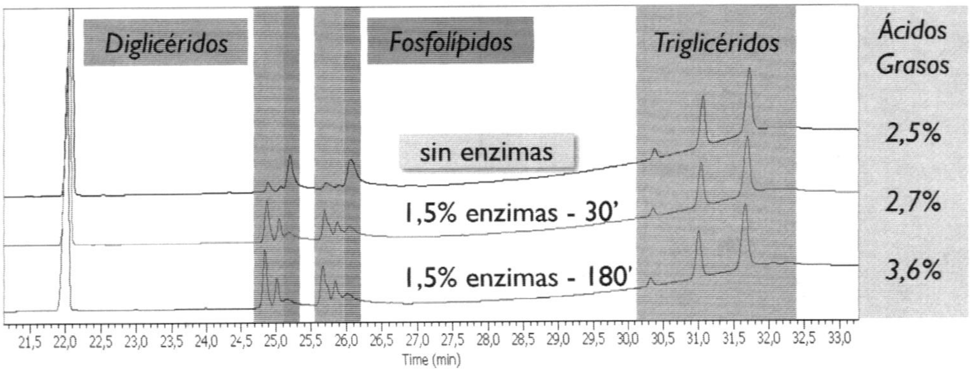

Resultados y conclusiones

- ✓ Se verifica la transformación de fosfolípidos en diglicéridos (GC) y ácidos grasos (químico).
  - PLC: 45% conversión a Diglicéridos (*semicuantitativo*)
  - PLA: 13% conversión a Ácidos grasos
- ✓ Se trabaja en las otras determinaciones analíticas:
  - HPLC: se realizaron curvas de calibrado. Se observa linealidad, pero los valores cambian en el tiempo. Los valores son relativos al día de la determinación.
  - FRX: se realizan curvas de calibrado. Los análisis de las mezclas de reacción difieren de los valores analizados con otras técnicas.

Resultados y conclusiones

- ✓ Se trabaja en las otras determinaciones analíticas:
  - *Material insoluble en acetona*
    es una medida directa de los fosfolípidos

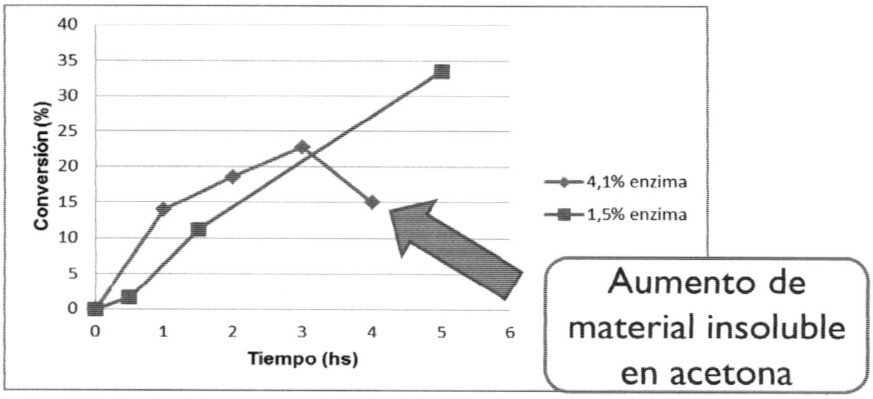

Algunos productos de reacción también son insoluble en acetona

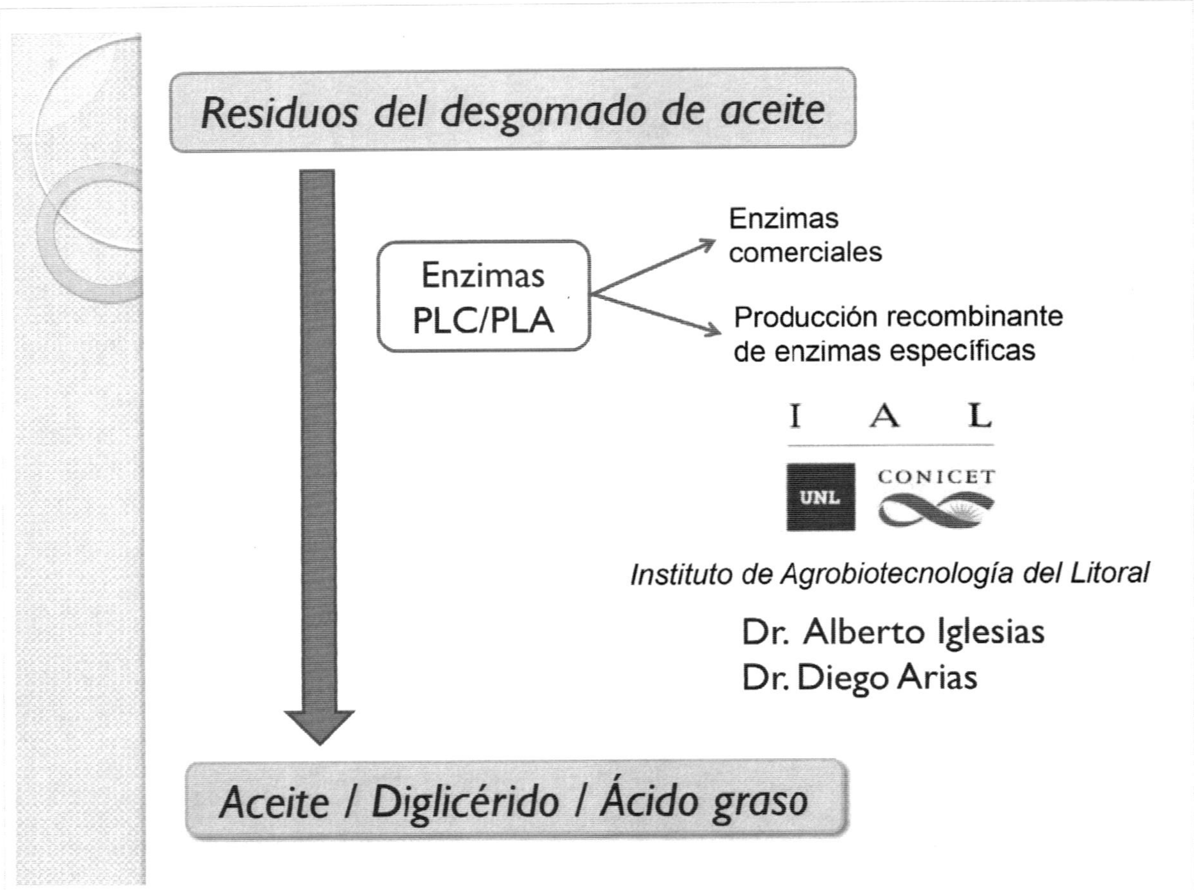

Residuos del desgomado de aceite
Enzimas PLC/PLA
Enzimas comerciales
Producción recombinante de enzimas específicas
I A L
UNL
CONICET
Instituto de Agrobiotecnología del Litoral
Dr. Alberto Iglesias
Dr. Diego Arias
Aceite / Diglicérido / Ácido graso

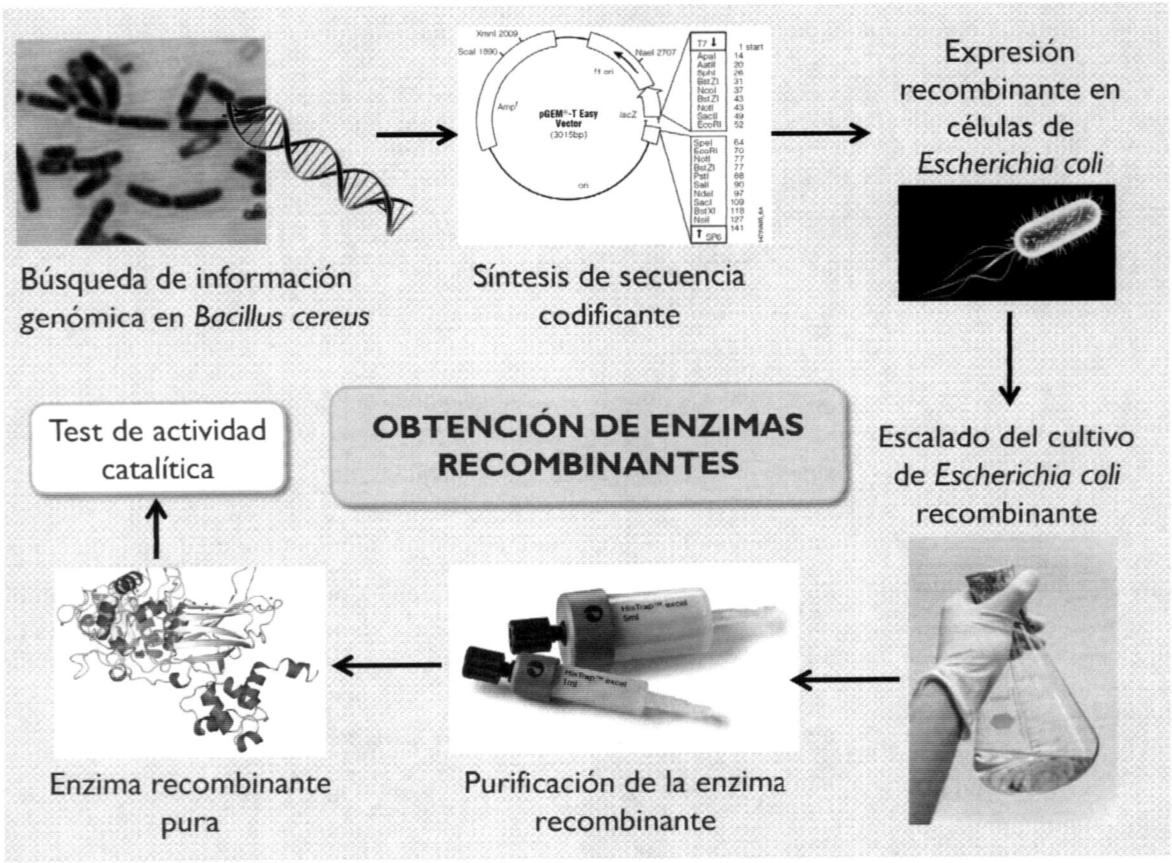

Búsqueda de información genómica en Bacillus cereus
pGEM®-T Easy Vector
Síntesis de secuencia codificante
Expresión recombinante en células de Escherichia coli
Escalado del cultivo de Escherichia coli recombinante
OBTENCIÓN DE ENZIMAS RECOMBINANTES
Test de actividad catalítica
Purificación de la enzima recombinante
Enzima recombinante pura

Test de actividad catalítica

La actividad PLC fue ensayada por seguimiento de la hidrólisis de O-(4-nitrofenilfosforil)colina (NPPC), monitoreando la absorbancia.

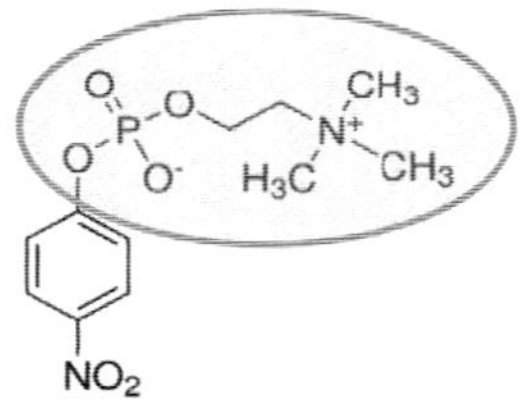

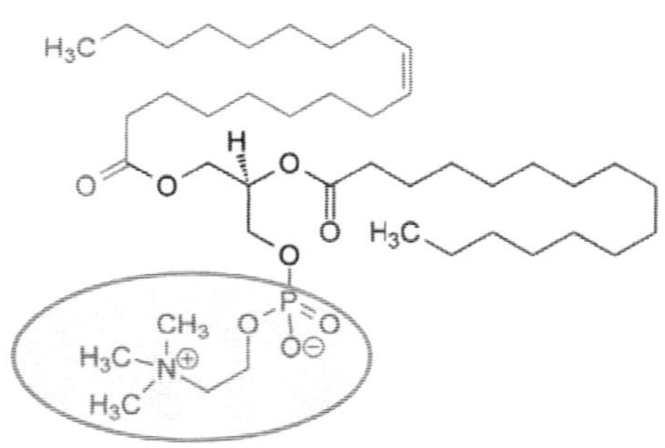

Una unidad de actividad (U) se define como la cantidad de enzima que cataliza la conversión de 1 μmol de sustrato por min bajo las condiciones especificadas.

Test de actividad catalítica

Fosfolipasas C de Bacillus cereus (BcePLC) - dos secuencias aminoacídicas (1, 2)

I A L

UNL CONICET

Enzima	Rendimiento (mg de proteína/500 ml de cultivo)	Actividad enzimática específica (U/mg de proteína)
BcePLC1	**36**	**0,054**
BcePLC2	**50**	**0,63**

Muchas gracias por su atención

mlpisa@fiq.unl.edu.ar

Facultad de
Ciencias Agrarias
UNIVERSIDAD NACIONAL DE ROSARIO
ZAVALLA - SANTA FE
Campo Experimental "Villarino"
Casilla de Correo Nº 14 -
S2125ZAA
Zavalla - Santa Fe – Argentina
www.fcagr.unr.edu.ar –
Pyrolysis of Lignocellulosic Biomass: The Case Study of Spartina argentinensis
Emiliano Jozami, Melisa Bertero, Ulises Sedran, Bárbara M. Civit and Susana R. Feldman
ejozami@unr.edu.ar

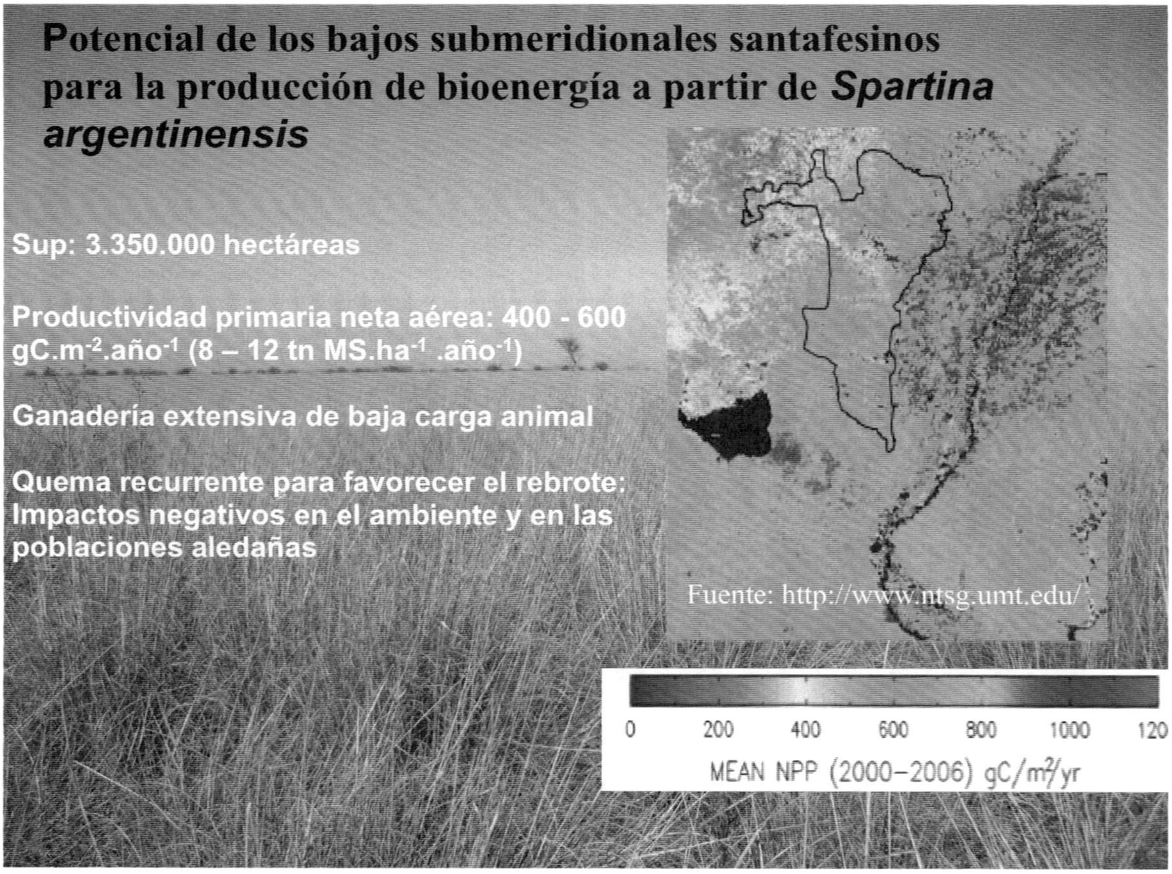
Potencial de los bajos submeridionales santafesinos para la producción de bioenergía a partir de Spartina argentinensis
Sup: 3.350.000 hectáreas
Productividad primaria neta aérea: 400 - 600 gC.m-2.año-1 (8 – 12 tn MS.ha-1 .año-1)
Ganadería extensiva de baja carga animal
Quema recurrente para favorecer el rebrote: Impactos negativos en el ambiente y en las poblaciones aledañas
Fuente: http://www.ntsg.umt.edu/
0
200
400
600
800
1000
120
MEAN NPP (2000–2006) gC/m2/yr

Pathways of lignocellulosic biomass conversion processes for bioenergy obtention

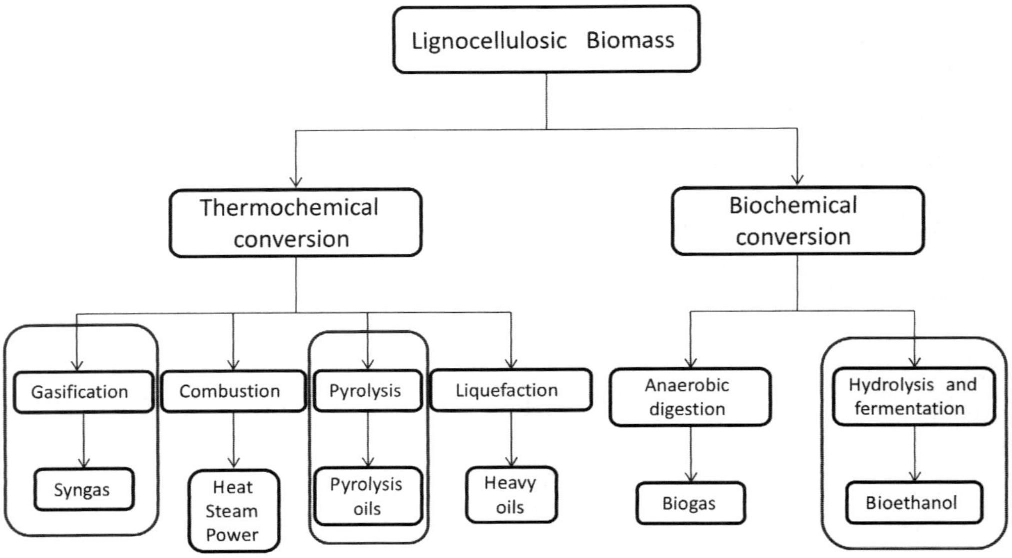

Facultad de Ciencias Agrarias
UNIVERSIDAD NACIONAL DE ROSARIO

TABLE 1. BIOETHANOL YIELD POTENTIAL ACCORDING TO DIFFERENT HYDROLYSIS EFFICIENCIES

Hydrolysis efficiency (%)	Bioethanol yield	
	L*Mg^{-1}	L*ha^{-1}
100	402.38	2816.66
90	362.14	2534.98
60	241.43	1690.01

TABLE 2. PERCENTAGE OF ARGENTINEAN AND SANTA FE GASOLINE DEMAND

Bioethanol (PJ year^{-1})	Percentage of Argentinean gasoline demand	Percentage of Santa Fe gasoline demand
19.76	8.81	96.3
29.63	13.22	144.4

Facultad de Ciencias Agrarias
UNIVERSIDAD NACIONAL DE ROSARIO

Pretreatment method	Processes
Physical pretreatments	Milling
Physicochemical and chemical pretreatments	Ammonia fiber explosion (AFEX)
	Dilute acid: Sulfuric acid Hydrochloric acid
	Sodium hydroxide
	Ammonia
	Lime
	Ozonolysis
Biological pretreatments	Fungi

El-Naggar *et al.* (2014), Biotechnology

Facultad de Cienc
UNIVERSIDAD NACIONAL DE ROSARI

Aphyllophorales (fungi without gills)

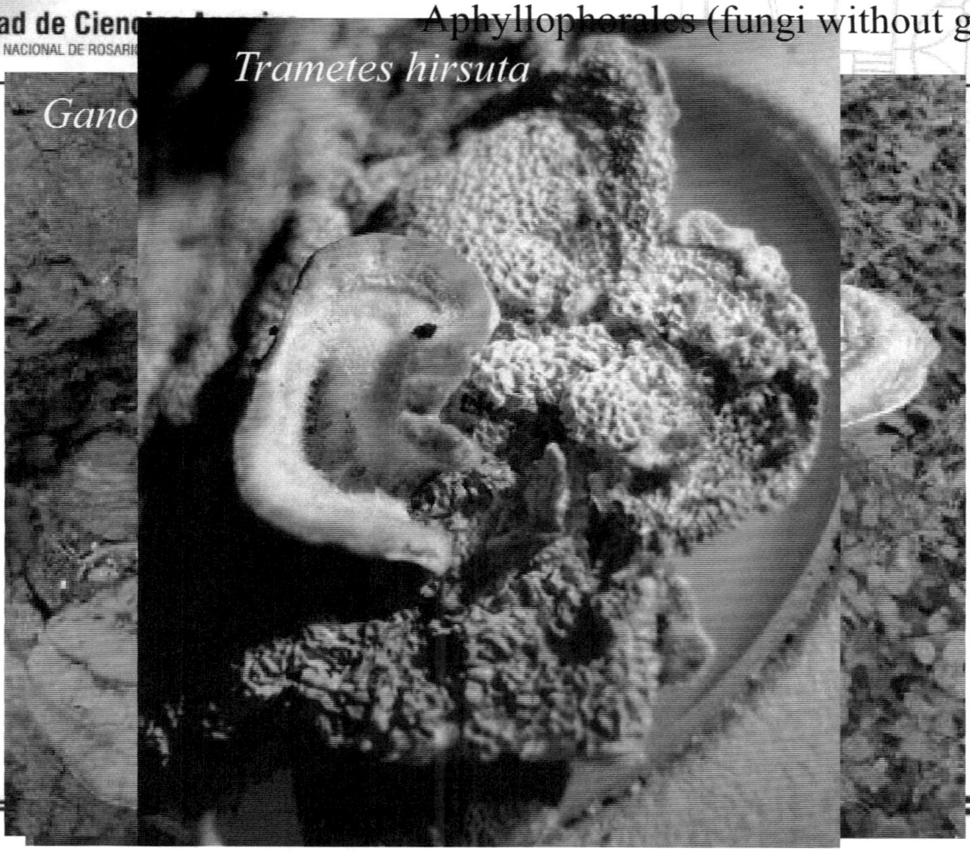

Facultad de Ciencias Agrarias
UNIVERSIDAD NACIONAL DE ROSARIO

Pretreatment	% Of hydrolysed cellulose
H_3PO_4	4.74
H_3PO_4 + laccase 0.06 U/tube	2.99
Laccase 0.06 U/tube	5.69
Laccase 0.005 U/tube	2.89
MnP 0.005 U/tube	4.85
LiP 0.005 U/tube	4.66
MnP–LiP–laccase	6.53
MnP–LiP	6.45
MnP–laccase	5.99
LiP–laccase	6.01
Trametes hirsuta supernatant	27.85
P. sanguineus supernatant	56.84

Facultad de Ciencias Agrarias
UNIVERSIDAD NACIONAL DE ROSARIO

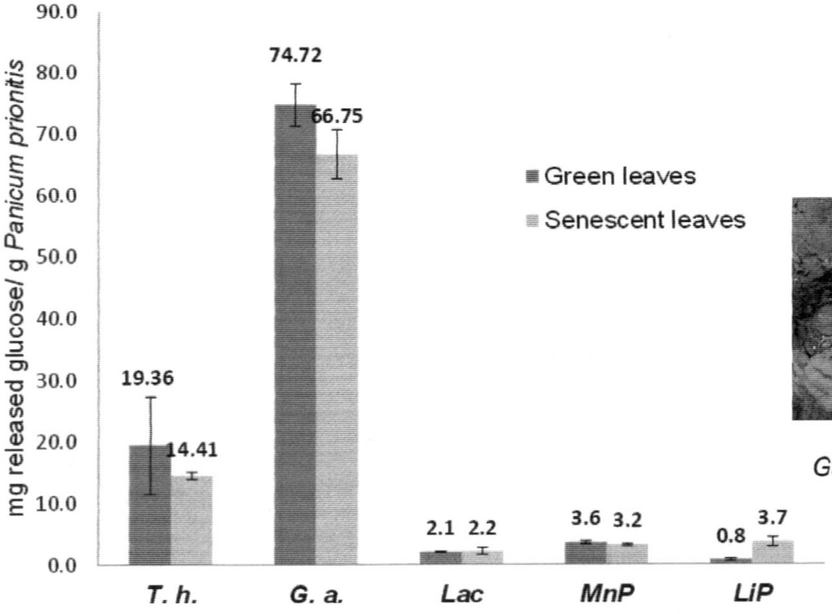

Ganoderma applanatum

Facultad de Ciencias Agrarias
UNIVERSIDAD NACIONAL DE ROSARIO

Facultad de Ciencias Agrarias
UNIVERSIDAD NACIONAL DE ROSARIO

Facultad de Ciencias Agrarias
UNIVERSIDAD NACIONAL DE ROSARIO

ENARCIV 2015

Procesos analizados para el balance energético

Campo/Field

- Disqueado
- segadora
- enrolladora

transporte 30 km

Planta industrial
Industrial Plant
INTI

- secado
- picado
- molido
- peleteado
- equipamiento auxiliar

Facultad de Ciencias Agrarias
UNIVERSIDAD NACIONAL DE ROSARIO

ENARCIV 2015

Recepción		1ra. Transformación: picado grueso	2da. Transformación: picado con malla		3ra. Transformación: Molido fino	
		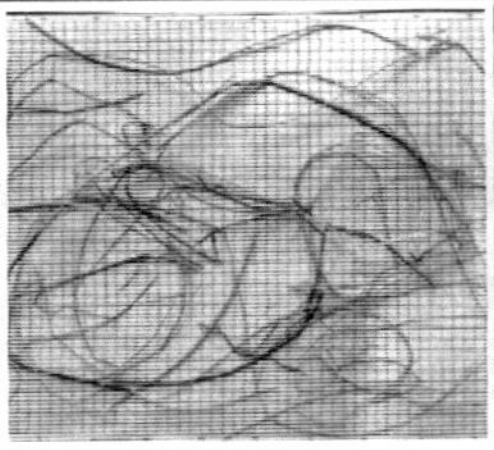			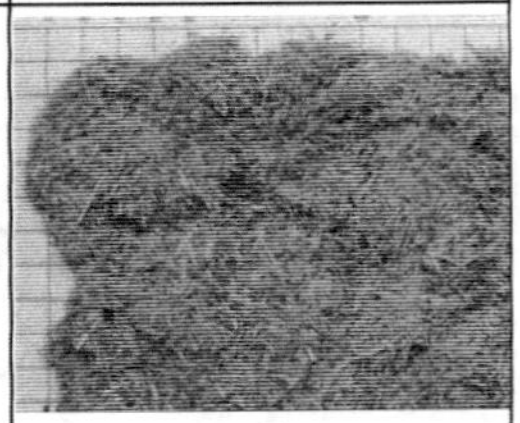	
Densidad: ~29 kg/m³ (en rollo)	Tamaño: ~700 mm	Invariable, menos entrelazado	Densidad: 66,9 [kg/m³]	Tamaño: ~25-50 mm	Densidad: 165 [kg/m³]	Tamaño: 0,2 a 0,5 mm

Obtención de Pellets de *Spartina argentinensis*

	valor real	Enplus-A1
Diámetro/Diameter (mm)	6,4	6 ó 8
Longitud/Length (mm)	13,9	≥3,15; ≤ 40
Humedad/Moisture (%)	10	≤ 10
Densidad/Density (kg/m3)	798,3	≥ 600
Durabilidad Mecánica /Mechanical durability (%)	99,6	≥ 97,5
Finos/Fines <3,15 mm (%)	0,5	<1

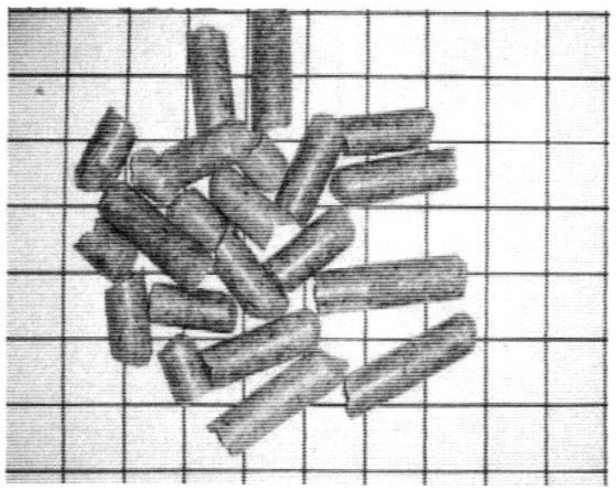

Facultad de Ciencias Agrarias
UNIVERSIDAD NACIONAL DE ROSARIO
ENARCIV 2015

Costo energético de aprovechamiento (CEA)

$$CEA = \frac{CEtotales}{PCI} = \frac{CEcorte + CEenrollado + CEtransporte + CEacondicionamiento + CEpeletizado}{PCI\ de\ una\ tonelada\ de\ Spartina\ argentinensis}$$

Tasa de retorno energético (TRE)

$$TRE = \frac{ENERGÍA\ OBTENIDA/KG\ DE\ PELLET}{SUMATORIA\ DE\ ENERGÍA\ INVERTIDA\ EN\ CADA\ ETAPA/KG\ DE\ PELLET}$$

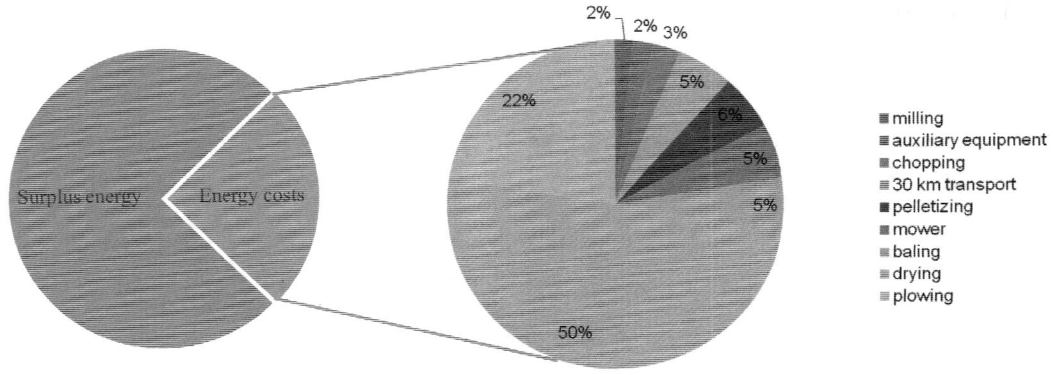

CEA =32%

TRE=2,6

- Consumo de gasoil por ha para corte y enrollado 7,5 mayor a datos bibliográficos de pasturas
- Se cosecharon sólo 800 kg de materia seca por ha
- Los rollos no presentaron la densidad adecuada por lo cual se humedecieron por las lluvias generando elevados costos energéticos para el secado

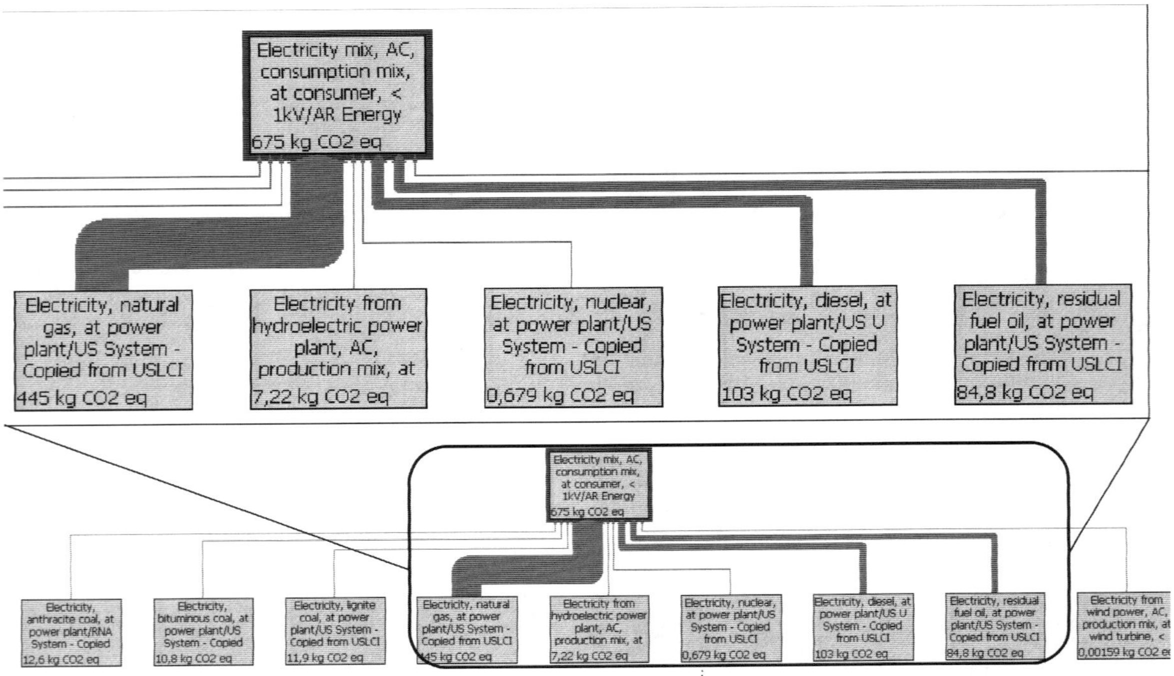

Diagrama de redes de los escenarios analizados mostrando los procesos principales

Facultad de Ciencias Agrarias
UNIVERSIDAD NACIONAL DE ROSARIO

Pathways of lignocellulosic biomass conversion processes for bioenergy obtention

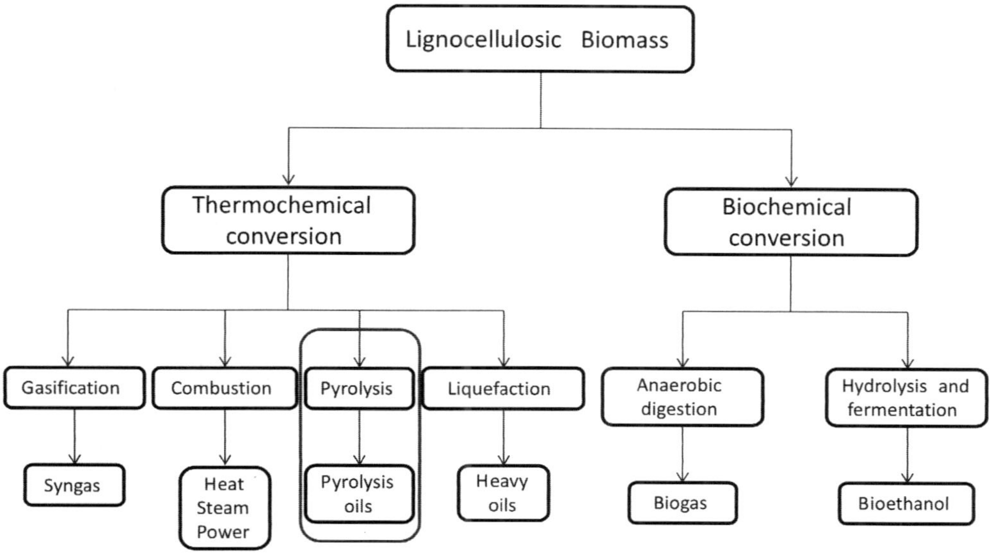

Facultad de Ciencias Agrarias
UNIVERSIDAD NACIONAL DE ROSARIO

Pirólisis

- La pirólisis es un proceso que permite producir combustibles líquidos a partir de biomasa lignocelulósica.

- Los bio-oils son mezclas de compuestos derivados de las reacciones de despolimerización y fragmentación de los tres componentes principales de la biomasa: celulosa, hemicelulosa y lignina.

- Los bio-oils pueden convertirse en combustibles de mayor valor agregado (gasolina, diesel) sobre catalizadores sólidos ácidos.

Pirólisis

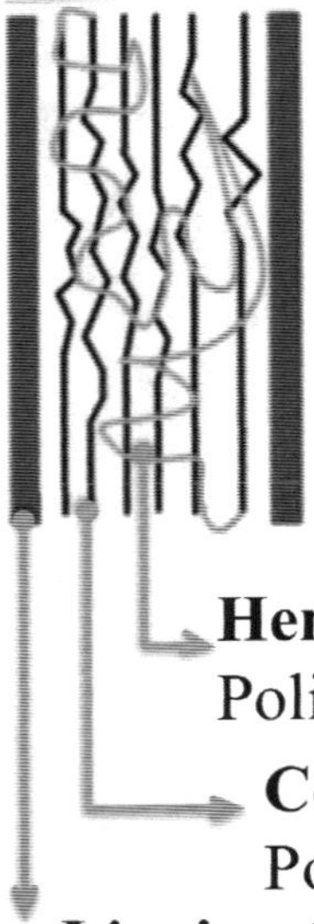

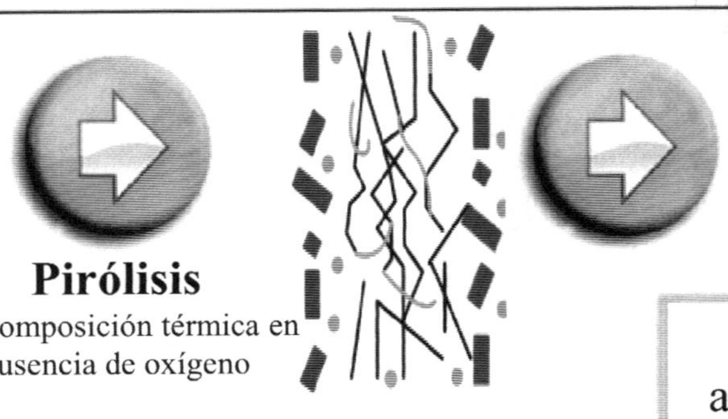

Pirólisis
Descomposición térmica en ausencia de oxígeno

Producto líquido altamente oxigenado que puede separarse fácilmente en dos fracciones, según su solubilidad en agua

Hemicelulosa (25 y 40 %)
Polisacárido (C5 y C6)

Celulosa (42 y 47 %)
Polisacárido de glucosa

Lignina (23 y 33 %)
Polímero heterogéneo
Derivados de fenoles ramificados

Bio-oil, soluble en agua

Tar, insoluble en agua

Equipamiento y análisis de productos

Reactor de pirólisis

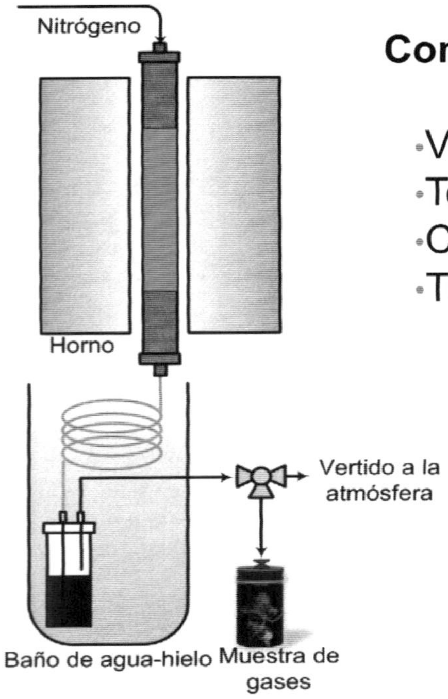

Condiciones de reacción

- Velocidad de calentamiento: 20 °C/min
- Temperatura: 500 °C
- Caudal de nitrógeno: 30 ml/min
- Tiempo a 500 °C: 30 min

Pirólisis

Distribución de productos

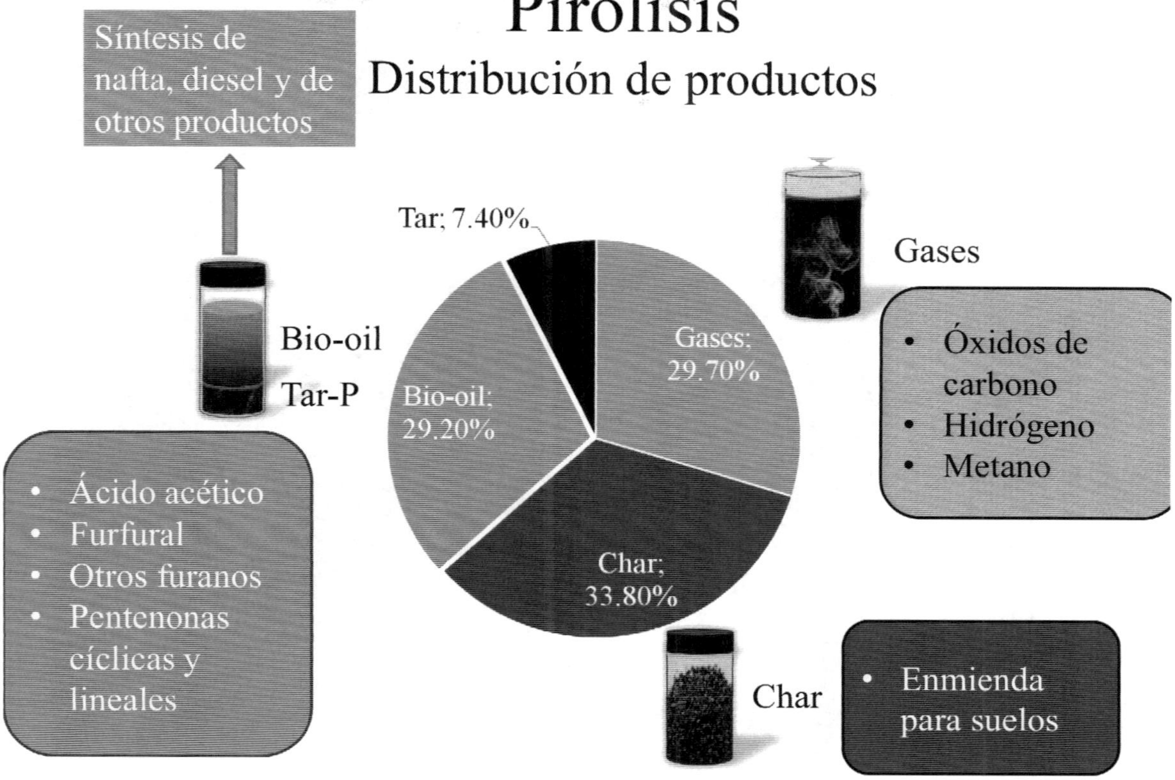

Facultad de Ciencias Agrarias
UNIVERSIDAD NACIONAL DE ROSARIO

Gracias por su atención

Thanks for your attention

ejozami@unr.edu.ar

IOWA STATE UNIVERSITY
Bioeconomy Institute

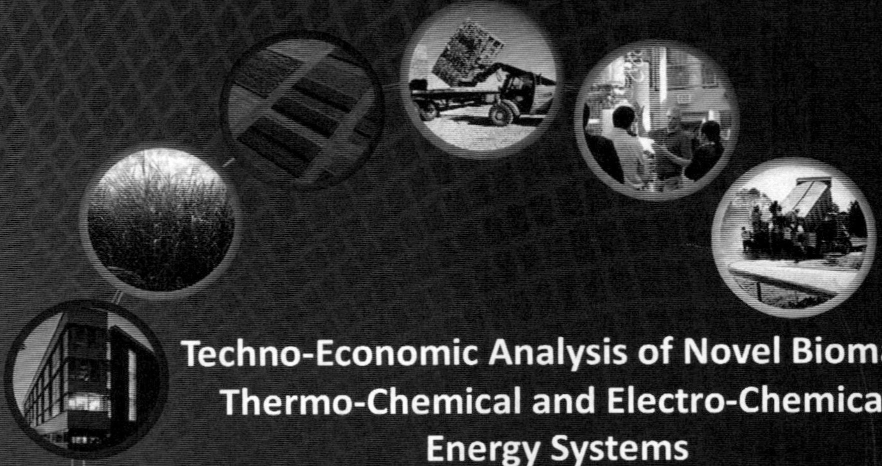

Techno-Economic Analysis of Novel Biomass Thermo-Chemical and Electro-Chemical Energy Systems

Qi Dang and Mark Mba Wright

September 14, 2016

Outline

- **Biofuels and bioproducts from an integrated pyrolysis refinery**
  - Process description
  - Techno-economics
- **Integrating fuel cells into a fast pyrolysis biorefinery**
  - Cost analysis method
  - Preliminary results

- **Biofuels and bioproducts from an integrated pyrolysis refinery**

- Integrating fuel cells into a fast pyrolysis biorefinery

Fast pyrolysis platform

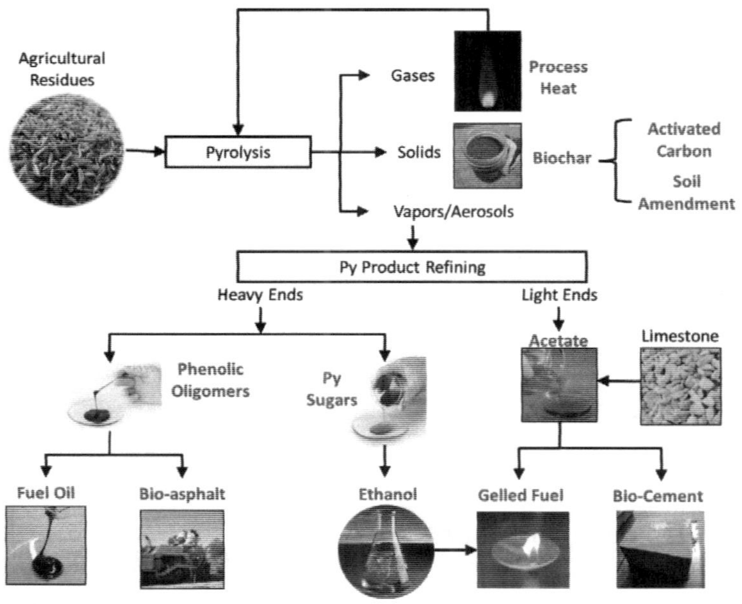

Integrated biorefinery

Three product portfolios

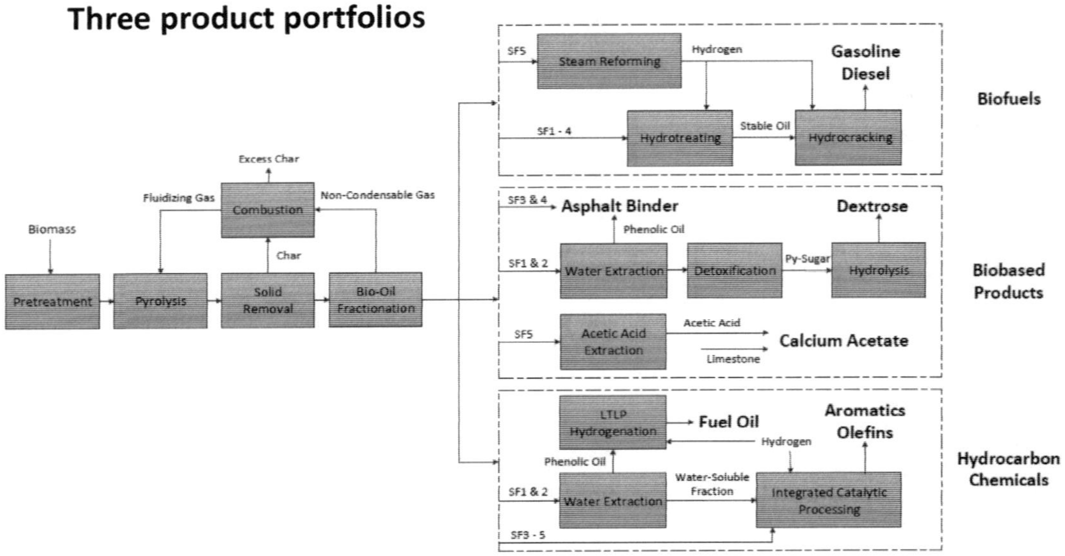

Methodology

- We apply and adapt the methodology our previous study
- Based on a conceptual level process design: biomass (red oak) processing capacity: 2000 metric tons per day (MTPD)
- Integrating data from various sources such as Aspen Plus process model, published literature, and historical data into the economic model
- Internal rate of return is chosen for economic evaluation
- Developing a new method to explore all possible IRRs for various combinations of three product portfolios under historical market product prices (low, average, and high)
- Sensitivity and uncertainty analysis

Main assumptions

Parameter	Assumption	Parameter	Assumption
Plant life	30 years	Total purchased equipment cost (TPEC)	1*TPEC
Equity	40%	Total installed cost (TIC)	3.02*TPEC
Loan interest	7.5%	Indirect cost (IC)	0.89*TPEC
Loan term	10 years	Project contingency	0.78*TPEC
Income tax rate	39%	Fixed capital investment (FCI)	4.69*TPEC
		Working capital (WC)	0.15*TPI
		Land use	0.06* TPEC
		Total project investment (TPI) with land	5.60*TPEC

IOWA STATE UNIVERSITY
Bioeconomy Institute

Capital costs for specific product portfolios

	Biofuels ($MM)	Biobased products ($MM)	Hydrocarbon chemicals ($MM)
Total purchased equipment cost	67.54	35.47	47.06
Pretreatment and pyrolysis	27.97	27.97	27.97
Fractionation	2.45	2.45	2.45
Upgrading	**12.12**	**1.00**	**12.59**
Hydrogen generation	**20.96**	**0**	**0**
Co-generation	1.41	1.41	1.41
Cooling plant	0.92	0.92	0.92
Balance of plant	1.72	1.72	1.72
Fixed capital investment	**316.90**	**166.41**	**220.79**

IOWA STATE UNIVERSITY
Bioeconomy Institute

Material prices and usage rates

Materials/ Energy	Price	Unit	Biofuels	Bio—based products	Hydrocarbon chemicals	Unit (per day)
Red Oak	80	$/MT	2000	2000	2000	MT
Hydrogen	1.5	$/kg	0	0	75100	kg
Natural Gas	0.26	$/kg	230500	0	0	
Biochar	-20	$/MT	126	126	126	MT
Electricity	6.16	Cents/kWh	801800	553900	561800	kWh
Sulfuric Acid	0.08	$/kg	0	38900	0	kg

Material prices and usage rates

Material/energy	Price	Unit	Ref.	Usage			Unit
				Biofuels	Bio-based products	Hydrocarbon chemicals	
Red oak	80	$/MT		2000	2000	2000	MT
Purchased hydrogen	1.5	$/kg	28	0	0	75096.40	kg
Natural gas	0.26	$/kg	15	230523	0	0	lb
Hydrotreating catalyst (Mo-based)	18.25	$/lb	21	1218.43	0	0	lb
Hydrocracking catalyst (Mo-based)	18.25	$/lb	21	852.90	0	0	lb
Hydrotreating catalyst (Ru-based)	2.52	$/lb	14	0	0	545.59	lb
Hydrotreating catalyst (Pt-based)	25.33	$/lb	14	0	0	545.59	lb
Hydrogenation catalyst (Pd-based)	25.33	$/lb	14	0	0	31.60	lb
Hydrocracking catalyst (Zeolite)	0.72	$/lb	14	0	0	186.55	lb
Sulfuric acid	0.08	$/kg	15	0	38893.80	0	kg
Limestone	17	$/MT	15	0	38.47	0	MT
Methanol	400	$/MT	15	0	0	38.51	MT
Biochar	–20	$/MT	15	125.83	125.83	125.83	MT
Hydrogen plant catalyst	4.24	¢/100 kscf H_2	21	34803.50	0	0	kscf
Hydrogen plant process water	195.26	¢/1000 gallon	29	142296	0	0	gallon
Boiler chemicals	1.65	$/lb	30	120	120	120	lb
Cooling tower chemicals	1.18	$/lb	30	17.14	17.14	17.14	lb
Cooling tower makeup	195.26	¢/1000 gallon	29	154008	154621	154621	gallon
Boiling feedwater makeup	195.26	¢/1000 gallon	29	809208	809208	809208	gallon
Sand and ash disposal	0.014	$/lb	28	712138	712138	712138	lb
Waste-water treatment	0.11	$/kg COD	31	794	625	625	kg
Extractor water	195.26	¢/1000 gallon	29	0	180307	180307	gallon
Electricity	6.16	¢/kWh	9	801782	553923	561782	kWh

Product yields

Input	Output			Unit (MTPD)
Biomass (2000 MTPD)	Biochar			260.8
	Non-condensable gas			421.73
	Bio-oil	SF1		274.25
		SF2		408.29
		SF3		86.82
		SF4		33.01
		SF5		515.1
Bio-oil (1317.47 MTPD)	Biofuels only	Gasoline		266
		Diesel		245
	Biobased products only	Asphalt binder		493.68
		Dextrose		197.62
		Calcium acetate		60.84
	Hydrocarbon chemicals only	Aromatics	Benzene	27.77
			Toluene	51.26
			Xylene	20.02
			Ethylbenzene	2.41
			Indene	0.51
			Naphthalene	0.51
			Styrene	1.44
		Olefins	Ethylene	83.3
			Propylene	140.07
			Butylene	36.9
		Fuel Oil		287.44

Historical product price distributions

Commodity	Probability distribution and parameters	Low (10th percentile)	Average (mean)	High (90th percentile)
Gasoline ($/MT)	Lognormal $[\mu, \sigma]$ μ =6.7607, σ =0.3207	572	909	1302
Diesel ($/MT)	Lognormal $[\mu, \sigma]$ μ =6.6439, σ =0.3598	484	820	1218
Asphalt binder ($/MT)	Logistic $[\mu, \beta]$ μ =953.651, β =244.734	416	950	1491
Dextrose ($/MT)	Logistic $[\mu, \beta]$ μ =569.491, β =76.34	402	564	737
Cement ($/MT)	Cauchy $[a, b]$ a =239.562, b =5.6184	222	240	257
Benzene ($/MT)	Lognormal $[\mu, \sigma]$ μ =7.2091, σ =0.07856	1222	1356	1495
Toluene ($/MT)	Lognormal $[\mu, \sigma]$ μ =7.0753, σ =0.07856	1069	1186	1308
Xylene ($/MT)	Lognormal $[\mu, \sigma]$ μ =7.2440, σ =0.07856	1266	1404	1548
Heating oil ($/MT)	Lognormal $[\mu, \sigma]$ μ =5.8318, σ =0.5358	172	392	678

IOWA STATE UNIVERSITY
Bioeconomy Institute

Maximum IRRs

Scenario	Prices			Bio-oil fractions			Maximum IRR
	Biofuels	Biobased products	Hydrocarbon chemicals	Biofuels	Biobased products	Hydrocarbon chemicals	
1	Low	Low	Low	0	0	1	9.10%
2	Low	Low	Average	0	0	1	17.50%
3	Low	Low	High	0	0	1	27.00%
4	Low	Average	Low	0	1	0	42.50%
5	Low	Average	Average	0	1	0	42.50%
6	Low	Average	High	0	1	0	41.70%
7	Low	High	Low	0	1	0	67.90%
8	Low	High	Average	0	1	0	67.90%
9	Low	High	High	0	1	0	67.50%
10	Average	Low	Low	0	0	1	9.10%
11	Average	Low	Average	0	0	1	17.40%
12	Average	Low	High	0	0	1	26.80%
13	Average	Average	Low	0	1	0	42.40%
14	Average	Average	Average	0	1	0	42.40%
15	Average	Average	High	0	1	0	40.60%
16	Average	High	Low	0	1	0	67.70%
17	Average	High	Average	0	1	0	67.80%
18	Average	High	High	0	1	0	67.80%
19	High	Low	Low	1	0	0	18.50%
20	High	Low	Average	1	0	0	18.50%
21	High	Low	High	0	0	1	27.00%
22	High	Average	Low	0	1	0	40.30%
23	High	Average	Average	0	1	0	42.50%
24	High	Average	High	0	1	0	42.40%
25	High	High	Low	0	1	0	67.90%
26	High	High	Average	0	1	0	67.80%
27	High	High	High	0	1	0	67.80%

Maximum IRRs

- Among the investigated scenarios, most scenarios (Scenarios 4–9, 13–18, and 22–27) tend to **maximize IRRs when bio-oil fractions are used for bio-based products** production as long as the biorefinery sells the bio-based products at their historical average price or higher
- When switching bio-based product prices to their historical low prices, seven scenarios (Scenarios 1–3, 10–13, and 21) that produce hydrocarbon chemicals generate higher IRRs
- **With historical low price level of biobased product**s, we also found two scenarios (Scenarios 19 and 20) where producing **biofuels with their high selling prices**

We determined internal rate of returns for different types of biorefineries.

Is an Integrated Biorefinery a better platform?

Triangle contour plots

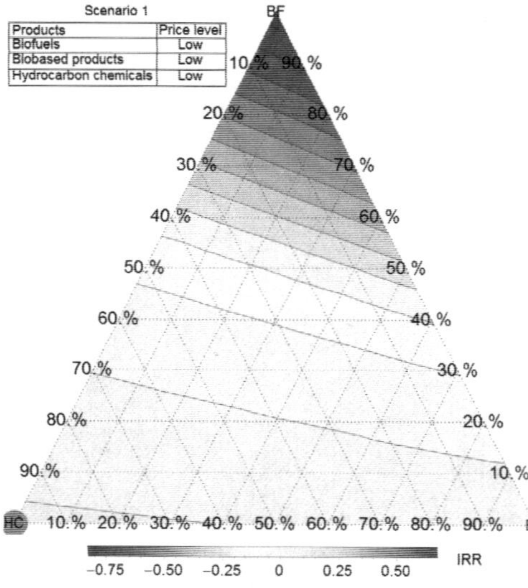

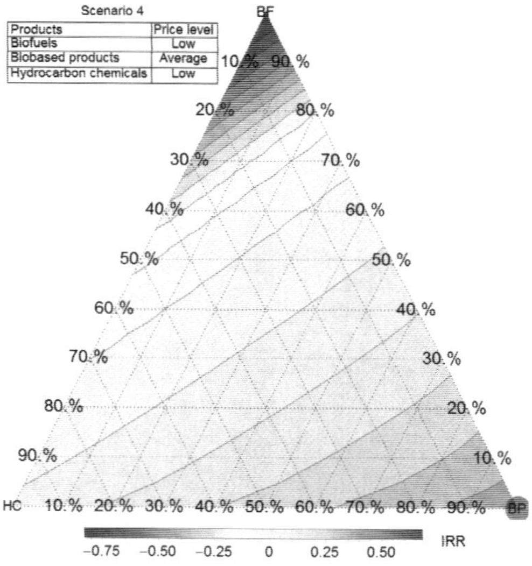

	Scenario 1	Scenario 4
Fixed capital cost ($MM)	220.8	166.5
Annual operating cost ($MM)	129.0	80.6
Maximum IRR	9.1%	42.5%
Product	Hydrocarbon chemicals	Biobased products

Triangle contour plots

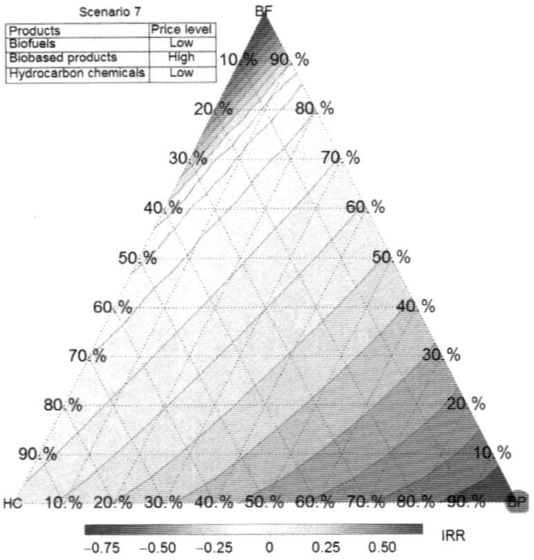

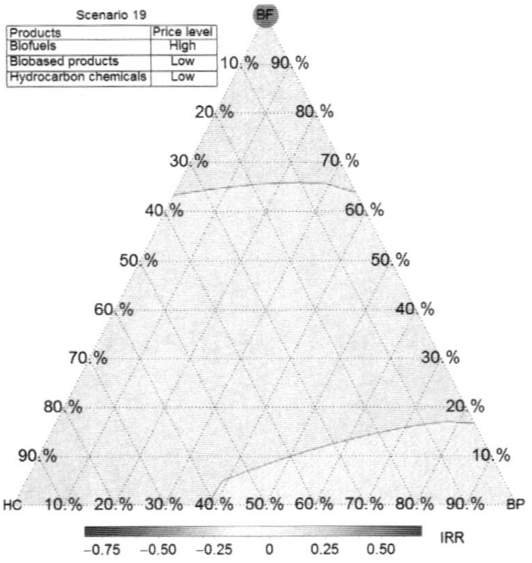

	Scenario 7	Scenario 19
Fixed capital cost ($MM)	166.5	316.9
Annual operating cost ($MM)	80.6	122.6
Maximum IRR	67.9%	18.5%
Product	Biobased products	Biofuels

IOWA STA
Bioeconomy In

16/21

Uncertainty Analysis

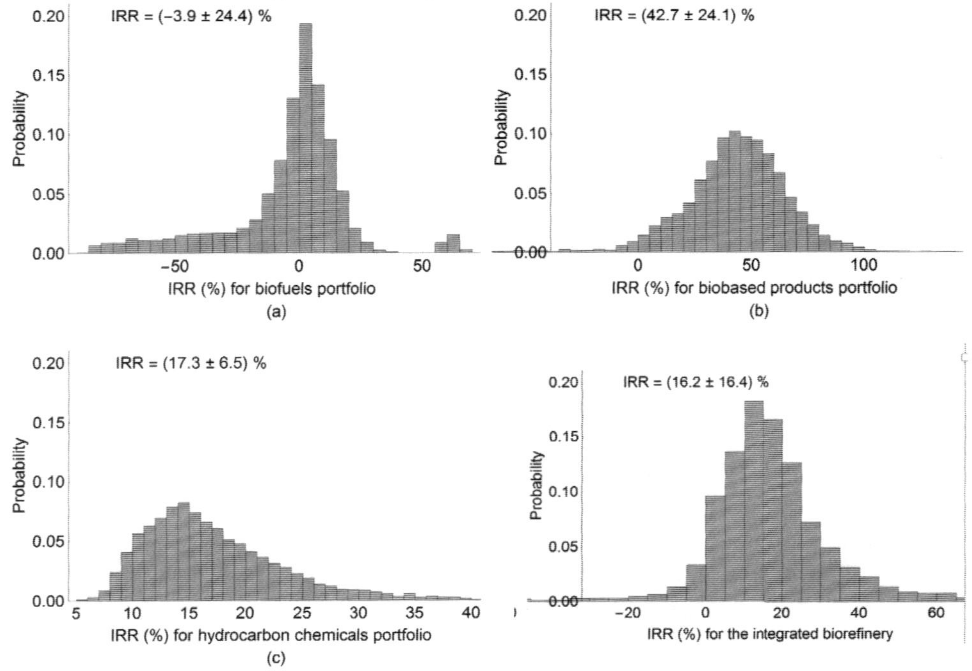

17/21

Sensitivity Analysis

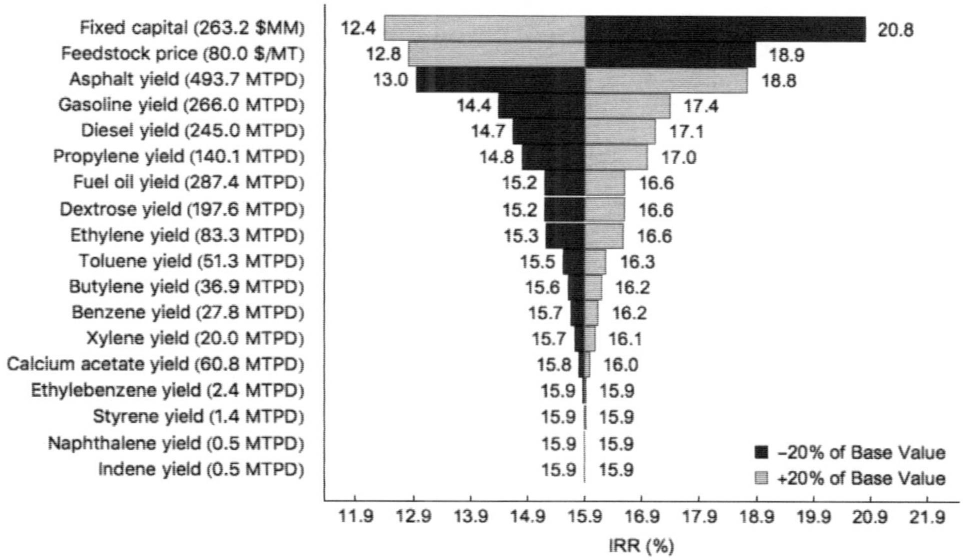

Conclusions for Part 1

- **The biorefinery has higher chances to gain profits by producing bio-based products followed by hydrocarbon chemicals and biofuels**
- **There are potential economic advantages in optimizing the product allocations of an integrated biorefinery based on market prices**
- **These findings serve as guidance for product selection to reduce the investment risk with known market conditions**

- Biofuels and bioproducts from an integrated pyrolysis refinery

- **Integrating fuel cells into a fast pyrolysis biorefinery**

Hybrid biomass energy system

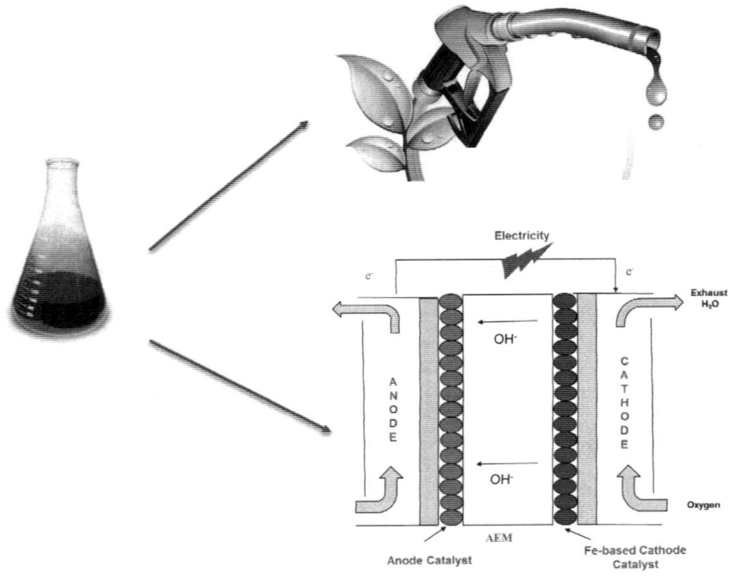

Cost analysis for battery

The price of a flow battery system

Darling et al. 2014

$$P_0 = ca\,A + (c_{add} + cbop)E_d t_d^{-1} + \sum_i c_{m,i} m_i$$

Electrode cost   Auxiliary cost   Electrolyte cost

P_0 : the initial system price in \$

c_a: the reactor cost per unit area in \$ m^{-2}

A: the electrode area in m^2

c_{bop}: the addition to the capital cost to reach the system price in \$ kW^{-1}

E_d: the energy discharged by the battery system in kWh or J

t_d: the discharge time of the battery in h

$c_{m,i}$: the cost per unit mass of electrolyte species I in \$ kg^{-1}

m_i: the required mass of species i in kg

Cost analysis for hybrid system

$$Unit\ cost\ of\ hybrid\ system = \frac{cost\ of\ biorefinery + cost\ of\ battery\ system}{biofuel\ from\ biorefinery + electricity\ from\ battery\ system}$$

Cost analysis for battery

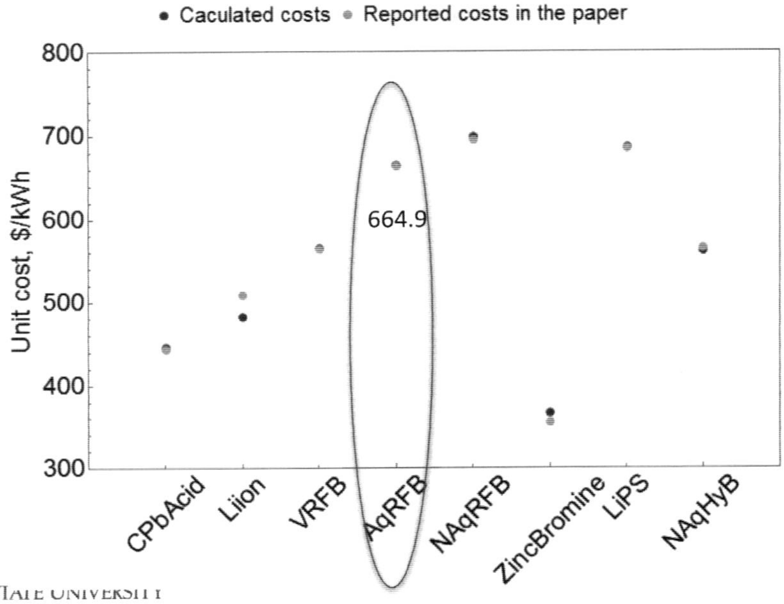

IOWA STATE UNIVERSITY
Bioeconomy Institute

Unit cost for hybrid system

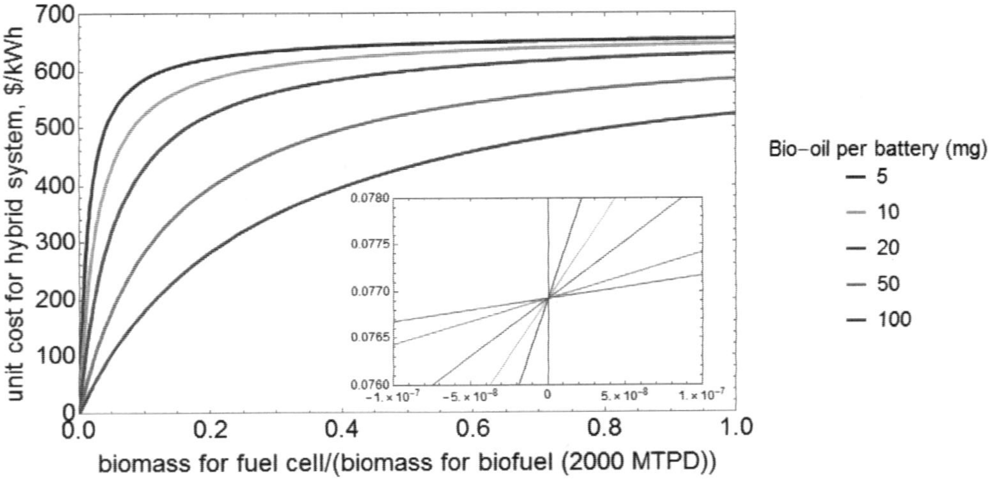

IOWA STATE UNIVERSITY
Bioeconomy Institute

Conclusions for Part 2

- This analysis aims at providing a new methodology of estimating the hybrid biomass thermo-chemical and electro-chemical energy systems
- Various hybrid system pathways based on fast pyrolysis will be compared in future work

IOWA STATE UNIVERSITY
Bioeconomy Institute

26/21

Acknowledgements

- EPSCoR funding from National Science Foundation
- Grant from Iowa Energy Center
- Dr. Brown and Dr. Li
- Students and staff from BEI

IOWA STATE UNIVERSITY
Bioeconomy Institute

27/21

Questions?

Mark Mba Wright: markmw@iastate.edu

Qi Dang: qdang@iastate.edu

Bioeconomy Institute, Iowa State University

Back-up Slides

Uncertainty Analysis

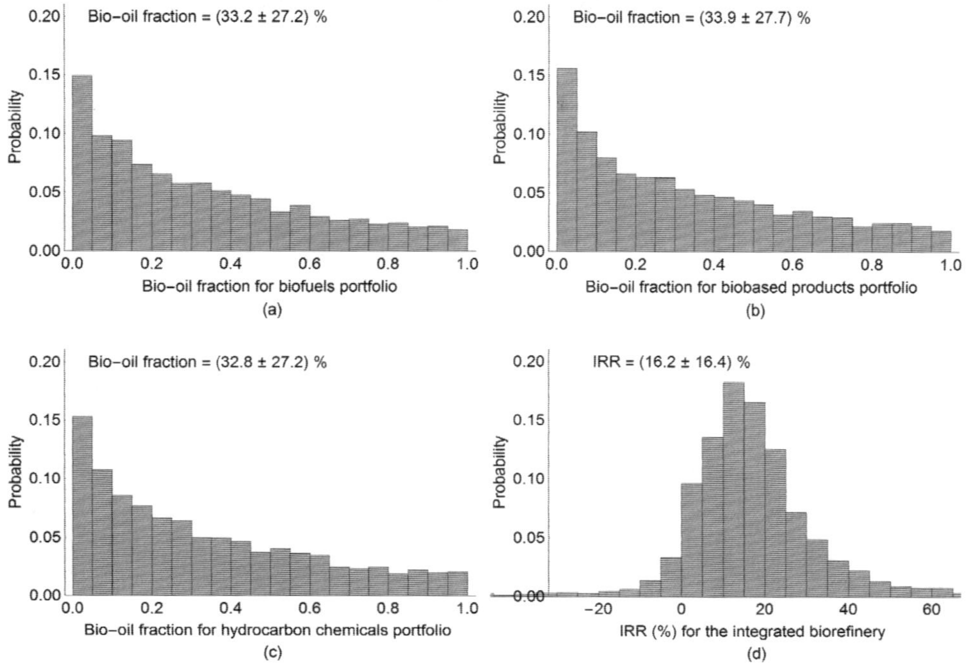

30/21

Cost analysis for battery

The price of a flow battery system

$$P_0 = ca\,A + (c_{add} + cbop)E_d t_d^{-1} + \sum_i c_{m,i} m_i$$

$$A = \frac{E_d}{\varepsilon_{sys,d}\, i_d V_d t_d}$$

$$\sum_i c_{m,i} m_i = \frac{E_d}{\varepsilon_{sys,d}\,\varepsilon_{q,rt}\, n_e F\, Vd}\left(\frac{s_+ M_+}{x_+}\left(c_{m,+} + \frac{c_{m,e+}}{S_+}\right) + \frac{s_- M_-}{x_-}\left(c_{m,-} + \frac{c_{m,e-}}{S_-}\right)\right)$$

$\varepsilon_{sys,d}$: an efficiency accounting for losses associated with auxiliary equipment

i_d: current density; V_d : voltage; $i_d{}^* V_d$: power density in W cm^{-2} or kW m^{-2}

$\varepsilon_{q,rt}$: the round-trip coulombic efficiency

m: the mass of the active species; M: the molecular weight

s: the stoichiometry of the active species in the energy storage reaction

n_e: the number of electrons; F: Faraday constant

X: the allowable state of charge range (SOC)

IOWA STATE UNIVERSITY
Bioeconomy Institute

31

Cost analysis for battery

The total system price for useable energy

$\varepsilon_{v,d} = V_d/U$

$V_d = U - i_d R$

$$\frac{P_0}{E_d} = \frac{c_a R}{\varepsilon_{sys,d} U^2 \varepsilon_{v,d} (1 - \varepsilon_{v,d}) t_d} + \frac{1}{\varepsilon_{sys,d} \varepsilon_{q,rt} n_e F \varepsilon_{v,d} U}$$

$$\times \left(\frac{S_+ M_+}{x_+} \left(c_{m,+} + \frac{c_{me+}}{S_+} \right) + \frac{S\_M\_}{x\_} \left(c_{m,-} + \frac{c_{me-}}{S\_} \right) \right) + \frac{c_{add} + cbop}{t_d}$$

$\frac{P_0}{E_d}$: unit price in $ per kWh

$\varepsilon_{v,d}$: discharge voltage efficiency

+: the positive active species

-: the negative active species

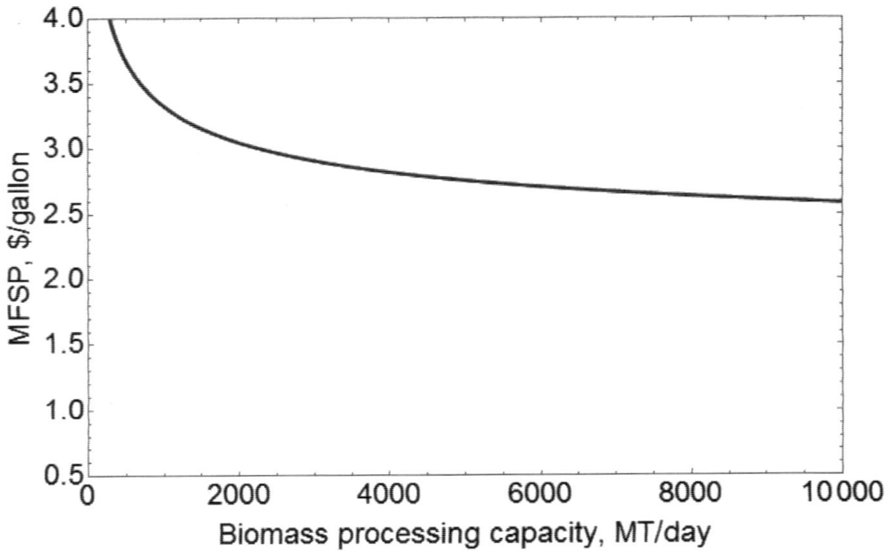

Unit cost for hybrid system

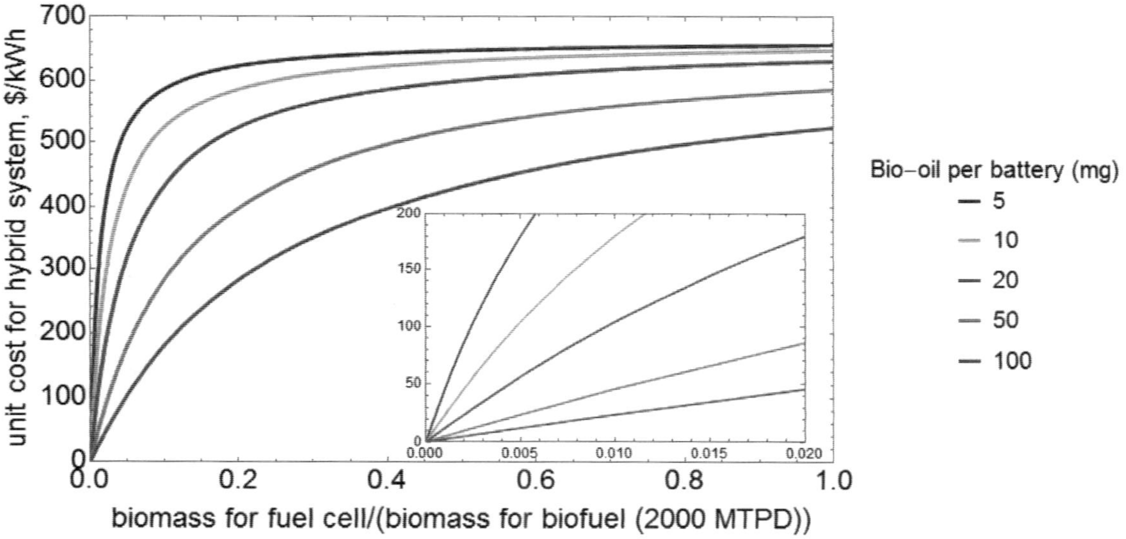

IOWA STATE UNIVERSITY
Bioeconomy Institute

34

Financial feasibility of short rotation woody crops intended for bioenergy production

Conference: RCN Conference on Pan American Biofuel and Bioenergy Sustainability

Year: 2016

Proceeding: 2nd RCN Conference on Pan American Biofuels & Bioenergy Sustainability

Group: Posters

Session: Rapid Fire Poster Session

Time: Wednesday, September 14, 2016 - 4:30pm-4:31pm

Authors:

Acuña, E., Universidad de Concepción
Rubilar, R., Universidad de Concepcion
Cancino, J.
Mena, P. A., Universidad de Concepcion

Decision makers often do not adapt as fast as it should be, in order to expect for the overt changing economic conditions. This is also the case of the conversion of agricultural land into short rotation woody crops. Several explanatory approaches are currently discussed in order to explain this behavior. A relatively known explanatory approach is given by the Monte Carlo simulation. This approach utilizes a stochastic comprehensive dynamic model that combines the uncertainty of the investments through probability distribution functions of variables which are important. By means of site-specific conditions and different scenarios, a full economic assessment for short rotation plantations for energy is provided. The financial feasibility for plantations of Eucalyptus globulus stablished in densities of 5000, 7500, 10000 and 15000 tress ha^{-1} was determined in four sites. Results indicate that woody crops became profitable in one sites, and only under intensive silviculture. The most profitable species was E. globulus with 5,000 trees ha^{-1} in sandy soils. If the conditions for short rotation commercial plantations were improved, it is recommended to incorporate enough and efficient subsidies to this type of woody crops.

Key words: short rotation woody crops, bioenergy, economic evaluation, Monte Carlo, uncertainty.

Using Innovative Automation Solutions in Biofuels – Technology Features, Applications and Economics

- Conference: RCN Conference on Pan American Biofuel and Bioenergy Sustainability
- Year: 2016
- Group: Posters
- Time:
  Wednesday, September 14, 2016 - 4:31pm-4:32pm

Authors: Lorenz, O. Sr., Siemens AG
Imbrogno, S.

Abstract

Since decades biofuels plants are automated with programmable logic controllers (PLC) or distributed control systems (DCS) and continuous improved by operational excellence programs. However, highly automated plants are still small and the authors observe that there is an ongoing drift between the advanced operation techniques and practice. Innovative automation solutions, such as model predictive control (MPC), Smith Predictor Control, Control Performance Monitoring (CPM), PID Analysis & Tuning and Softsensors are promise increased plant throughput, improvements of operability and energy efficiency, reduction of plant start-up time as well as reduced engineering cost and standardization in biofuels plants. Innovative automation solutions could in the opinion of the authors - be implemented much more in practice and this contribution aims to show the possibilities from an application point of view and stimulating a dialogue between academia, suppliers and users.Specialized suppliers offer tools and support with sophisticated functionalities, independently from the distributed control system (DCS). Furthermore DCS suppliers offer separate solutions which can be linked via OPC to their own automation system. Alternatives to the established separate solutions, several DCS suppliers have developed integrated advanced process control solutions in the recent years. This opens up potential new application fields. General aspects related to the selection of attractive applications in bioethanol process units
(slurry solids, fermentation, distillation, drying) are summarized. Actual status, advantages and disadvantages of the different approaches are shown to provide orientation and decision factors for user in the biofuels industry. The most common used technology in industrial practice Model Predictive Control is illustrated with respect to tools, suppliers, degree of utilization, hurdles for implementation, engineering, standardization and cost-benefit in investment and maintenance.

Non-Technical Barriers for the Development of a Bioenergy Industry in Argentina

- Conference: RCN Conference on Pan American Biofuel and Bioenergy Sustainability
- Year: 2016
- Group: Posters
- Time:
  Wednesday, September 14, 2016 - 4:33pm-4:34pm

Authors: Castelao Caruana, M. E., CESOT (UBA)- CONICET

The bioenergy industry in Argentina is a developing sector that, following many years of unfavorable public policies and economic conditions, it finds nowadays numerous opportunities for domestic growth (new energy prices, the opening of public tenders for renewable energy projects, access to international funding, etc.). However, the sector still faces significant economic, political and social challenges associated with the development of productive factor markets, economic uncertainty, limited awareness of bioenergy throughout society, and project implementation issues.
The barriers that hinder the industry´s development need to be identified and addressed in order to design innovative policy approaches and private strategies that help realize its potential. For this, involvement of various stakeholders in the identification of barriers is extremely important, as their perceptions on barriers and barrier-removal measures differ. The harmonization of the policies and strategies of public and private stakeholders is critical to ensure their effectiveness.

Barrier is a metaphor to refer to those constraints that influence to a certain extent the implementation and development of a bioenergy system (McCormick & Kåberger, 2007)[1]. Most of the available research on barriers has been developed regarding countries of the European Union and the USA, reflecting the perspective of one or a few stakeholders and focusing specially on technical and economic issues.

The proposed poster will present the results of a research on the barriers that hinder the development of a bioenergy industry in Argentina. This research contributes to the identification, analysis, and discussion of barriers and barrier-removal strategies behind bioenergy development from both a production structure and a market structure perspective, focusing on the dynamic forces on the bioenergy market rather than on isolated, individual bioenergy projects.

The research builds on past international studies and personal in depth interviews with various stakeholders such as wholesale consumers, equipment manufacturers, bioenergy project developers, technology agencies and policy makers to find out their perspective on the barriers to the development of this sector. The survey first gauged the respondentsâ perceptions as to the most significant barriers to expanding bioenergy from a list of major factors previously identified in the literature (e.g., transaction costs, policy regulations, market prices, know how, etc.). The survey then delves into each of these factors to determine what aspects posed the greatest obstacle to its developments, and how strongly each particular barrier would hinder the penetration of bioenergy.

The barriers identified are analyzed considering their influence on productivity, production costs and transaction costs within the business, and their importance to stakeholders. This analysis contemplates the different roles stakeholders have in the value chain of bioenergy, their levels of influence and their relations in the market structure

[1] McCormick, K. y KÃ¥berger, T. (2007). Key barriers for bioenergy in Europe: Economic conditions, know-how and institutional capacity, and supply chain co-ordination, *Biomass and Bioenergy,* 31, pp. 443â452.

Production of Biodiesel from an Olive Oil Industry Waste

- Conference: RCN Conference on Pan American Biofuel and Bioenergy Sustainability
- Year: 2016
- Group: Posters
- Time:
  Wednesday, September 14, 2016 - 4:34pm-4:35pm

Authors: Alvarez Serafini, M. S., CONICET
Reinoso, D. M., CONICET
Tonetto, G. M., Planta Piloto de Ingeniería Química

With the growth of global energy demand and high levels of pollution, sustainable fuel alternatives have become a high priority for many countries. Biodiesel is one of these sustainable fuels obtained from vegetable oils or animal fats. Consists of alkyl esters derived from either the transesterification of triglycerides or the esterification of free fatty acids with short-chained alcohols. Biodiesel comes with many advantages; low emissions of CO2 compared to petroleum diesel, biodegradable, non-toxic. Argentina is currently one of the major biodiesel producer worldwide, using soybean oil as raw material. In the country, the diesel must contain 10% biodiesel since 2014. Biofuel production programs have several social and environmental impacts and affected food markets through diverting grain away from food for fuel. Therefore, utilizing biomass residues for biodiesel production will reduce its contribution to food price. Also, the use of cheap low quality feed stocks will help in improving the economic feasibility of biodiesel.

Biomass (such as agricultural residues, animal wastes, forestry and wood processing residues, municipal or industrial wastes) is seen as one of the most promising, renewable energy sources because it can be converted to other usable forms of energy, like biofuels. It is relatively cheap energy resource to provide the initial feedstock in bioenergy industry. This would mean the elimination of a waste stream and gives biofuel producer a chance to reduce costs.

Alperujo is a waste product from the process of olive oil extraction (two-phase extraction process). It is a blend made of water, oil, cellulose, lignin, proteins, carbohydrates, small portions of active phenol compounds and other derivatives. It is not environment-friendly and it leads to serious ecological problems once alperujo is scattered on the soil. It has a phototoxic activity that does not permit plants to germinate. Olive plantations in Argentina are rapidly expanding, therefore alperujo is a waste to the environment that needs to be treated.

In the industry is used an alkaline homogeneous catalyst. One of the main problems of these catalysts is the difficulty in using low-cost raw material, such as the formation of soap due to the neutralization of free fatty acids present in low-quality oils, or the saponification of triglycerides in the presence of water. The use of a heterogeneous catalyst would simplify the separation and purification steps. There is a growing interest in finding a solid catalyst that is active in the transesterification of triglycerides, while it is selective with raw materials of low quality, that is to say a catalyst that is stable in the presence of FFA and water. Given that it will be used in âgreenâ processes, it should also be non-pollutant and non-toxic.

In this study, alperujo has been successfully used as feedstock to obtain biodiesel using zinc stearate as a new clean catalyst for an environmentally-friendly biodiesel production. The catalyst was able to catalyze simultaneously the triglyceride transesterification and the fatty acid esterification reactions with high activity and selectivity under moderate operating conditions.

Simultaneous Saccharification and Fermentation of Bagasse for Ethanol Production: Preliminary Design

- Conference: RCN Conference on Pan American Biofuel and Bioenergy Sustainability
- Year: 2016
- Group: Posters
- Time:
  Wednesday, September 14, 2016 - 4:35pm-4:36pm

Authors:
Albernas Carvajal, Y., Universidad Central Marta Abreu de las Villas
González Suárez, E., Universidad Central Marta Abreu de las Villas

The interest of ethanol production from agro-industry residues as raw material has deserved great attention since these feedstocks may constitute an alternative to fossil fuels as well as represent beneficial effects both from economic and environmental point of view. The sugar cane bagasse (SCB) is presented as an available and cheap opportunity of raw material. Although the SCB is usually burn at the sugar mill for steam and electricity production, there still remains certain amount of residues available for bioethanol production, according with (Albernas et al., 2014).
Fuel ethanol from SCB can be obtained through pretreatment, enzymatic hydrolysis and fermentation process stages. Once the pretreatment is performed, there are several ways for carrying out saccharification and fermentation. The simultaneous saccharification and fermentation (SSF) is one option, which as its name indicates, has the advantage of carrying out simultaneously the enzymatic hydrolysis and fermentation stages on the same processing unit, obtaining directly ethanol as a product. Then, it represents a promising separation technology because end-product inhibition is minimized (as glucose is converted as it is released), and therefore higher solid levels are allowed (Mesa et al., 2011). On the other hand, the number of process equipment is reduced and consequently the investment cost. Also, less energy is required and therefore production costs are also decreased and more sustainable process is achieved.

In this work a preliminary ethanol plant design is proposed. A mixed integer linear programming (MILP) model is formulated in order to obtain the optimal number of parallel units in each stage and their sizes (plant configuration and capacity). The production planning along the time horizon of one year is determined, given by the number of batches to be processed and its size. A formulation based on fixed sizes factors and times for discontinuous stages is proposed, which were obtained by experimental results of Albernas (2014) and Albernas et al. (2015). The model is implemented and solved in GAMS (General Algebric Modeling System). Different production sustainable scenarios are analyzed, for which the investment cost is minimized, and economic profitability indicators are calculated.

References

Albernas, Y., Corsano, G., Morales, M., González, M., Santos, R., & González, E. (2014) Optimal design for an ethanol plant combining first and second-generation technologies. CT&F - Ciencia, Tecnología y Futuro, Vol. 5, No. 5, 97-120.

Mesa, L.; González, E.; Romero, I.; Roiz, E.; Cara, C.; Castro, E. Comparison of process configurations for ethanol production from two-steps pretreatment sugar cane bagasse. Chemical Engineering Journal 2011, 175, 185-191.

Albernas, Y. (2014) Procedimiento para la síntesis y el diseño óptimo de plantas discontinuas de obtención de bioetanol empleando bagazo de caña de azúcar. Tesis en Opción al Grado Científico de Doctor en Ciencias Técnicas, Universidad Central Marta Abreu de Las Villas, Departamento de Ingeniería Química, Santa Clara.

Albernas, Y., Corsano, G., Mesa, L., Santos, R. & González, E. (2015) Estudio de la cinética de la hidrólisis enzimática del bagazo pretratado. Afinidad, Vol. LXXII, No. 570, Abril-Junio, 127-132.

Synthesis and Analysis of Integrated Process Schemes for Sugarcane Conversion

Conference: RCN Conference on Pan American Biofuel and Bioenergy Sustainability

Year: 2016

Proceeding: 2nd RCN Conference on Pan American Biofuels & Bioenergy Sustainability

Group: Posters

Session: Rapid Fire Poster Session

Time: Wednesday, September 14, 2016 - 4:36pm-4:37pm

Authors:

Albernas Carvajal, Y., Central University of Las Villas
Corsano, G., CONICET
González Cortés, M., Central University Marta Abreu of Las Villas
Herrera Álvarez, S., Central University of Las Villas

In this paper it is carried out the synthesis and analysis of technological schemes for sugar and derivative process integration. Several technological process such as first and second generation ethanol, torula yeast, furfural, lignin and CO2 are considered in the analysis. Sugar cane is the principal raw matter; parameters such as price of the products in the market are considered.
Process simulations in ASPEN PLUS were performed to evaluate mass and energy balances and evaluate 6 sceneries with variations in products price, plants outside of operation, scenario for time of no harvest, among other parameters.
Some improvements are proposed, including process thermal integration, adopting the pre-treatment of sugarcane bagasse by steam explosion, and including an integrated cogeneration system. It was assumed that sugarcane trash and lignin cake, a hydrolysis residue, are available as supplementary fuel.
The exergies of streams involved in the process were calculated, along with their exergetic cost. For the conventional process, in which ethanol and surplus electricity are the major products, a second-law efficiency of 28% was found. In the integrated process, with furfural, CO2 and lignin as an additional product, a higher performance in resources management was achieved, with values in the range of 35 -37%

Fourth Generation Biofuels: Metabolic Modelling of Synechocystis sp. PCC 6803 for Ethanol Production

- Conference: RCN Conference on Pan American Biofuel and Bioenergy Sustainability
- Year: 2016
- Group: Posters
- Time:
  Wednesday, September 14, 2016 - 4:37pm-4:38pm

Authors:
Lasry Testa, R., PLAPIQUI, CONICET, UNS
Delpino, C., Planta Piloto de Ingeniería Química (PLAPIQUI), Universidad Nacional del Sur-CONICET
Estrada, V., Planta Piloto de Ingenieria Quimica - Universidad Nacional del Sur

Cyanobacteria are autotrophic unicellular microorganisms that are gaining significance as a medium to obtain fourth generation hydrocarbon biofuels, owing to their capacity to use solar energy, atmospheric carbon dioxide and water to grow. In this work, we analyze the metabolic network of the cyanobacterium Synechocystis PCC 6803 for the production of ethanol. A genomic scale metabolic model of the network, which has 523 metabolites and 661 reactions, is considered (Knoop *et al.* 2013). Firstly, the genetic modifications (i.e., knockout of genes, overexpression of genes) already made to the strain experimentally by different authors are considered. A Flux Balance Analysis (FBA) optimization technique is used to model these modifications and ascertain this way the best ethanol production rates that can obtained from these modified strains. The results are compared with the experimental results of the authors considered (Dexter and Fu 2009, Duhring *et al.* 2010, Gao *et al.* 2012, Dienst *et al.* 2014). Secondly, a bi-level optimization technique is used to determine what other modifications could be done to the cyanobacterium to improve its ethanol yield. For this, the maximization of biomass production is set as the inner problem and the maximization of ethanol production is set as the outer problem. Biomass and ethanol production pose a cellular objective and a biotechnological objective, respectively. Binary variables are added to model the possibility of a knockout. This gives rise, after reformulating the problem with available techniques, to a mixed integer linear problem (MILP). All the models formulated in this work were solved using GAMS. Numerical results obtained provide useful insights on the biofuel production of this strain within the context of genomic-scale cyanobacterial metabolism.

References

Dexter, J. and Fu, P. (2009). Metabolic engineering of cyanobacteria for ethanol production. Energy and Environmental Sciences, 2: 857-864.

Dienst, D., Georg, J., Abts, T., Jakorew, L., Kuchmina, E., Borner, T., Wilde, A., Duhring, U., Enke, H. and Hess, W. R. (2014). Transcriptomic response to prolonged ethanol production in the cyanobacterium *Synechocystis* sp. PCC 6803. Biotechnol Biofuels, 7: 1-21.

Duhring, U., Enke, H., Kramer, D., Smith, C. R., Woods, R. P., Baier, K. and Oesterhelt, C. (2010) Genetically modified photoautotrophic ethanol producing host cells, method for producing the host cells, constructs for the transformation of the host cells, method for testing a photoautotrophic strain for a desired growth property and method of producing ethanol using the host cells. Patent Publication Number: US20100297736A1.

Gao, Z., Zhao, H., Li, Z., Tan, X. and Lu, X. (2012) Photosynthetic production of ethanol from carbon dioxide in genetically engineered Cyanobacteria. Energy Environ Sci, 5: 9857â9865.

Knoop H., GrÃ¼ndel M., Zilliges Y., Lehmann R., Hoffmann S., Lockau W. and Steuer, R. (2013).Â Flux balance analysis of cyanobacterial metabolism: the metabolic network of *Synechocystis* sp. PCC 6803. PLoS Comput. Biol. 9:e1003081.10.1371/journal.pcbi.1003081.

Propuesta para Lograr una Produccion Sucro-Alcoholera más Sustentable

- Conference: RCN Conference on Pan American Biofuel and Bioenergy Sustainability
- Year: 2016
- Group: Posters
- Time:
  Wednesday, September 14, 2016 - 4:38pm-4:39pm

Authors:

Trabajo: *Propuesta para lograr un proceso sucro-alcoholero más sustentable*Director de Tesis: Lic. Alberto S. CAPRAAutor: Ing. Ricardo E. FOURCADE RESUMEN

La producción de azúcar en la provincia de Tucumán, Argentina, tiene alrededor de 3500 pequeños productores independientes que entregan su caña a 15 ingenios utilizando procedimientos de zafra que incluyen corte manual y quema de caña cortada y a consecuencia de los métodos de zafra inapropiados y la falta de una gestión mancomunada entre ellos y la industria, la materia prima que recibe el ingenio es variable en calidad y composición química dificultando el procesado y aumentando la exigencia ambiental.

El presente trabajo de Tesis plantea desde una visión holística una posible mejora a la problemática que permitiría al ingenio disponer de materia prima de composición más uniforme y a la vez ajustar los procesos de manera de obtener un producto final del ciclo Azúcar-BioEtanol, llamado vinaza, que pueda utilizarse en todo tipo de sustratos para fertirriego, aumentando sobremanera la sustentabilidad de la actividad.

Se propone también una posible modificación al proceso de fermentación que permitiría aumentar el rendimiento de obtención de alcohol, resultando un subproducto que es el mosto, cuya alta carga orgánica lo haría apropiado para continuar otro proceso biológico como podría ser la producción de BioGas u otros de interés.

Antecedentes del autor:

Ingeniero Industrial recibido en la Universidad Nacional de Sur año 1969.

Magister en Ingeniería de Procesos en la Universidad de Carabobo, Venezuela, año 2004.

Heat Transfer Model and Its Experimental Validation in a Small Scale Biodigestor

Conference: RCN Conference on Pan American Biofuel and Bioenergy Sustainability

Year: 2016

Proceeding: 2nd RCN Conference on Pan American Biofuels & Bioenergy Sustainability

Group: Posters

Session: Rapid Fire Poster Session

Time: Wednesday, September 14, 2016 - 4:39pm-4:40pm

Authors:

Adrover, E.
Rayes, M.
Madies, E.
Cotabarren, I.
Pedernera, M.

A small scale anaerobic digester was installed in a region of Argentina with important thermal amplitude. The digester operated in semi-batch mode and was fed with a mixture of rabbit manure, sorghum crops and water and operated with a hydraulic retention time of 45 days. The performance of the anaerobic digester was studied with special focus on thermal effects. A simple non- stationary model was developed in order to predict the slurry temperature and determine the challenges associated with biogas production in our region. Radiative, convective and conductive heat transfer phenomena were taken into account and the model has been validated with experimental data. Biogas stream was quantitatively evaluated in terms of volumetric total flowrate (bubble soap meter) and pH were measured over a long period of time. Biogas composition was determined by means of gas chromatography. Previously, the presence of sulfur in the biogas stream was analyzed using the silver nitrate tritation method through its reaction to yield silver sulfide (black precipitate). The produced digestate was also characterized by determining, in samples with different period of storage, the organic content and dry weight by gravimetric analysis, pH and conductivity, size distribution of suspended solids by laser diffraction, hydrogen, nitrogen and total carbon content by elemental analysis, potassium content by flame spectrophotometry and phosphorus content by the colorimetric method of Scheel.

Energy Use of Residual Biomass of Lemon Cultivation By Gasification in a Citrus Industry in Tucuman

- Conference: RCN Conference on Pan American Biofuel and Bioenergy Sustainability
- Year: 2016
- Group: Posters
- Time:
  Wednesday, September 14, 2016 - 4:40pm-4:41pm

Authors: Diaz, G. F.
Paz, D.

The citrus industry is of great importance in Tucuman, being in the top spots worldwide. Peel drying during lemon processing, has the highest consumption of natural gas, which comes into deficit at harvest time. Large amounts of biomass waste are generated from lemon cultivation, which can be exploited by gasification: a thermochemical conversion process of a solid or liquid fuel into a gas by incomplete combustion. The aim of this study was to evaluate technically and economically a gasification plant of biomass residues from lemon harvest to supply energy to the lemon peel drying of a factory processing 50 t/h located in Lules (Tucumán, Argentina).
Availability and energy potential of residual biomass were analyzed. Process was defined including: transport of biomass wastes to the gathering centers; preconditioning; transportation to the factory; drying of the chips; gasification and lean gas cleaning; poor gas combustion; mass and energy transfer in peel drying. The equipment required for the processing and transport of biomass were identified and quantified. A system of equations from mass and energy balances and sizing of the wood chip dryer was raised. For economic and financial analysis of the project, investment, operating costs (ranked as fixed and variable costs) were identified and quantified, and then the cash flow was calculated. From cash flows, the equivalent annual costs (EAC) of the project and the current situation of the factory were calculated for comparison. The project was analyzed in three scenarios: without bank credit, with bank credit, and selling carbon credits. The sensitivity of EAC of the project and the current situation was performed regarding the most important variables.

As result, the study determined the need to install two collection centers of 1.3 ha received 8.639 t / month of biomass with 50 % humidity, located at a distance of 5 km from the factory to the north and the south, respectively. The gasification plant attached to the factory process 3,960 kg / h of chips at 15% moisture, producing 12.657 kW of thermal energy with 9.900 Nm3 / h of poor gas. Sizing of the rotary dryer chip, energy integrated with peel driers, with moisture gradient of 30 % to 15%, results in a diameter of 3 m and a length of 11.8 m. The initial investment for the project is 49 million pesos, with total annual costs of about 6.5 million. The distance from the collection centers and fuel costs showed not to have significant impact on EAC. The EAC of the current situation is equal to the project with bank credit rate of 19% annual interest and selling carbon credits at 6 â¬ / tCO_2.

Integrated Microalgae-Based Biorefinery for Astaxanthin Production

Conference: RCN Conference on Pan American Biofuel and Bioenergy Sustainability

Year: 2016

Proceeding: 2nd RCN Conference on Pan American Biofuels & Bioenergy Sustainability

Group: Posters

Session: Rapid Fire Poster Session

Time: Wednesday, September 14, 2016 - 4:41pm-4:42pm

Authors:

Estrada, V., Planta Piloto de Ingenieria Quimica - Universidad Nacional del Sur

In this work, we address the optimization of an integrated microalgae-based biorefinery through the formulation of a nonlinear programming (NLP) model for the production of the natural ketocaroteinoid astaxanthin, a powerful antioxidant with application in nutraceuticals, pharmaceuticals, cosmetics and food industries andco-products. Biodiesel from microalgae and biogas from waste streams are considered to reduce the dependency of external energy sources. Biodiesel is produced from oil transesterification, producing glycerol as a by-product. Glycerol is mostly used for pharmaceutical products, food and cosmetics but be also used as raw material for higher added value products as polyhydroxibutirates (PHB). PHB is a biopolymer produced by microorganisms and it constitutes an alternative to fossil fuel based polymers as it has similar properties. In this work, PHBis used for packaging proposes. A combined heat and power cycle to transform biogas generated by the anaerobic digestion of waste streams(algal cake, waste paper, glycerol, sludge and cell residual material from PHB extraction) is also included. Anaerobic digestion model accounts for detailed composition of the different feed streams. Detailed cost models are formulated for process equipment. The proposed model is implemented in GAMS to maximize net present value and minimize energy consumption. Numerical results provide useful insights on integrated biorefineries that use carbon dioxide as carbon source, as well as quantitative information on green processes.

Studies on Oil Extraction, Fatty Acids Profile and Biodiesel Production from Syagrus Romanzoffiana

- Conference: RCN Conference on Pan American Biofuel and Bioenergy Sustainability
- Year: 2016
- Group: Posters
- Time:
  Wednesday, September 14, 2016 - 4:42pm-4:43pm

Authors: Sanguinetti Flores, P., Facultad de Agronomía
Mongelli, E.

Most of the fuels that are used worldwide (natural gas, gasoil, fueloil) comes from fossil sources. These are non renewable energy sources, since the production rate is very much lower than the consumption rate. On the other hand, biofuels are produced from renewable energy sources like agricultural production, livestock, forestry and urban and industrial wastes. They have environmental advantages and offer the possibility of diversifying the supply sources. The main attraction of biofuels such as biodiesel and bioethanol is that the uptake of carbon dioxide during their growth balances the release of carbon dioxide when is combusted. It is also important that the biofuel production is sustainable and does not negatively impact on environmental quality. The life cycle assessment of biodiesel from different sources shows a positive balance.
Nowadays, most of the worldwide biofuels production is made with plants that come from edible species. Soybean, palm and canola in the case of biodiesel, and sugar cane and corn for the bioethanol production. For this reason, the main criticism to liquid biofuels is that they are made from edible feedstocks, competing in this way with the food production.

Thus, the present tendency is the study of non-traditional species with the aim of diversifying the productive scene with particular interest in regions that are considered marginal for the traditional agriculture.

Syagrus romanzoffiana (Arecaceae) is a palm that grows in South America. It is known as pindó, ybá pitá, yerivá, coquito, guariroba. It is an species that resists frosts, water stress and thermal amplitude.

In this research, pindó seed oil was investigated as a promising non-traditional species for preparation of biodiesel fuel. The maximum extraction of oil using hexane in a continuos Soxhlet extractor was 42 (wt) %.

For the evaluation of the biodiesel production, an alkaline transesterification at 60 ° C was carried on, using different catalyst concentrations. The maximum yield production was 92 (wt) %. The gas chromatographyâmass spectrometry (GC-MS) analysis of the FAME showed the following fatty acid profile: Caprylic (8:0) 5 % Capric (10:0) 6 % Lauric (12:0) 35 % Myristic (14:0) 10 % Palmitic (16:0) 7 % Estearic (18:0) 5 % Oleic (18:1) 25% Linoleic (18:2) 7 %. Some properties of the biodiesel obtained from *Syagrus romanzoffiana* will be discussed based on this profile.

Increase in the H2 Yield Sustainability in APR of Biomass Derivates By Reduction of Pt Loading in Pt-Ni-Co Catalysts

- Conference: RCN Conference on Pan American Biofuel and Bioenergy Sustainability
- Year: 2016
- Group: Posters
- Time:
  Wednesday, September 14, 2016 - 4:43pm-4:44pm

Authors: Dosso, L.
Teves, M.
Tuler, N.
Grau, J.

The industry constantly searches for alternative to H_2 production to cover the deficit caused by high consumption in oil refineries. This is due to increased hydrotreating (HDT) of fuel to bring it to ecological quality and the handling of heavy cuts, with high contents of S, N and metals which is regulated by HDT. Conventional generation of H_2 is by reforming naphtha or methane (HC), non-renewable raw materials, which have high cost and high environmental impact. An alternative to cover the deficit is to increase H_2 production from renewable resources derived from digestion of waste lignocellulosic biomass (CHO). The incorporation of biomass in the Refinery contributes to its transformation into Bio refinery, future target for the generation of Biofuels.

The conversion method to transform CHO derived from biomass into H_2 is the catalytic reforming. By its higher thermal sensitivity and water solubility, CHO reforming is performed at lower temperatures than HC and in aqueous phase (APR). The catalysts used in APR are multimetallic Pt based and must have high loads and dispersions of the active metal.

Hydrolysis products of glucose such as alcohols can be transformed into H_2 with different performance on the same catalyst. As still it has not found a catalyst suitable properties and acceptable cost for scaling. In this work we have studied the behavior of Pt catalysts with the addition of Ni and Co supported on alumina, in APR of different Â alcohols. The objective of this work is increase the sustainability of this process to find the most selective to H_2 and stable catalyst modifying the metal charge, using it in APR of aqueous solutions of ethylene glycol, glycerol and sorbitol. Two supported metal catalysts were prepared: a reference one with 1% Pt and a trimetallic Pt-Ni-Co. The support was Î³-Al_2O_3 Sasol, thermally stabilized (Vp = 1.48 cm^3 / g, SG = 204 m^2 / g), ground to 35-80 mesh. To determine the appropriate ratio of Ni and Co to maximize hydrogen performance and reduce CH_4 selectivity, different bimetallic catalysts were prepared varying the Ni:Co ratio, with a total metal loading of 6%.

The metallic phase was characterized by ICP and CO chemisorption. The dehydrogenation capacity was measured with DHC. The APR experiments were performed with 10 wt% alcohols in aqueous solutions, with 0.5 g catalyst reduced in situ in H_2 1 h at 500°C. The catalyst deactivation by coke deposition was studied by temperature programmed oxidation (TPO).

The good yield to H_2 of Pt can be replaced by the Ni addition. The Ni is a high productor of H_2 and CH_4, which decreases the selectivity. Co helps maintain low levels of coke and methane, which is a good additive to compensate for the high activity to undesirable reactions of Ni and promote stability. The catalyst Pt(0.5)Ni(1.5)Co(4.5)/Al_2O_3 with 0.5 wt% Pt, 1.5 wt% Ni and 4.5 wt% Co, shows a suitable yield and selectivity to H_2. At higher molecular weight of the alcohol feeded, less catalyst stability.

Sustainable Biodiesel Production through an Integrated Perspective

- Conference: RCN Conference on Pan American Biofuel and Bioenergy Sustainability
- Year: 2016
- Group: Posters
- Time:
  Wednesday, September 14, 2016 - 4:44pm-4:45pm

Authors:
Bálsamo, N. F., Centro de Investigación y Tecnología Química (CITeQ), UTN-CONICET, Facultad Regional Córdoba, Universidad Tecnológica Nacional

Biofuels are renewable solutions to replace the environmentally pollutant fossil liquid fuels. They are produced from low cost sustainable feedstocks. In the la last years, biodiesel production had increased exponentially. This is generating high quantities of glycerol as by-product. Significant economic benefits to the industry of biodiesel, is to research in a green processes through enzyme immobilized and to add value of a byproduct generated in abundance, the glycerol.

One of the objective of this work is to develop a catalyst for its use in the biofuels production from renewable raw materials using an enzymatic immobilization on solid supports nano-structured. In fact, the main purpose is to immobilize the lipase of Pseudomonas fluorescens in the mesoporous material SBA-15 (Santa Barbara Amorphous) pure, synthesized by hydrothermal treatment. In addition, by comparing their activity for the biodiesel production, from vegetable oils, with the enzyme immobilized on materials SBA-15 modified with metals (Ca, Na, Fe, Cu, Al, K) by the wet impregnation method. The physicochemical properties of the synthesized mesoporous materials were determined by Small-angle X-ray scattering (SAXS), Transmission electron microscopy (TEM) and UV-visible. The enzyme was immobilized by mixing each mesoporous material with an enzyme solution to achieve immobilization by physical adsorption. The effective incorporation of the enzyme in the materials was confirmed determining the amount of protein in the soluble fraction after immobilization by the Bradford method. The optimal conditions for the catalytic activity of the biocatalyst were determined: oil / ethanol ratio,% water, amount of immobilized enzyme / mg of SBA-15 materials, reaction time and activity of the biocatalyst respect to the metal impregnated in the solid used .

Simultaneously, glycerol value-added products such as monoglycerides were produced via transesterification reaction of methyl stearate. Monoglycerides derived from glycerol are compounds employed as emulsifiers in food, cosmetic and pharmaceutical products. This green and selective process involved catalysis from hydrotalcite-like compounds. Nano-structured materials from layered double hydroxides are industrially used in heterogeneous catalysis processes. They offer great advantages compared to homogeneous catalysis, since not only favors the environment, but also allows the use of separation operations less complex.

The proposal is to use derivatives nano-clays mixed metal oxides with different metals incorporated (Li, K or Cs) to improve their basic properties. The presence of metal ions, hydroxyl groups on the surfaces of the sheets and anions interchangeable makes them ideal compounds for preparing functionalization products allowing the design of new nanoscale materials. The materials physicochemical properties were characterized. The correlation between their properties and the catalytic activity in the transesterification reaction was investigated. The highest activity was found for the catalyst with Li incorporated by the co-precipitation method with yield values around 80% for monoglycerides. The catalyst with Li incorporated also showed the greatest Lewis medium basic sites density, a 31% more than the rest of the materials.

The production of biodiesel and the revalorization of glycerol to minimized costs, through selective heterogeneous catalytic way, which encourages the use of energy and renewable raw materials with less environmental cost, shows the road to sustainability.

Assessment Biogas Production in Argentina from Co-Digestion of Sludge and Municipal Solid Waste

Conference: RCN Conference on Pan American Biofuel and Bioenergy Sustainability

Year: 2016

Proceeding: 2nd RCN Conference on Pan American Biofuels & Bioenergy Sustainability

Group: Posters

Session: Rapid Fire Poster Session

Time: Wednesday, September 14, 2016 - 4:45pm-4:46pm

Authors:

Morero, B., INTEC - CONICET
Vicentin, R.
Campanella, E. A.

In Argentina there is an important potential to utilize organic waste as a source of bioenergy generation. This work analyzes the environmental impacts, energetic and economic requirements of biogas production by digestion of the sludge (SS) produced in a wastewater treatment plant of a medium city of Argentina. The SS is co-digested with the organic fraction of municipal solid waste (OFMSW) and the basis of the study is the life cycle assessment (LCA). The LCA is performed according to ISO 14040-44 using the SimaPro simulator. In a first stage, the transport of the different raw materials to the biogas plant was defined. Then, the co-digestion and the biogas treatment for final use were evaluated. The co-digestion was improved with glycerol and the generation of biogas was estimated with the GPS-X software. Two alternatives of biogas end use were considered: combined heat and power (CHP) and biomethane generation. For the first option H2S is removed from the raw biogas stream, while the second option also removes CO2. The H2S removal process was simulated in the SuperPro software by anaerobic biofiltration. The same software was used to simulate the removal of CO2 absorption-desorption with water as solvent. Finally the environmental impacts related to the end use of biogas (CHP and biomethane) were evaluated. The environmental analysis showed that the use of organic waste as an energy source, solving problems related to waste disposal, generates less environmental impact and lower costs than conventional disposal in a landfill. The feed to the digester needs to be big enough for a more economic attractive biogas production; then the disposal of organic waste (OFMSW + SS) of a medium-sized city should be complemented with waste from nearby towns or agricultural/livestock waste. The energy balance is highly positive with the use of energy for the overall process (Feedstock collection and transport, Anaerobic Digestion, Digestate Management and Biogas Treatment Plant) between 30 to 39% of the energy generated.

Life CYCLE Assessment of Sugarcane-Based Ethanol Production in the Province of Tucuman (ARGENTINA)

- Conference: RCN Conference on Pan American Biofuel and Bioenergy Sustainability
- Year: 2016
- Group: Posters
- Time:
  Wednesday, September 14, 2016 - 4:46pm-4:47pm

Authors: Garolera De Nucci, P.
Tonatto, J.
Romero, E., EEAOC
Cárdenas, G.

Sugarcane production, as well as bioethanol from sugarcane, are part of the most important activities in Tucuman and North Western (NW) region of Argentina. This study shows the application of life cycle assessment (LCA) to such activity. LCA is an appropriate tool to estimate the environmental burdens associated with a product or process during their entire life cycle from cradle (extraction of raw materials) to grave (disposal). This technique allows identifying critical process steps susceptible of improvements and is an important contribution in decision-making. LCA is performed according to the procedures included in the ISO 14040 standards.
The aim of this paper is to present the environmental profile of bioethanol from sugarcane in the province of Tucuman (NW region), evaluating a set of relevant environmental impacts through LCA technique.

The boundaries of the sugarcane system included biomass, sugar and alcohol production. The functional unit is defined in 1 kg of bioethanol obtained at the gate of the distillery. The data used for making inventories have been mostly collected directly from the field, mill and distillery, through interviews with experts from agribusiness and Estación Experimental Agroindustrial Obispo Colombres (EEAOC) as well as available publications. Material and energy balances were also performed and international databases were used (eg. Ecoinvent v3). The LCA study is conducted with the help of software SimaPro® v8.1.1.16 (PRé Consultants), a software tool compatible with ISO 14040, while resorting to method of ReCiPe Midpoint V1.12 for impact assessment.

The study revealed that all impact categories are dominated by the contribution of the production of sugar and molasses, which embeds the mill and sugarcane environmental impact too.

As a result of this study, the environmental profile of bioethanol produced in Tucumán was obtained. It has to be noted that in terms of climate change category, a positive environmental impact prevails due to the absorption of CO_2 during sugarcane growth.

This study also makes available information of sugar and ethanol production in other impact categories, which contributes to its sustainability definition as a biofuel. This depends largely on the actions and technological developments implemented to mitigate the environmental impact of biofuel production stages.

Desarrollo De Biocombustibles En La Provincia De Salta

- Conference: RCN Conference on Pan American Biofuel and Bioenergy Sustainability
- Year: 2016
- Group: Posters
- Time:
  Wednesday, September 14, 2016 - 4:47pm-4:48pm

Authors: Cadenas, A. J., Asme
Jimenez, L. A., Consejo Profesional de Ingenieria Industrial

Desarrollo de Biocombustibles en la Provincia de Salta[1]

En los últimos años, el sector de biocombustibles ha crecido rápidamente en buena medida ayudado por las señales auspiciosas del mercado internacional y por el establecimiento de un marco regulatorio específico.

Las políticas para el desarrollo de los biocombustibles se relacionan con los beneficios de diversificar y aumentar el valor agregado de la oferta exportable de base agropecuaria; con su potencial para modificar la matriz energética frente a la caída en las reservas de hidrocarburos, y con la posibilidad de reducir las emisiones atmosféricas asociadas al transporte.

Por su ubicación, a la región de la Provincia de Salta le corresponde un clima subtropical. El sistema montañoso ubicado al oeste de las principales zonas productivas de grano, ejerce una marcada influencia sobre las precipitaciones y las temperaturas, generando una gran diversidad de ambientes. Las lluvias anuales disminuyen de oeste a este desde 1.000 mm hasta menos de 550 mm.

Se observa una gran variedad de residuos que pueden ser aprovechados para su uso directo como ser transformados a biocombustibles. Se destacan entre ellos los residuos del monte, sean estos de montes nativos como de montes implantados. Sobresale también la disponibilidad de residuos de la actividad frutal, especialmente cítricos y banana. Adicionalmente, se ha detectado una oferta interesante de residuos de la actividad ganadera intensiva (feedlots, faena y frigoríficos y tambos).

En este contexto, el objetivo de este estudio es la identificación y determinación del potencial de los biocombustibles como fuente de energía renovable y eficiente para la Provincia de Salta. Para ello se identifican los distintos cultivos actuales o existentes en la región y otros recursos susceptibles de ser aprovechados como recursos energéticos. Se analiza su oferta directa e indirecta, se realiza una estimación y distribución geográfica de las reservas, se elaboran mapas de aptitud agroecológica y ambiental de los cultivos con fines bioenergéticos y se determinan las características climáticas y de suelo necesarias para este tipo de cultivos. Asimismo, en este trabajo se identifican y analizan las barreras regulatorias, legales y tributarias para la promoción y fomento de los biocombustibles en la Provincia; se identificación las barreras técnicas y de infraestructura para la generación de los mismos y se presentan los criterios de sostenibilidad ambiental, económica y social para proyectos de esta índole. Finalmente, y en función de los recursos relevados en este trabajo, se resumen las acciones que pueden sugerirse con el fin de desarrollar los biocombustibles en la Provincia de Salta.

[1] Trabajo realizada para el "Programa de Desarrollo Integral de Energias Alternativas en las Provincias de Buenos Aires y Salta Financiado por el BID (préstamo: ATN/OC-11500-AR, Apoyo al Desarrollo de Biocombustibles en la Provincia de Salta)."

ATLAS BioenergÉTica De Ecuador: Punto De Partida DEL PLAN De DiversificaciÓN De La Matriz EnergÉTica Nacional

- Conference: RCN Conference on Pan American Biofuel and Bioenergy Sustainability
- Year: 2016
- Group: Posters
- Time:
  Wednesday, September 14, 2016 - 4:48pm-4:49pm

Authors: Fazzini, C. A.
Cadenas, C. V.

ATLAS BIOENERGÉTICA DE ECUADOR: PUNTO DE PARTIDA DEL PLAN DE DIVERSIFICACIÓN DE LA MATRIZ ENERGÉTICA NACIONAL[1]

El acelerado incremento del consumo energético a nivel mundial en los últimos años, el interés por diversificar la matriz energética, reducir la dependencia del petróleo y la necesidad de disminuir los problemas ambientales asociados con el uso de los combustibles fósiles, han motivado la realización de múltiples esfuerzos por parte de los organismos relacionados con el quehacer energético, hacia la utilización de potenciales fuentes alternativas y renovables de energía.

Desde la óptica de desarrollo sostenible, actualmente el Ecuador apunta a la diversificación de la matriz energética con la incorporación de energías renovables, entre ellas, aquella proveniente de la biomasa, con la finalidad de reemplazar el uso de combustibles fósiles altamente contaminantes y garantizar su autonomía energética.

El Ministerio de Electricidad y Energía Renovable, el Ministerio Coordinador de Producción, Empleo y Competitividad y el Instituto Nacional de Preinversión, impulsaron en el 2014 la elaboración del Altas Bioenergético, cuya finalidad es cuantificar y ubicar geográficamente la biomasa residual con potencial energético, resultante de la actividad agropecuaria y agroindustrial a nivel cantonal, en la República del Ecuador, así como también aquella generada por la explotación de bosques implantados y de la industria forestal consecuente.

La elaboración de estudios de preinversión asegura una adecuada ejecución de la inversión pública porque permite planificar los proyectos desde su etapa inicial y optar por las mejores alternativas técnicas, económicas, sociales, ambientales y de financiamiento, tanto en su diseño como en su construcción. Contar con información referencial, precisa, detallada y técnicamente probada de cuál es la alternativa más viable para ejecutar un proyecto al menor costo y tiempo, con el mínimo de impactos posibles garantiza, además, que se realicen obras de calidad, con el consecuente ahorro de recursos para el Estado.

Los estudios desarrollados tienen como propósito apuntalar la transformación de la matriz productiva, garantizar el derecho al acceso al agua, promover el desarrollo de la ciencia y la investigación, proveer de una infraestructura vial moderna y segura, usar nuevas TIC´s para la innovación y, por sobre todo, mejorar la calidad de vida de los ecuatorianos.

La disponibilidad de biomasa utilizable con fines energéticos en el Ecuador, se ha plasmado en mapas que facilitan el acceso a su conocimiento y análisis de una manera clara y simple.

Este Atlas brindará información para el desarrollo de proyectos de generación eléctrica, térmica y cogeneración, así como para la implementación de políticas de promoción del aprovechamiento de la biomasa procedente de residuos agropecuarios en el paÃs, contribuyendo al progreso tecnolÃ³gico del Ecuador y de esta forma a la consecuciÃ³n de los objetivos nacionales para el Buen Vivir.

[1] Prestamos BID 2585/OC-EC â SP-INP-MEER-2013-002

Formulation and Development of an Integrated Pervaporation Module: Application to the Process Synthesis and Optimization of Bio-Ethanol Production from Sugar-Cane

- Conference: RCN Conference on Pan American Biofuel and Bioenergy Sustainability
- Year: 2016
- Group: Posters
- Time:
  Wednesday, September 14, 2016 - 4:49pm-4:50pm

Authors: Colombo, M., Universidad Nacional de Tucuman
Gatica, J. E., Cleveland State University
Mele, F. D., Universidad Nacional de Tucuman
Hernández, M. R., Universidad Nacional de Tucuman

Todayâs process synthesis has been greatly facilitated by the versatility of modern day process simulators. Indeed, these tools are currently integrated in the R&D of most industries. One of the advantages is the ability to provide rapid assessment of new technologies by integrating process development with its economic evaluation and energetic integration.
Unfortunately, in order to take advantage of rapid processing platforms and rigorous thermodynamics databases available with modern process simulators, rigorous models for critical units are needed. A common area of lacking in most commercial simulators are chemical reactors, reactive distillations, and separation processes not based on equilibrium.

Â One such an example is the rising popularity of the industry of bio-renewables and other sustainable technologies.Â A rapidly growing regional market is the bio-ethanol production, due to the possible uses of the anhydrous product in energy generation. This production however, is weighed down by the energetic demands and the environmental effects of the distillation-dehydration steps needed to to reach fuel-grade specifications.

This work focus its attention on the process synthesis of the dehydration step, a separation stage where most conventional separation processes are hindered by the presence of an azeotrope in the ethanol-water mixture generated in the distillation train. The alternatives examined here are based on the formulation and use of a pervaporation module to complement the production of fuel-grade bio-ethanol in distillery installations integrated with sugar mills.

Reaction Engineering Routes to Waste Gasification for Sustainable Living Environments

- Conference: RCN Conference on Pan American Biofuel and Bioenergy Sustainability
- Year: 2016
- Group: Posters
- Time:
  Wednesday, September 14, 2016 - 4:50pm-4:51pm

Authors: Colombo, M., Universidad Nacional de Tucuman
Lange, E. M., Cleveland State University
Barbutti, A. D.
DeMattia, B., Cleveland State University
Moten, J.
Hernández, M. R., Universidad Nacional de Tucuman
Gatica, J. E., Cleveland State University

Concern over âgreen,â or environmentally friendly, technologies has risen considerably over the past decade. Thus, over the past 50 years, metropolitan areas have realized a need for waste disposal alternatives as landfills approach capacity and take up valuable land space. Although newer technology has made incinerators more efficient, there is an increasing interest in formulating more âgreenâ gasification options as alternatives to incinerators.
Gasification converts organic and carbonaceous materials into a combination of gaseous products known as âsyngas,â or synthetic gas. This research focuses on advancing the current knowledge of a catalytic gasification process as a potential in-situ resource utilization and waste management system. With significance in a variety of engineering applications, this research is of particular relevance towards municipal waste management and as in-situ resource utilization alternative for advancing space exploration.

The catalytic gasification reaction mechanism evolves through a multi-phase reaction mechanism. The process is initiated by a liquid phase oxidation reactions of long chain polymers. The liquid phase oxidation reactions produce carbon monoxide, carbon dioxide, and water. Current long chain polymers being studied in the catalytic gasification mechanism are focused on cellulose and polyethylene as model substrates. The oxidation reactions are complemented by two gas phase reactions: the Water Gas Shift and the Sabatier reaction, the main stages in the pathway of producing hydrogen and methane.

Gasification experiments examine gasification dynamics for different substrates promoted by different catalysts. It utilizes a high pressure batch scale reactor which allows examining controlled multi-phase catalytic isothermal oxidation of solid and melting substrates in the presence high purity air.

The studied substrates, Polyethylene and Cellulose, are both long chain organic polymers, and make up a substantial portion of both space and municipal waste composition. Although similar in nature, these substrates exhibit marked differences as it pertains to gasification, and were therefore selected as model substrates.

Analysis of the liquid and solid reaction residues, along with the gaseous products, enabled the characterization of the gasification kinetic pathway. The results are used in conjunction with a kinetic model for the gas-phase to formulate a phenomenological model for the gasification of solid waste.

Preliminary results suggest that catalytic gasification provides a potential route for control of the liquid-phase oxidations. Under appropriate conditions, oxidation was found to follow the incomplete oxidation pathway; which resulted in significant yields of hydrogen and methane. Attention to these products was focused as they are potentially high value products that can be used for energy recovery.

The results will be presented along with a preliminary extension for a continuous gasification process, which clearly illustrate the potential of catalytic gasification as a sustainable waste management alternative.

Optimal Design of Grain Sorghum Based Ethanol Plant

- Conference: RCN Conference on Pan American Biofuel and Bioenergy Sustainability
- Year: 2016
- Group: Posters
- Time:
  Wednesday, September 14, 2016 - 4:51pm-4:52pm

Authors: Estrada, V., Planta Piloto de Ingenieria Quimica - Universidad Nacional del Sur
Montain, M. E., PLAPIQUI (UNS-CONICET)
Borio, D., PLAPIQUI (UNS-CONICET)

The increasing interest in supplement or substitute products such as fuels, chemicals, plastics, etc. derived from fossil resources has driven the study and development of different processes that could use a wide variety of renewable raw materials. In particular, bioethanol production can improve energy security, reduce carbon dioxide emissions, and decrease urban air pollution.
In the present work, we carry out a study for an ethanol production plant based on grain sorghum. This cereal has a low water requirement making it suitable for cultivation in drier areas and it is not use for human consumption. The produced ethanol could be used to produce ethanol-blended fuels or as raw material for ethylene production. We formulate mass and energy balances for the dry milling process, which is the most widely used for grain-based ethanol production. Byproducts are carbon dioxide and DDGS, distilled dried grains with solubles Carbon dioxide can be used, after ethanol removal, in soft drinks plants, as supercritical solvent in many extraction processes, as raw material in algal-based biorefineries, etc. DDGS are used for animal feed due to their high nutritional value. We perform heat integration and optimization of water consumption on the proposed ethanol process. An economic evaluation is also carried out for the two scenarios (optimized and non-optimized (base case) plant).

In this work we consider a grain sorghum based ethanol plant production of 310 ton ethanol/day. Water consumption is a key issue in this type of plants, being an acceptable value 3 l water/l ethanol. In our case study, numerical results show that water consumption could decrease from 4 l water/l ethanol for the base case to 1.3 l water/l ethanol for the optimized case. On the other hand, a reduction of 24% on energy consumption is achieved in the optimized plant with respect to the base case. The economic analysis reveals an increase of 23% in the net present value for the optimized plant for a project useful life of 10 years. Another important result obtained from the costs evaluation is that raw material cost constitutes about 75% of ethanol production cost, which emphasizes the advantages of grain sorghum against corn, due its lower price with respect to corn.

Energy Impact of the Feed Composition to the Dehydration Column in the Production of Bioethanol

- Conference: RCN Conference on Pan American Biofuel and Bioenergy Sustainability
- Year: 2016
- Group: Posters
- Time:
  Wednesday, September 14, 2016 - 4:52pm-4:53pm

Authors: Cuezzo, A. M.
Araujo, P. Z.
Hernández, M. R., Universidad Nacional de Tucuman
Colombo, M., Universidad Nacional de Tucuman
Gatica, J. E., Cleveland State University

The high price and the environmental harmfulness of petroleum fuels have highlighted the need for alternative energy sources, being bioethanol one of the most important source. Bioethanol is a co-product of high profitability and great opportunities for development, its use as biofuel improves the quality of gasoline in an environmentally friendly level.

In 2006, Argentina sanctioned Law 26.093 of Promotion of Biofuels, setting a minimum percentage of 5% of ethanol in gasoline. In April 2016, the mandatory percentage of bioethanol in gasoline was increased to 12% by volume, in the framework of Laws 26,093 and 26,334 for the Regulation and Promotion of Biofuel Production and according to the consolidation of Belgrano Plan. These regulations triggered a significant increase in the production of bioethanol in the province of Tucuman.

The bioethanol production process in Argentina is mainly from molasses fermentation and / or sugar cane juice. The mixture of alcohol and water produced by simple distillation, enters a subsequent separation process for dehydration purposes. Through distillation conventional methods, it is only possible to obtain ethanol concentrations close to 89% mole fraction. Therefore, as the anhydrous ethanol production processes are now a priority environmental, technological and economic; energy efficient separation methods that break the azeotrope, such as extractive distillation are needed.

This study examines the steps involved in simple distillation and the following dehydration step using glycerol as an extractive agent. For this, an extractive column leading anhydrous ethanol and a distillation column for recycling glycerol are needed. Alternative schemes are proposed to assess the energetic impact of the composition of the feed stream to the dehydration section through process simulation, which saves consuming time, costs and experimentation in the design of bioprocesses.
In the conventional process the feed stream to the dehydration step is a mixture of ethanol-water at the azeotropic point. In the proposed alternative processes, we studied the effect of different ethanol-water mixtures obtained as product in the simple distillation column and its effect on the dehydration step by extractive distillation.

Conversion of Cellulose to Polyols with Ag Catalysts Supported on Mesoporous Mordenite

- Conference: RCN Conference on Pan American Biofuel and Bioenergy Sustainability
- Year: 2016
- Group: Posters
- Time:
  Wednesday, September 14, 2016 - 4:53pm-4:54pm

Authors: Aspromonte, S. G.
Romero, A.
Alonso, E.
Boix, A. V.

Actually, the use of biomass as feedstock for the production of fuel, energy and chemicals represents one of the future technologies that can help reduce global warming and reduce dependence on fossil fuels. Specifically, cellulose is one of the most abundant components of biomass and, therefore, represents one of the promising materials in selective hydrolytic processes to generate fuels and chemicals of high added value [1].
Here, supercritical water (SWC) is recognized to possess good ability to depolymerize cellulose. Although more than 96% of the cellulose is solubilized, the presence of oligomers remains high (> 60% C) [2]. A possible mechanism to solve this problem is complete the hydrolysis of cellulose using heterogeneous acid catalysts (sulfonated mesoporous silica, sulfated zirconia, sulfonated activated carbon) [3]. Furthermore, Fukuoka and Dhepe [4] and Luo et al. [5] reported heterogeneous catalytic processes for the conversion of cellulose to sugars and microcrystalline alcohols, by combining hydrolysis hydrogenation over supported noble metal catalysts, such as Pt / Al2O3 and Ru/C [6]. Therefore, parameters such as porosity, specific area, active phase and the acidity of the catalysts are key factors in the development of these processes.

Microporous materials (pore diameter<2 nm) as zeolites offer high acidity, thermal and chemical stability at elevated temperature. On the other hand, mesoporous materials MCM and SBA type have larger pores with narrow size distribution which promotes steric selectivity for the conversion of bulky molecules such as sugar. However, the acidic properties are limited to the incorporation of another element such as Al for the synthesis of siliceous structure.

Therefore, the main objective of this work is the synthesis of an active for cellulose hydrolysis catalyst, the mordenite zeolite using as a source of Si and Al, and a structuring agent to expand the pore size. In addition, it is proposed to study the effect of adding Ag on the conversion of cellulose to glucose.

The synthesized mesoporous zeolitic support showed a significant increase in the pore volume and size thereof.

The incorporation of Ag does not significantly change the pore diameter, allowing oligomers bulky

Consequential LCA (C-LCA), Agri Industrial Symbiosis and Waste Biorrefinery. a Literature Review

Conference: RCN Conference on Pan American Biofuel and Bioenergy Sustainability

Year: 2016

Proceeding: 2nd RCN Conference on Pan American Biofuels & Bioenergy Sustainability

Group: Posters

Session: Rapid Fire Poster Session

Time: Wednesday, September 14, 2016 - 4:54pm-4:55pm

Authors:

Schein, L., INAHE CONICET
Civit, B. M., INAHE - CONICET

This article is a review of published papers in various journals and conference proceedings on the topics of bioenergy and biorefinery, waste to energy, waste to wealth and agri industrial symbiosis, between January 2006 and June 2016. Around 135 papers are identified. The article intends to build a map of concepts and its evolution, in the areas of agri industrial symbiosis and biorefinery schemes from a consequential LCA approach. The literature is analyzed by theoretical and methodological advances as well as case studies applications. Through the development of a set of criteria to value each articleâs approach, a ranking of importance is settled, from which different key elements are identified as potential components of a model, that ultimately aims to fulfil complementary objectives: help building value on waste biomass by efficiently managing agricultural waste streams.

The late emergence of these topics, blooming along the circular economy approach, have been observed, and some peculiar overlapping of ideas has been identified. These many perspectives over a complex issue inspired a phD thesis on how to efficiently model waste biomass as a resource, in the agri industrial conglomerate, considering its territorial expression. This project is being undertaken and it's expected to be ready within 4 years.

Study Nonisothermal Drying Kinetic By Thermogravimetric Analysis

- Conference: RCN Conference on Pan American Biofuel and Bioenergy Sustainability
- Year: 2016
- Group: Posters
- Time:
  Wednesday, September 14, 2016 - 4:55pm-4:56pm

Authors: Fernandez, A.
Echegaray, M.
Palacios, C.
Mazza, G., Universidad Nacional del Comahue
Rodriguez, R., Facultad de Ingeniería

Biomass, particularly agro-industrial wastes, is a renewable fuel and the fourth largest following coal, oil and natural gas. Compared with these fossil fuels, it has the advantages of being neutral concerning the emissions of the greenhouse gas carbon dioxide, as this participates in biomass growth through the photosynthesis reactions, and reducing pollutant species generation, given the low sulfur and nitrogen contents. Bio-wastes can create major environmental problems if not properly handled.
On the other hand, one of the most important economic activities in the Cuyo Region, Argentina, is the agroindustry, this sector produces a significant environmental impact in specific geographical areas. These wastes generated have significant amounts of lignocellulosic materials.

The recent Argentine political, are strongly encouraging the use of biomass for energy purposes, mainly owing to three targets economic and social development of countryside, elimination of wastes and reduction of CO_2 emissions.

The more developed technologies for obtaining energy or fuels from biomass are those based on thermochemical treatments, namely, pyrolysis, gasification and combustion.

The moisture content of biomass wastes is an important property in the thermal treatment since the moisture levels may affect its performance and reliability. As is well known, the first step of pyrolysis and combustion corresponds to water evaporation, which requires energy and reduces overall process efficiency.

After the forced drying at 378K until constant weight, for example, the moisture content of biomass can decrease between 2 at 5%. Numerous pyrolysis and combustion studies have used this dry-basis biomass for experimental research and kinetic analysis.

High moisture content has a much lower net energy density by mass, owing to the weight of the water, but also by volume due to the energy required to evaporate the water. Consequently, drying pretreatment of biomass to thermal treatment is usually desirable and in some cases essential, in order to minimize the effect of moisture on the treatment.

The kinetic modeling of the drying stage is crucial for an accurate prediction of the materialâs behavior under different working conditions. Investigation of this stage kinetics and its parameters contributes to further information of heat and mass transfer.

As biomass is linear heating in the thermogravimetric analyzer, the moisture continuously absorbs energy and then evaporates. However, not all water molecules can be turned to vapor and released. Only the water molecules that have obtained sufficient energy for gasification can undergo the transition. This level of gasification energy can be regarded as the activation energy required for biomass to dry. The higher the drying activation energy the more heat requirement in the drying process.

The aim of this study is the determination of nonisothermal drying kinetics of marc, stalk and peach pits from Argentine agro-industry. Different models were applied at obtained TG data of water evaporation under inert and oxidative atmosphere. The activation energy was determined using integral and differential methods applied at the model with better fit.

Bio-Oil Upgrading. Catalytic Hydrodeoxygenation of Furfural

Conference: RCN Conference on Pan American Biofuel and Bioenergy Sustainability

Year: 2016

Proceeding: 2nd RCN Conference on Pan American Biofuels & Bioenergy Sustainability

Group: Posters

Session: Rapid Fire Poster Session

Time: Wednesday, September 14, 2016 - 4:56pm-4:57pm

Authors:

Zanuttini, M. S., Instituto de Investigaciones en catálisis y petroquímica
Querini, C., Instituto de Investigaciones en Catálisis y Petroquímica (INCAPE - UNL/CONICET)
Gross, M. S.
Zuttión, A. M., Universidad Nacional del Litoral

Fast pyrolysis of lignocellulosic wastes has become an interesting process to generate fuel. TheÂ liquid obtained known as bio-oil, contains a variety of oxygenated compounds, including water.Â The energy potential is very important. However, for industrial purposes a pretreatment isÂ required. Catalytic hydrodeoxygenation (HDO) is one of the technologies currently studied toÂ improve the quality of bio-oils.Â Due to the complexity of the liquid, initial studies are based on model compounds or synthetic mixtures. In previous works, the hydrodeoxygenation of model compounds such as m-cresol, anisole, phenol, and others was studied. Various catalysts, such as Pt on Î³-Al_2O_3, SiO_2Â or proton beta zeolite exchanged with sodium or lanthanum, Pd on Î³-Al_2O_3, SiO_2, and Î±-C, Fe or Ni/SiO_2, were evaluated. In this work, deoxygenation of a model compound of bio-oil, furfural, was studied at atmospheric pressure in presence of hydrogen, using platinum or iron supported on silica as catalysts. Furfural was chosen as model molecule because it the main compound observed in bio-oils. Monometalic (Pt or Fe) and bimetallic (Fe-Pt) catalysts were evaluated. The loading of the metals were varied. The catalysts were characterized by XPS, XRD, BET, TPR, and TEM. The main products obtained were furan and 2-methylfuran, the latter being the most desirable in terms of overall system performance and fuel properties. Possible reaction pathways were proposed. Influence of the electronic nature of the supported metals and the reaction temperature on products distribution was analyzed. Due to its electronic configuration, iron atoms have selective attraction to the oxygen atom present in the C=O group of the furfural molecule and repulsion to the aromatic ring. This allows high selectivity to 2-methylfuran. However, it stability is much lower than the one of platinum catalyst. A catalyst regeneration study of Fe/SiO_2 revealed that the carbonaceous deposits are removed by treatment in air at 300 °C. Moreover, it was observed that the incorporation of 0.05% of platinum to iron catalyst improves its stability.

Sustentabilidad DEL Biodiesel Utilizando EL Glicerol Como Materia PRIMA PARA Biorefinería

- Conference: RCN Conference on Pan American Biofuel and Bioenergy Sustainability
- Year: 2016
- Group: Posters
- Time:
  Wednesday, September 14, 2016 - 4:57pm-4:58pm

Authors:
Comelli, R. A., INCAPE (FIQ-UNL, CONICET)

The process to produce biodiesel generates glycerol as by-product, interesting its transformation into added-values compounds and/or energetic ones to improve the sustainability of the overall process. In the province of Santa Fe, in the influencing area of Greater Rosario, a capacity of 2.5 million tons annual of biodiesel is installed, generating consequently 250,000 tons/year of glycerol. Glycerol is an intermediate to synthesis of large amount of compounds employed to industries, but the problem related to its increasing availability linked to the increased production of biodiesel is not solved by adopting a single-product strategy, because it would become the new problem; consequently, solution should have a diversified approach such as applying the concept of biorefinery, which is the functional unit that converts biomass into added-value products and other ones capable to supply energy. In this context, catalytic processes were developed to transform sustainability glycerol to compounds with industrial applications, such as dihydroxyacetone (DHA), acetol, propyleneglycol (PG), 1,3-propanediol, lactic acid, and methanol, as well as compounds with energetic use such as hydrogen, synthesis gas, and ethers of glycerol. Reactions of selective oxidation and reduction, steam reforming, and hydrogenolysis of glycerol were studied, showing the possible integration of processes in an environment of biorefinery. Selective oxidation to DHA occurred on Pt/K-FER, improving with Pt-Bi/K-FER, reaching 75.9% conversion and 93.9% selectivity to DHA. Selective reduction to PG in gas phase took place on Cu-containing catalysts, reaching the best performance Cu-Ce/Alumina, with 99.8% conversion and 83.2% selectivity to PG, being the main byproduct acetol (14.3%), which is the reaction intermediate. Oxidation to lactic acid, which demand is increasing, occurred on Cu/Alumina with Cu loadings between 6 and 40%w/w, at 240ºC and 1.4 MPa, reaching the best performance the material with the largest loading. The steam reforming produced mainly hydrogen, carbon oxides, and methane, depending of the catalyst design and operating conditions that favor either the formation of hydrogen, an energy vector considered the future fuel, or the production of synthesis gas, a mixture of hydrogen and carbon monoxide, which can be used for methanol and Fisher-Tropsch synthesis or as a fuel mixture; catalysts of nickel supported on alumina, modified with Zr, Mg, Ce, and Co, were prepared, characterized, and evaluated. Moreover, the hydrogenolysis of glycerol in liquid phase to methanol, which is raw material for biodiesel production, allowing it the biodiesel process as a fully sustainable one; Ru-containing catalysts showed promising results. Consequently, only feeding glycerol and integrating processes is possible to obtain DHA, PG, acetol, lactic acid, hydrogen, synthesis gas, carbon dioxide, methane, and methanol. Two catalysts, for selective oxidation to DHA and selective reduction to PG, were developed and patented; moreover, a funding to scaling up a pilot plant to produce 100 tn of PG per year was obtained, while another funding was solicited for scaling up the steam reforming process to a pilot plant to produce hydrogen, which is raw material into the selective reduction.

Carbon Footprint and Energy Analyses of Biomethane from Anaerobic Digestion of a Food Waste / Manure Mixture in Colorado, USA

Conference: RCN Conference on Pan American Biofuel and Bioenergy Sustainability

Year: 2016

Proceeding: 2nd RCN Conference on Pan American Biofuels & Bioenergy Sustainability

Group: Posters

Session: Rapid Fire Poster Session

Time: Wednesday, September 14, 2016 - 4:58pm-4:59pm

Authors:

Shonnard, D. R., Michigan Technological University
Ankathi, S., Michigan Technological University
Parra-Alvarez, N.
Potter, J., AgEnergy USA LLC

This study determined the life cycle greenhouse gas (GHG) emissions and Net Energy Return on Investment of bio-methane produced from anaerobic digestion of a mixture of food waste and cow manure in a facility near LaSalle, Colorado. The analysis compares a business-as-usual (BAU) case with an anaerobic digestion (AD) case. The BAU case includes the composting of food waste with manure plus the production and use of fossil natural gas. The AD case includes production of bioCH_4 through an advanced AD process as well as production and use of peat moss in an amount that represents the difference in compost production between the two cases. Included in the system boundary for this AD analysis are long distance semi-truck transport of collected food waste from the Denver metropolitan area, short haul transport of manure waste from local dairy farms, AD process and biogas purification, distribution of the bio-methane into the natural gas pipeline grid, combustion of bio-methane, and distribution of digestate from the AD process to local dairy farm fields for provide both irrigation water and mineral nutrients. The carbon accounting included both fossil and biogenic carbon. The effects of co-products solid compost sold to retail gardening market and nutrient rich liquid digestate distributed to local farm fields were accounted for in the analysis.

After including emission credits for avoiding landfill emissions of CO_2 and CH_4, bio-methane emits over its life cycle -3.5 kg CO_2 equivalents / kg bio-methane, thus reducing global warming effect with each unit of production of bio-methane. This favorable emission intensity is compared to that of fossil methane of between 3.3 to 5.1 kg CO_2 equivalents / kg fossil methane. Effects of life cycle system parameters in the model on GHG results are investigated in scenario analyses and also in time dependent analysis of landfill emissions.

Is Biodiesel from Microalgae a Real Energy Option? Preliminary Life Cycle Assessment (LCA) of a Multigeneration System

- Conference: RCN Conference on Pan American Biofuel and Bioenergy Sustainability
- Year: 2016
- Group: Posters
- Time:
  Wednesday, September 14, 2016 - 4:59pm-5:00pm

Authors:
Rodríguez, P. D., Grupo CLIOPE "Energía, Ambiente y Desarrollo Sustentable"-Universidad Tecnológica Nacional, Facultad Regional Mendoza-CONICET
Arena, A. P., Universidad Tecnológica Nacional
Piastrellini, R., CLIOPE. Universidad Tecnológica Nacional. Facultad Regional Mendoza - CONICET

Global predictions have indicated that the world population will need 50% more fuel, 50% more water and a 50-80% reduction in carbon dioxide (CO_2) emissions by the year 2050. In this context, renewable energies will have an important role replacing fossil fuels in critical issues, especially biofuels (biodiesel and bioethanol), since they can be used in transportation and electricity generation and they can store energy. Many authors have pointed out the microalgae as the major raw material for biofuels production, principally biodiesel. This is possible because the microalgae have higher productivities than crops, they can accumulate large amounts of fatty acids (the feedstock for biodiesel production), they consume CO_2, they do not compromise the production of food and they can grow in wastewater. Therefore, the production of biodiesel from microalgae is a well explored topic. Nevertheless, it is facing numerous economic and technological challenges that need to be solved before biodiesel becomes a real alternative to fossil fuels usage. Also, many energetic and environmental analysis have been conducted and they exhibit very divergent results that put the sustainability of microalgae-based biodiesel into doubt. Consequently, biodiesel from microalgae production is an issue that needs to be fully explored. The aim of this study is to evaluate the environmental impacts in the life cycle of a multigeneration system for biofuels and co-products production through microalgae cultivation in compact tubular photobioreactors (PBR). For this purpose, the Life Cycle Assessment (LCA) based on the recommendations of the ISO 14040 is performed. The results show that the production of biomass from microalgae has a large amount of energy inputs, which involve another environmental impacts, like greenhouse gases (GHG) emissions. For that reason, it is necessary to take into account the production of various co-products to enhance the environmental performance of the microalgae-derived biodiesel. Multigeneration strategies exhibit a big potential to produce energy and multiple products from microalgae. As a consequence, biodiesel from microalgae will become an alternative source of energy only if it is produced into multigeneration systems. This study also demonstrates the importance of LCA to evaluate different scenarios into multigeneration systems and to indicate critic points in the biofuels production chain. In addition, the system under study presents some challenges to solve with LCA, such as the allocation between multiproducts, the CO_2 consumption by microalgae, the use of wastewater as nutrient for microalgae and the inclusion of the waste treatment into the system.

A Mixed Integer Linear Programming Model for Optimizing the Design of the Forest Supply Chain in Argentina

- Conference: RCN Conference on Pan American Biofuel and Bioenergy Sustainability
- Year: 2016
- Group: Posters
- Time:
  Wednesday, September 14, 2016 - 5:00pm-5:01pm

Authors: Alasino, N.

Forest industry faces many challenges in the present. The industrial network is formed by the wood manufacturing and the pulp and paper industries. In the particular case of Argentina, a strategic national plan is required in order to promote specific developments in the involved sectors. Regarding forest provision, there is a scarce provision of eucalyptus and salicaceae species (hardwood) and an oversupply of pine (softwood) is located in one province: Corrientes. In addition, investments are required for new sawmills given the limited installed capacity in the country. The pulp and paper sector is a capital intensive industry and incentives are needed in order to attract local and foreign investments guaranteeing a sustainable industrialization and import substitution. In this sense, Argentina produces 980 thousand tons of pulps per year while other neighbors such as Chile, Uruguay and Brazil produce 4.5, 1.2 and 13.5 millions of tons, respectively. Furthermore, the commercial balance presents deficits in some specific type of papers. The lack of research and development activities in the involved companies turn out into low innovation regarding the development of new products as well as in the ways of doing business. The forest supply chain (FSC) is then driven to expand its limits, increase the installed capacities and develop a product portfolio with higher added value. One interesting alternative is to integrate ethanol bio-refineries into the FSC. Forest resources represent 80% of the global biomass and are widely distributed. The abundance of these sources as well as the fact that they do not directly compete with food needs, make these an attractive raw material for the bio-ethanol industry. Indeed, in order to remain competitive, the FSC must increase installed capacities and develop a product portfolio with higher added value. With this objective, the integration of ethanol bio-refineries into the FSC is evaluated in this work. The FSC involves forest sites that can be expanded, installed and potential pulp and paper plants, timber mills, new cellulosic bio-ethanol plants and demand nodes. Different types of wood species and residues from the trees felling and the industrial processes are considered to select the most convenient. Operating nodes can be expanded and new ones installed. Four processes are proposed for the bio-ethanol plants, i.e. simultaneous saccharification and fermentation; simultaneous saccharification and co-fermentation; acid hydrolysis and separate fermentation and enzymatic hydrolysis and separate fermentation; each of them with different operational and investment costs as well as production efficiencies for each raw material. The objective is to determine the optimal location of nodes, links, technologies and feedstock minimizing total costs. The problem is formulated as an MILP and implemented in GAMS 24.1 system.

An Integrated Sustainability Evaluation of a Biorefinery Based on Jatropha curcas Fruits in Yucatan Mexico

Conference: RCN Conference on Pan American Biofuel and Bioenergy Sustainability

Year: 2016

Proceeding: 2nd RCN Conference on Pan American Biofuels & Bioenergy Sustainability

Group: Posters

Session: Rapid Fire Poster Session

Time: Wednesday, September 14, 2016 - 5:01pm-5:02pm

Authors:

Sacramento-Rivero, J. C., Universidad Autonoma de Yucatan
Navarro-Pineda, F. S., Universidad Autonoma de Yucatan

A jatropha-based whole-crop biorefinery was designed and assessed under the sustainability framework developed by Sacramento-Rivero (2012, http://dx.doi.org/10.1002/bbb.335). This framework involves techno-economic analysis, along with environmental life cycle assessment, and non-renewable energy and water consumption, into a set of indicators whose values are normalized and range between cero to infinity. Indicators that range between 0 and 1 are considered sustainable. The indicators are gathered into 5 categories including renewability, economics, fossil-based products displacement, biodiversity protection, environmental mitigation, and community development. Information regarding the cultivation of jatropha was collected from local jatropha plantations and two potential scenarios to harness the jatropha biomass were identified, simulated, and assessed. The first scenario (BPP) refers to a standalone biodiesel production process which does not harness the lignocellulosic residual biomass, while the second (IBP-2) uses it for producing bio-oil, bio-char, and the internal heat and power requirements. Results indicate that the biorefinery scenario (IBP-2) performs better than the standalone biodiesel production (BPP) due to a more diverse portfolio of products allowing for a better sustainability performance. However, the use of pesticides and fertilizers, along with an intense water consumption for the jatropha crops, make the system unsustainable in the fossil-based products displacement and environmental-mitigation categories. In fact, the agricultural stage also contributes the most (77%) to the total production cost, environmental load (more than 60%), non-renewable energy consumption (66%), and water consumption (99%).

Efficiency Assessment of Gasification Waste from the Pistachios Processing

- Conference: RCN Conference on Pan American Biofuel and Bioenergy Sustainability
- Year: 2016
- Group: Posters
- Time:
  Wednesday, September 14, 2016 - 5:02pm-5:03pm

Authors:
Zalazar García, D. I. Y., Instituto de Ingeniería Química. Facultad de Ingeniería. Universidad Nacional de San Juan
Romera, S. B., Instituto de Ingeniería Química. Facultad de Ingeniería. Universidad Nacional de San Juan, San Juan, Argentina,
Feresin, G. E., Instituto de Biotecnología. Facultad de Ingeniería. Universidad Nacional de San Juan
In San Juan pistachio production has grown in recent years due to the great interest in the consumption of pistachio nut, which has health benefits. Production waste from industrialization represents a problem, because it is accumulated and stored in acceptable ways occupying space and generating unpleasant odors among others. Gasification of waste for energy represents advantages from the economic point because it gives added value to these materials, on the one hand and on the other, environmentally can say that the balance of C is zero because it is residual biomass. This process involves the conversion of waste, in a substoichiometric atmosphere in a fuel gas, which can be used as fuel for electricity power generation or as a chemical for the manufacture of methanol, oils, etc. The aim of this work was to study the performance of the gasification process residue pistachio steam, as it improves yields syngas production, based on thermodynamic analysis to improve the gas composition using an equilibrium model to predict the gas composition based on the minimization of Gibbs free energy. The simulation was performed using two types of residues pistachio stems, leaves, fleshy exocarp (R1) on one side and the lignocellulosic mesocarp (R2) on the other. The prediction model was developed using Mathcad software. The thermodynamic equilibrium model applied to predict the composition process, allowed syngas comprising CO, CO_2, H_2, CH_4 and N_2, and the process efficiency. Performance was evaluated taking into account the process operating conditions such as temperature, humidity of each residue, ER (actual equivalent ratio moles oxygen / mol biomass fed to the reactor) and SBR (waste steam / biomass ratio). One exergo-economic analysis was also performed. Values energy costs depending on the generated power were allocated. The analysis was applied to different operating conditions, whose values were obtained from an equilibrium model in residue R1 and R2. The results indicated that the higher the temperature, the performance of gasification is decreased by 2% every 50°C rise in the range between 450 and 700°C for R1 and 1% between 450 and 1000 ° C for R2 respectively, which also involve lower costs of exergy. The maximum yield was 89.2% at 450°C and 72.7% at least 1000°C for R1

and 89.4% and 78.1% for R2. Regarding the process performance to changes in equivalent ratios (ER:, values between 0.1 and 0.9 were tested, implying better performance of the process near 0.1 for both residues. Furthermore, the process behavior with respect to moisture variation no significant changes of its efficiency in both residues. According to the changes molar proportions of gasifying agent (superheated steam), a corresponding optimum value at steam / biomass ratio = 4 (SBR) was obtained, although substantially constant for both waste. Economic analysis suggests that the gasification process is desirable because it provides solution to the problem of environmental pollution with low operating costs.

Wastewater Characterization and Treatment: Influence in Bio-Hydrogen and Methane Production

- Conference: RCN Conference on Pan American Biofuel and Bioenergy Sustainability
- Year: 2016
- Group: Posters
- Time:
  Wednesday, September 14, 2016 - 5:04pm-5:05pm

Authors: Fuentes Mora, M., INGAR (CONICET-UTN)
Aguirre, P. A.

Environmental conservation issues increasingly grow in importance and bring about changes in the way of addressing and optimizing industrial processes from their very conception, weighing environmental conditions and sustainability aspects. In particular, this paper deals with the issue of wastewater treatment including biohydrogen and methane recovery, and lays emphasis on effluent characterization and treatment system selection. In this context, complex mathematical models are developed whose solution correspond to optimal treatment systems configurations, minimizing costs and serving restrictions of dumping against scenarios of theoretical and practical interest. These models perform a superstructure that enables the description of reactors, kinetics of removing pollutants (C, N), and operational issues and design considerations. The importance of selecting a suitable scheme of degradation is made clear in previous works. It is also relevant to include the description of inhibitions factors; i.e. that physicochemical and biochemical process kinetics becomes decisive, irrespective of the reactor type to be designed. On the other hand, effluent characterization is based on parameters which measure the concentration of contaminants: chemical and biological oxygen demands, and volatile suspended solids, described mainly as volatile fatty acids, carbohydrates, and protein percentages.
The objective is to calculate optimal solutions that ponder biofuel recovery in wastewater treatment, taking into account: (a) different levels of pollution and (b) different alternatives using simple or combined methods of biological treatment. The objective function is net present value (NPV) minimization, which considers investment and operation costs, thus satisfying allowed discharge limits. For the studied cases, alternatives that involve biohydrogen production in an initial anaerobic stage (at low pH) in effluents with high organic load and a proper C/N ratio are attractive. Depending on methaneâs tentative price (as a gas substitute), some alternatives including this recovery before an aerobic treatment stage are feasible. So, it is essential to identify the biofuel production cost component in each case. When effluent complexity increases, the concentration of suspended solids in the treated effluent decreases, resulting in a high NPV. The models were implemented in a (general Process Modeling System, gPROMS) software that allows simulating several scenarios and becomes an essential tool for decision-making.

Revaluation of the Natural Sponge Luffa Cylindrica As Adsorbent for Lead and Raw Material for Paperboard

Housam Ankah[1], Gladys Rincón[2,3], Alejandra Meza[1], Narciso Pérez[3], José Guillermo Cardenas[4], Sandra Peña[4]

[1]Escuela de Ingeniería Química, Universidad Central de Venezuela, Carcas, Venezuela

[2]Becario Ateneo-Senescyt, Facultad de Ingeniería Química, Universidad de Guayaquil, Guayaquil, Ecuador.

[3]Laboratorio de Carbón y Residuales, Universidad Simón Bolívar, Carcas, Venezuela.

[4]Facultad de Ingeniería Química, Universidad de Guayaquil, Guayaquil, Ecuador.

ABSTRAC

Adsorption process is of utmost importance in most industrial processes, mainly from the point of view of residual treatment. In recent times, International organizations such as the World Health Organization (WHO) and the International Labour Organization (ILO) are paying a lot of attention to water effluents, and are asking for much more stronger restrictions on the conditions and concentrations of water effluents. Due to its extensive usage, lead (Pb) is found in high concentrations in waters dedicated to human consumption, this brings in severe damage to public health. WHO suggest a Pb concentration below 0.01 mg/l for drinking water, this is a quite low and difficult to achieve limit, which makes necessary either to improve the existing water purification technologies or developing new ones. Any material of biological origin able to adsorb metallic or metalloid species is referred to as a bio-adsorbent. Typical examples of bio-adsorbents are bacteria, fungi, aquatic algae and some woody lignocellulosics materials. In this work we study lead adsorption by Lignin obtained from the natural sponge *luffa cylindrica*. The lignin is extracted from the sponge by Kraft's method with NaOH solutions of 1, 3, 5, 7, 10 and 15% concentration, we find 10% to be optimal. The cellulose we get corresponds to 30.44% of the initial mass of the sponge. It is important to avoid refining or bleaching the cellulose when manufacturing paperboard. We obtained the bioadsorbent through the following procedure, the black liquor was acidify to a pH 4, this ensures that the lignin precipitates, the precipitation was assisted by microwaving, we got a â24% mass recovery. The lignin so obtained was put in contact with the leaded effluent with Ph varying between 5 and 9; Ph 6 showed to be the best value. A similar procedure was used to test for contact times (we tried 10-40 min), we found that from 20 min on there is a significant rate of lead adsorption. The mass to volume ratio range is 0.01-2 g/l, the maximum adsorption been obtained at 0.01 g/l. We did also test for temperature variations in the range 24 y 60Â°C and found that no significant increase in lead retention was observed, we therefore propose to work at environmental temperatures. By using the method of Langmuir and Freundlich isotherms we were able to determine that the observed phenomenon was indeed chemical adsorption, kinetic studies show that for the concentration range we tried, the reaction may be adjusted to either a first or second order one, we recommend further studies.

REVALORIZACIÓN DE LA ESPONJA NATURAL, LUFFA CYLINDRICA, COMO ADSORBENTE DE PLOMO Y MATERIA PRIMA PARA CARTÓN

Housam Ankah1, Gladys Rincón2,3, Alejandra Meza1, Narciso Pérez3, José Guillermo Cardenas4, Sandra Peña4

1Escuela de Ingeniería Química, Universidad Central de Venezuela, Carcas, Venezuela

2 Becario Ateneo-Senescyt, Facultad de Ingeniería Química, Universidad de Guayaquil, Guayaquil, Ecuador.

3 Laboratorio de Carbón y Residuales, Universidad Simón Bolívar, Carcas, Venezuela.

4 Facultad de Ingeniería Química, Universidad de Guayaquil, Guayaquil, Ecuador.

RESUMEN

El proceso de adsorción es muy importancia en la mayoría de los procesos industriales, principalmente desde la perspectiva ambiental para el tratamiento de residuales industriales. En los últimos años, organismos internacionales como la Organización Mundial de la Salud (OMS) y la OIT (Organización Internacional del Trabajo) están poniendo especial interés en los efluentes acuáticos, proponiendo cada vez restricciones más rigurosas respecto a las condiciones y concentraciones de efluentes en aguas. El plomo (Pb) por su uso generalizado se encuentra en concentraciones importantes en el agua de consumo humano, causando graves problemas de salud pública. La OMS sugiere que la concentración de Pb no supere los 0.01 mg/l para agua potable: limite muy baja y difícil de alcanzar, por lo que se hace necesario el desarrollo de nuevas tecnologías y/o la optimización de las existentes con el fin de garantizar agua saludable para la población. El término biosorbente corresponde a cualquier material de origen biológico que tenga la capacidad de adsorber especies metálicas o metaloides. Principalmente se utilizan los biosorbentes de estructura bacteriana, estructura fúngica, algas marinas y materiales leñosos lignocelulosos. En esta investigación se realiza la bioadsorción de plomo con ligninas obtenidas a partir de la esponja natural *luffa cylindrica*. Para lo cual, primero, se extrae la lignina de la esponja por el método Kraft, con NaOH concentrado a diferentes niveles desde 1 hasta 15% de concentración. La celulosa obtenida como subproducto corresponde al 30,44% de la masa inicial de esponja por lo que se recomienda no refinar ni blanquear la celulosa para utilizarla en la fabricación de cartón. Para la obtención del bioadsorbente, primero, se acidificó el licor negro hasta llevarlo a pH 4, condición en la que precipita la lignina. La precipitaciÃ³n se asiste por medio de un microondas para obtener mayor cantidad de precipitado: lignina recuperada â24% de masa inicial de la esponja. La lignina obtenida se pone en contacto con las muestras del efluente con plomo y se varían las condiciones de pH entre 5 y 9: el pH 6 resultó ser el mejor valor. Se aplica un procedimiento equivalente para establecer el tiempo de contacto (10 - 40 min), obteniendo que a partir de 20 minutos no se adsorben cantidades significativas de plomo. La relación masa de adsorbente entre volumen de solución se varía entre 0,01- 2 g/l, obteniéndose la máxima adsorción para 0,01g/l. El rango de temperaturas estudiado se encuentra entre 24 y 60ºC, resultando que el aumento de temperatura no representaba un aumento significativo en la retención de plomo: se propone operar a temperatura ambiente. Por medio de las isotermas de Langmuir y Freundlich se determina que el fenómeno presente es adsorción química y el estudio cinético determina que, en el rango de concentraciones empleado, la reacción se puede ajustar a una de primero como de segundo orden. Se recomienda seguir evaluando la factibilidad de industrialización de este proceso.

Valorization of By-Products from Biomass Processing through Bio-Catalytic Oxidation

Conference: RCN Conference on Pan American Biofuel and Bioenergy Sustainability

Year: 2016

Proceeding: 2nd RCN Conference on Pan American Biofuels & Bioenergy Sustainability

Group: Posters

Session: Rapid Fire Poster Session

Time: Wednesday, September 14, 2016 - 5:06pm-5:07pm

Authors:

Parodi, A., Instituto de Investigación y Desarrollo en Ingeniería de Procesos y Química Aplicada
Sappia, P.
Falco, M.
Ferreira, M. L.
Zygadlo, J.
Magario, I.

The oxidation of biomass-derived hydrocarbons or aliphatic alcohols into ketones, aldehydes and carboxylic acids may confer enhanced bioactivity properties as e.g. bactericide, antifungal or insecticides. Due to the increase of the biodiesel industry, based on sustainability politics for renewable resources utilization, glycerol as the main by-product is being supplied at levels much higher than demanded thus being poorly commercially exploited. On the other side, terpenes are renewable natural hydrocarbons, which are readily available in plants. Among them, limonene is nowadays one of the most abundant and easily available since it is contained in citrus peels, which are the waste product of juice industries. With the exception of di-hydroxyacetone, the market of the oxidation products of glycerol but also of limonene was not developed because of high production costs although all of them play as very valuable intermediates in fine chemistry. This study focalizes in the valorisation of glycerol and limonene by catalytic oxidation and the characterization of the oxidized products in terms of its insecticide action. Not only pure standards of both substances but also raw samples of industries, from which are by-produced are considered to be tested for the oxidation reaction. Special attention is cared to the use of environmental-friendly processing free of toxic solvents and oxidants without the need of cofactors. In addition to their potential high selectivity and simplicity, two types of catalytic systems were pre-selected for screening studies: pure oxidases and metallic complexes based on Mn and Fe. Preliminary results relates to the use of a natural iron porphyrin ânamed as hematin- as the oxidazing catalyst in first oxidation trials, optimization of reaction kinetics and analytics establishments. Hematin possess the ability to mimicking the action of peroxidases and catalases thus co-catalazing the decomposition of the oxidant, H_2O_2. The radicals formed by this route may cooperate in the oxidation of organic matter. Therefore a careful selection of reaction conditions in terms of solvent mixture, pH, temperature and molar ratio of reducing substrate to oxidant was required in order to afford a complete activity check.

Application of a Process-Based Model for Predicting Aboveground Biomass of Eucalyptus Globulus Bioenergy Plantations in Southern Chile

- Conference: RCN Conference on Pan American Biofuel and Bioenergy Sustainability
- Year: 2016
- Group: Posters
- Time:
  Wednesday, September 14, 2016 - 5:07pm-5:08pm

Authors: Mena, P. A., Universidad de Concepcion
Acuña, E., Universidad de Concepcion
Rubilar, R., Universidad de Concepcion

An alternative to reduce the impact of climate change is to reduce emissions of greenhouse gases through the use of alternative renewable and low cost energy as those based on biomass transformation, specifically dendroenergy Plantations. Operational crops from this kind are still scarce due to the lack of relevant information in terms of expected returns and silvicultural activities under different management schemes and weather conditions Ecophysiological process-based models for predicting productivity of forest plantations from the efficient use of solar radiation, determined by specific parameters for each species and weather information. The 3-PG model (Physiological Principles Predicting Growth) has proven accurate estimates on plantations of genus Eucalyptus, however, are scarce experiences of using this model in dendroenergetic plantations of high density, its parameterization could allow to estimate more certainty productivity in this type of site-specific plantations under different management schemes level. The general objective of this study was to evaluate the efficiency of 3-PG model in aboveground biomass prediction of *E. globulus* of a dentroenergetic plantation undergoing intensive silviculture in order to identify the full potential of biomass production. Two contrasting sites were examined from south central Chile (soils of volcanic sands and recent volcanic ash) and three levels of density were evaluated (5000, 10000 y15000 trees ha^{-1}) The 3PG model was successfully parameterized and contrasted with field observations, a high level of efficiency (97% average) for the variables evaluated (total, stem and foliar biomass,) was observed. It was identified that the highest cumulative aboveground biomass occurred in establishment density of 10000 and 15000 trees ha^{-1}, independent of the year in which decides culminate the rotation. In both densities, it is observed that sandy soils achieve a maximum monthly mean increase in aboveground biomass before two years (3-5 Mg ha^{-1} $month^{-1}$), in contrast to volcanic ash soils achieve maximum mean monthly aboveground biomass increment at 4 years (1.5-2.5 Mg ha-1 month-1). Finally, the maximum accumulation in aboveground biomass for a 4-year rotation occurs in soils of sandy type with an initial density of establishment of 10,000 trees ha-1 for a rotation of 5 or 6 years the maximum accumulation takes place in volcanic ash soils established at densities of 15,000 trees ha^{-1}.

Use of Eucalyptus Bark for Bioenergy

Conference: RCN Conference on Pan American Biofuel and Bioenergy Sustainability

Year: 2016

Proceeding: 2nd RCN Conference on Pan American Biofuels & Bioenergy Sustainability

Group: Posters

Session: Rapid Fire Poster Session

Time: Wednesday, September 14, 2016 - 5:08pm-5:09pm

Authors:

Hansted, A. L., UFSCar
Belini, G., UFSCar
Nakashima, G., UFSCar
Yamaji, F. M., UFSCar
de Morais, L. C., UNESP
Caraschi, J. C., UNESP

The use of biomass resources in order to generate energy has been increasing as a renewable alternative, foremost because of their appealing properties. In Brazil, one of the most important crops that is cultivating is sugarcane, mainly in SÃ£o Paulo State. Recently, Law 11,241/02 prohibits the utilization of fire to eliminate sugarcane straw during harvest process. The sugarcane straw has been searchingÂ toÂ apply in the energy production. The present study is aimed on the influence of storage time in the chemical characterization of sugarcane straw to determine its potential to be applied as a sustainable fuel. Three treatments were analyzed: T1 âÂ no storage time; T2 â 1 year storaged; T3 â 2 years storaged. The material was separated according to particle size (<425Âµm). The analyses involved particle size distribution, density determination, proximate analysis and chemical analysis to estimate extractives, lignin, holocellulose, Î±-cellulose and hemicellulose content. For sugarcane straw, it was obtained density of 0,08603 g/cm^{3}before the mill process. The redults for proximate analysis were 74,77% (T1), 72,11% (T2)Â and 73,39% (T3). The ash content was 8,98% (T1), 7,93% (T2) and 7,02% (T3). The highÂ ash content in T1 (8,98%)Â resulted in lowerÂ fixed carbon 16,25%. The total extractivesÂ were 18,02% (T1), 22,23% (T2) andÂ 23,43% (T3). Klason lignin content were 32,06% (T1), 39,65% (T2) and 40,01% (T3). The hollocelulose content is essential to obtain Î±-cellulose and hemicellulose concentration. The hollocelulose content was 74,41% (T1), 72,45% (T2) and 73,91% (T3). The Î±-cellulose and hemicellulose content were 62,23% (T1), 55,58% (T2) and 59,53% (T3) for Î±-celluloseÂ and 12,7% (T1), 16,82% (T2) and 14,61% (T3) for hemicellulose. It was possible to realize that storage time influences in the chemical properties of sugarcane straw, mainly ash content, which is related to minerals content in the sugarcane strawÂ and contaminants from harvest process.

The Particle Size Influence in the High Heating Value (HHV) of Sugarcane Trash

Conference: RCN Conference on Pan American Biofuel and Bioenergy Sustainability

Year: 2016

Proceeding: 2nd RCN Conference on Pan American Biofuels & Bioenergy Sustainability

Group: Posters

Session: Rapid Fire Poster Session

Time: Wednesday, September 14, 2016 - 5:09pm-5:10pm

Authors:

Nakashima, G., UFSCar
Hansted, A. L., UFSCar
Belini, G., UFSCar
Cruz, E.
Sette Júnior, C. R., UFG
Yamaji, F. M., UFSCar

Biomass is a resource type that is present worldwide. The demand for sugarcane coupled with soil, climate and territorial conditions make Brazil the largest producer of sugarcane in the world. Then, there is a large amount of waste from this culture, approximately 14 to 18% of sugarcane straw. The waste of sugarcane culture includes green leaves, dried leaves and tops. The aim of the study was to investigate the influence of storage time and particle size in sugarcane trash. It was analyzed the ash content and high heating value (HHV) for energy purposes. The treatments were based on: the position of the bale (inner and surface), the storage time (0, 1 and 2 years) and the particle size (> 0.420mm, 0.250-0.420mm and < 0.250mm). The results showed a behavior that the higher amount of ashes were observed in smaller particle size. The values for ash content varied from 2.16% (particles >0,420mm) to 12.72% (particles <0,250mm). These results were considered high for solid biofuel production. It was expected that the higher ash content resulted the lower HHV. The results for HHV were 15.93 $MJ.kg^{-1}$ (particles <0,250mm) to 18.32 $MJ.kg^{-1}$ (particles >0,420mm) for high heating value. The results for HHV did not showed statistical difference between the treatments. The particle size affects the results and should be chosen carefully according to the test purpose.

Maize (Zea mays L.) Genotype Selection with Ability to Production Lignocellulosic Bioenergy and Grain Yield

- Conference: RCN Conference on Pan American Biofuel and Bioenergy Sustainability
- Year: 2016
- Group: Posters
- Time:
  Wednesday, September 14, 2016 - 5:10pm-5:11pm

Authors: García Stepien, L. E., CONICET
Aulicino, M. B., UNLZ
Bertoia, L. M., UNLZ

Stover maize (*Zea mays* L.) can be converted into fermentable sugars to produce ethanol by similar processes they are subjected forage in rumen. The efficiency of these processes depends on the stover cell wall structure. Some authors have suggested that selection for grain yield and resistance to stalk lodging could reduce forage quality, and as a result, it potential ability to produce ethanol from stover. Others authors found evidence that would increase stover yield by selected genotypes with more plant height, stalk size and density, leaf area, and delayed senescence. Therefore it is important to find sources of genetic variability for silage production and lignocellulosic bioethanol without compromising grain yield. During the years 2012, 2013 and 2014, 144 genetically divergent genotypes were evaluated, including landraces (100 of them below to INTA Pergamino), commercial hybrids (12 from 10 seed companies), synthetic composites (10 of them below to DAS company), and experimental hybrids (22 from 3 seed companies), with different improvement level, cycle length, grain type, and genotypes with bmr genes. DK72-10VT3P (grain type, temperate germplasm), DK390VT3P (grain type, tropical germplasm) and BMR126HX (bmr, temperate germplasm) hybrids were included as checks in an augmented block design. The experimental unit consisted of 2 rows of 4 m × 0.5 m with a density of 80,000 plants ha^{-1} at harvest time. The harvest was made at physiological post-maturity. Agronomical performance was evaluated on Stover Dry Matter (SDMY, kg DM ha^{-1}), Grain Yield (GY, kg ha^{-1}); Stalk broken (SB), and Lodging (SL) and nutritional variables estimated by NIRS calibration: Neutral detergent fiber (NDF, %), Acid detergent fiber (ADF, %), and Acid detergent lignin (ADL, %). Two combined traits were calculated: Cellulose (CEL, %), as NDF-ADL; and Hemicellulose (HCEL, %) as FDN-FDA. Due to the dual propose (Stover and grain), selection indexes (SI) for each genotype and year were constructed using a nonparametric rank-sum index (Kang, 1988) that gave equal weights to high CEL, HCEL, SDMY, GY, and low ADL, SB, and SL. A SI Total (TSI) was calculated by the sum of the SI for each environment. As consequence, indexes allowed to identify the superior genotypes for both, their bio-energetic potential and stability. The ranking of the total sum of the indices grouped the best 15 genotypes (10% of the population), represented by six synthetic compounds (4 for silage and 2 tropical), 3 forage experimental hybrids, 4 landraces and 2 commercial hybrids. All selected genotypes were late-maturity cycle and tend to produce more biomass, both vegetative as reproductive. In contrast, the lower index group (11 landraces and 4 experimental hybrids) included early-maturity and no-breeding genotypes. Commercial hybrids demonstrated the highest stability across the year. The 4 selected landraces could be incorporated into breeding programs to obtain genotypes with the dual ability, bio-energy and grain yield.

Biomass for Bioethanol Production: Characterization of Two Varieties of Elephant Grass (Pennisetum purpureum Schumacher)

- Conference: RCN Conference on Pan American Biofuel and Bioenergy Sustainability
- Year: 2016
- Group: Posters
- Time:
  Wednesday, September 14, 2016 - 5:12pm-5:13pm

Authors: García Stepien, L. E., CONICET
Rossi, C. A., UNLZ
Borlandelli, M. S.
Arizaga Inda, M. A.
Rovegno, M. S.
Olmeda, M. F.

Elephant grass (*Pennisetum purpureum* Schumacher) is a species of C4 perennial tropical grass native to the African grasslands. It has a vigorous growth and high biomass production, which is commonly used for feeding cattle in subtropical areas. It has low water and nutrient requirements, and therefore can make use of otherwise uncultivated lands. Currently it is evaluating the potential for production of second generation bioethanol. The biomass energy fitness can be assessed using the same parameters used for determining the nutritional value of forages, due to the similarity that both processes occur. The bioenergetic potential of the biomass of two varieties of elephant grass in the grounds of the FCA-UNLZ was studied. Var. Mott (dwarf) and Var. Panamá (High) were implanted in individual plots since 1996. These showed a great adaptation to the soil and climate of the Pampeana region for 20 years. DBCA was used with 3 replications. The experimental unit consisted of 2 rows of 4m×1m (density at harvest 10000 plants ha^{-1}) on which dry matter yield was determined (DMY, kg ha^{-1}), plant height (Height, m), proportion of leaf (Leaf, %). The manually harvest took place on 1 April 2016 (31.07 %DM average). The quality parameters were determinate on a fraction of stem+leaves of the middle section of the plant. Total solids content (B, °Brix) was determined in fresh. Later, the samples were dried to determine the content of dry matter (DM,%) and then were ground to 1 mm to evaluate the following variables quality laboratory: Neutral detergent fiber (NDF,%); Acid detergent fiber (ADF,%); Acid detergent lignin (ADL,%) and ash (Ash,%). Lignin (L,%) was calculated as LDA-C, cellulose (CEL,%) as FDA-CEL, and hemicellulose (HCEL,%) as FDN-FDA. In turn, the quality variables were evaluated in the leaf fraction and stem separately. The result of ANVA when comparing both varieties showed that only significant differences for variables Height (2.49m vs. 1.78m) and DMY (23543kg ha^{-1} vs. 6226 kg ha^{-1}) for Panamá. Both varieties had similar values for quality parameters. However, Mott had a trend to higher levels of bioenergy quality. When analyzing stem vs. leaves fraction, no significant differences were found between varieties. Leaves presented a lower content of L than stem, with no significant differences in the other variables analyzed. In conclusion, Panamá presented a better performance to be used as resource to produced biomass to bioenergy because of its greater DMY and similar quality.

Use of Eucalyptus Residue for Bioenergy

Conference: RCN Conference on Pan American Biofuel and Bioenergy Sustainability

Year: 2016

Proceeding: 2nd RCN Conference on Pan American Biofuels & Bioenergy Sustainability

Group: Posters

Session: Rapid Fire Poster Session

Time: Wednesday, September 14, 2016 - 5:13pm-5:14pm

Authors:

Hansted, A. L., UFSCar
Nakashima, G., UFSCar
Belini, G., UFSCar
Caraschi, J. C., UNESP
de Morais, L. C., UNESP
de Barros, J. L., UNESP
Yamaji, F. M., UFSCar

The biomass obtained from eucalyptus forestry residues can be used as fuel for industries. It is a renewable energy and has advantages such as high amount of material and low cost. The aim of this research was to perform the physico-chemical characterization of eucalyptus bark used for power generation in a panel industry, and identify the existence and the components of the contaminants in the bark. Three treatments were performed for particle sizes of peel: T1 (850μm and 425μm), T2 (250μm) and T3 (<150μm). The material was assessed for moisture content and the three treatments were subjected to chemical analysis. Also, the high heating value was calculated. The ashes were analyzed under scaning electronic microscope (SEM) and energy dispersive spectroscopy (EDS) in order to identify the contaminants. All treatments showed contaminants in their ashes, with the presence of Ca, Fe and Si. The data obtained in this study were statistically analyzed using the software R. The material presented moisture content of 70% on a dry basis, considered high for use in bioenergy. The ash content and volatile content presented significant difference between treatments, and T1 showed lower value for ash (2.63%) and higher value for volatile (80.23%). The treatment T1 was considered best for use in energy purposes.

Chemical Characterization of Sugarcane Straw for Bioenergy Purposes

Conference: RCN Conference on Pan American Biofuel and Bioenergy Sustainability

Year: 2016

Proceeding: 2nd RCN Conference on Pan American Biofuels & Bioenergy Sustainability

Group: Posters

Session: Rapid Fire Poster Session

Time: Wednesday, September 14, 2016 - 5:14pm-5:15pm

Authors:

Belini, G., UFSCar
Teixeira, C., UFSCar
Nakashima, G., UFSCar
Hansted, A. L., UFSCar
Yamaji, F. M., UFSCar

The use of biomass resources in order to generate energy has been increasing as a renewable alternative, foremost because of their appealing properties. In Brazil, one of the most important crops that is cultivating is sugarcane, mainly in SÃ£o Paulo State. Recently, Law 11,241/02 prohibits the utilization of fire to eliminate sugarcane straw during harvest process. The sugarcane straw has been searching to apply in the energy production. The present study is aimed on the influence of storage time in the chemical characterization of sugarcane straw to determine its potential to be applied as a sustainable fuel. Three treatments were analyzed: T1 â no storage time; T2 â 1 year storaged; T3 â 2 years storaged. The material was separated according to particle size (<425Âµm). The analyses involved particle size distribution, density determination, proximate analysis and chemical analysis to estimate extractives, lignin, holocellulose, Î±-cellulose and hemicellulose content. For sugarcane straw, it was obtained density of 0,08603 g/cm3 before the mill process. The redults for proximate analysis were 74,77% (T1), 72,11% (T2) and 73,39% (T3). The ash content was 8,98% (T1), 7,93% (T2) and 7,02% (T3). The high ash content in T1 (8,98%) resulted in lower fixed carbon 16,25%. The total extractives were 18,02% (T1), 22,23% (T2) and 23,43% (T3). Klason lignin content were 32,06% (T1), 39,65% (T2) and 40,01% (T3). The hollocelulose content is essential to obtain Î±-cellulose and hemicellulose concentration. The hollocelulose content was 74,41% (T1), 72,45% (T2) and 73,91% (T3). The Î±-cellulose and hemicellulose content were 62,23% (T1), 55,58% (T2) and 59,53% (T3) for Î±-cellulose and 12,7% (T1), 16,82% (T2) and 14,61% (T3) for hemicellulose. It was possible to realize that storage time influences in the chemical properties of sugarcane straw, mainly ash content, which is related to minerals content in the sugarcane straw and contaminants from harvest process.

Carbon Footprint of Sugar Production in Mexico

Conference: RCN Conference on Pan American Biofuel and Bioenergy Sustainability

Year: 2016

Proceeding: 2nd RCN Conference on Pan American Biofuels & Bioenergy Sustainability

Group: Posters

Session: Rapid Fire Poster Session

Time: Wednesday, September 14, 2016 - 5:15pm-5:16pm

Authors:

García, C., Escuela Nacional de Estudios Superiores Unidad Morelia-UNAM

Global warming, caused mainly by increased worldwide emissions of greenhouse gases, is currently one of the greatest threats to the environment and human societies. Mexico has set an ambitious goal of reducing 30% of its greenhouse gases emissions by year 2020. The sugar agroindustry has been identified as one of the opportunities for mitigating emissions in this country. The aim of this work is to contribute towards identifying policy measures and practices for low-carbon sugar production by assessing the carbon footprint of sugar produced in four sugar mills in Mexico using a life cycle assessment method. System boundaries include agricultural practices, sugarcane harvesting, cane milling and sugar conversion. The results show that sugar production has carbon footprint values in the range of 0.45e0.63 kg CO2e/kg sugar. In these four cases, the agricultural stage contributes the majority of carbon emissions (59e74%). Most greenhouse gases emissions in the agricultural stage were from fertilizer production, nitrous oxide (N2O) emissions and biomass burning. The industrial stage contributed with 14-30% of total greenhouse gases emissions, mainly due to fossil fuel and bagasse use. The carbon footprint value is particularly sensitive to nitrogen fertilization, nitrous oxide emissions from the soil and sugarcane yields. Cogeneration in sugar mills could become an important way to reduce the carbon footprint of sugar and to produce electricity with low carbon emissions. We show the impact of different carbon footprint performance of sugar production process in Mexico. Data used on this manuscript came from real field measurements, and our results are accompanied by sensibility and uncertainty analyses. This is the first time that life cycle assessment has been used to estimate the carbon footprint of sugar production in Mexico including agricultural, industrial and transportation boundaries, to identify greenhouse gases mitigation opportunities. Studying techniques for improving sugar cane yield, making fertilizer use more efficient, minimizing cane burning and developing efficient cogeneration in sugar mills with bagasse as fuel is scientifically relevant. Applying concrete public policy measures to these areas of opportunity would allow production of low carbon sugar in Mexico. The results of this study may also be used as reference by other countries with similar sugar production conditions.

Energy Intensity of the Argentine Economy and Agriculture Sector: A Decomposition Exercise and Policy Evaluation

- Conference: RCN Conference on Pan American Biofuel and Bioenergy Sustainability
- Year: 2016
- Group: Posters
- Time:
  Wednesday, September 14, 2016 - 5:30pm-6:30pm

Authors: Ghersa, F.
Ferraro, D., Universidad de Buenos Aires
Moscovich, L.

Argentina is the seventh largest producer of agricultural products in the world, and as such its economy is highly dependent upon the relative health of its ecological systems to sustain agricultural production. Understanding energy intensity, the amount of energy used to produce one unit of output, is a key metric in order to gage the sustainability of a system. Considering that all physical systems are constrained by the laws of thermodynamics, the way they use energy to maintain structure and function tells us about their capacity to persist in time, what Ludwig Bolztmann stated as a âfight against entropy. In this thesis the economic energy intensity of the Argentine economy and agriculture sector between 1960 and 2013 is examined. This research project has two main objectives, the first is to decompose energy intensity and evaluate the weight of two contributing factors, the efficiency of energy use and the sectoral composition of the economy through Logarithmic Mean Divisia Index (LMDI I). The second objective is to provide an energy efficiency policy evaluation scheme under which we suggest certain proposals to future policymakers. Our results show that changes in technical energy efficiency have played a dominant role, relative to economic structure, in the trends of energy intensity. The general economy has a U-shaped curve in its intensity trend, such that between 1960 and 2013 there have been no real improvements in either energy consumption or energy efficiency. At the sectoral level the economic energy intensity trends have been either U-shaped or strictly increasing, which provides a partial explanation behind the patterns at the aggregate level. The agriculture sector has both increasing intensity, consumption and a high dependence on fossil fuels. Considering that we only take into account the energy used within the economic system (as opposed to all ecosystemic energy flows) our trends only show a partial view of the true situation. Thus, the evidence presented suggests that Argentina is treading down an unsustainable path from a systemic standpoint, in both its general economy and agriculture sector. Feeble energy efficiency oriented public policy and institutions, together with high dependence on fossil fuels, increasing consumption and either stable or increasing energy intensity propose serious difficulties to ensure the longevity of the productive forces behind the Argentine economy and agriculture sector. The incapacity of the productive systems to increase their efficiency could stem from the fact that the degradation of natural capital has exceeded that which technology and higher quality energy can compensate for.

An Integrated Approach to Sustainable Bioenergy/Biofuels Resources and Supply Chain Management: Modeling, Techno-Economic Analysis, Challenges, and Future Outlook

Conference: RCN Conference on Pan American Biofuel and Bioenergy Sustainability

Year: 2016

Proceeding: 2nd RCN Conference on Pan American Biofuels & Bioenergy Sustainability

Group: Abstracts

Session: Session 4: Integrated Biomass/Bioenergy Supply Logistics/ LogíStica Integrada De Suministro De Biomasa / Bioenergía

Time: Thursday, September 15, 2016 - 8:30am-8:50am

Authors:

Hosseini, M., The University of Texas Rio Grande Valley
Srivastava, A. K., The University of Texas Rio Grande Valley

Biomass supply chain and resource assessment seek to encourage the sustainable and cost-effective use of biomass energy. Biomass resource assessment includes estimating the required quantity of material, considering the constraints (i.e., environmental and technical), and evaluating the amount of recoverable material that can potentially be available for biomass energy use. Knowledge concerning the quantity and the type of biomass is essential because the success of bioenergy is critically dependent on having a large supply of low cost, high quality biomass which allows for the design of sustainable bioenergy chains and the appropriate selection of energy conversion technologies. Designing integrated and sustainable biomass supply chains that are highly efficient and cheaper allow bioenergy to be more competitive when compared to the energy generated from traditional sources. Herein, technological challenges for the developing and scaling up of integrated bioenergy/biofuels systems, improving bioconversion technologies to be more cost-effective, enhancing efforts for productivity of energy crops, establishing bioenergy/biofuels markets, and logistics are discussed. Furthermore, modeling and processing techniques for the bioconversion of biomass to bioenergy, planning for a self-sufficient bioenergy/biofuels chain, and a techno-economic evaluation of an integrated cost effective biomass to bioenergy supply chain are presented.

Optimal forest supply chain design including cluster formation

Nicolás A. Vanzetti, Jorge M. Montagna, Gabriela Corsano

Instituto de Desarrollo y Diseño (INGAR), Conicet

Santa Fe - Argentina

INGAR CONICET UTN

PAN AMERICAN BIOFUELS AND BIOENERGY SUSTAINABILITY
AN NSF RESEARCH COORDINATION NETWORK

Introduction

Supply chain (SC)

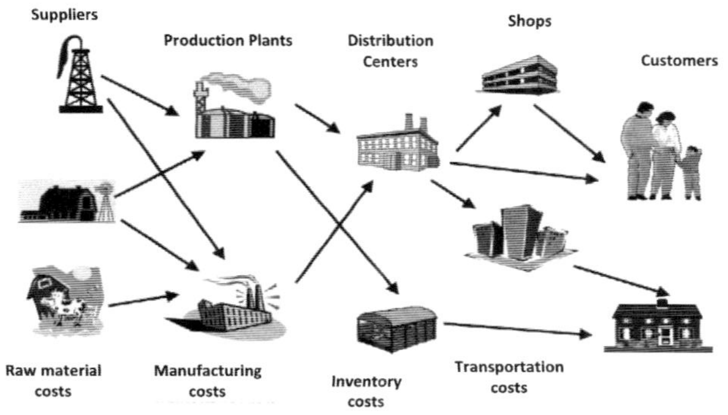

A SC is a network of firms and distribution channels organized to acquire raw materials, convert them to finished products, and distribute these products to customers.

2

Introduction

Industrial cluster

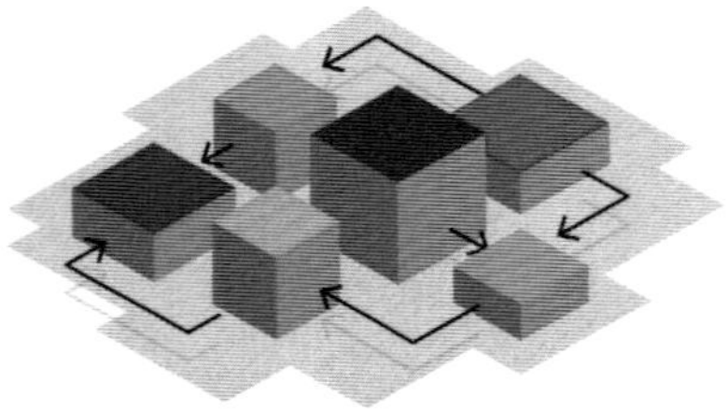

- An industrial cluster is formed by plants located in a given area.
- Advantages:
  - Access to inputs and infrastructure
  - Labor and human resources
  - Access to information and performance measures
  - Complementary products

3

Objective

- Development of a MILP model for the optimal design of a SC considering clusters

- Comparison:

  Individual Facilities vs Clusters

- Analysis of:

  → Production scale

  → Costs

4

Problem Definition

Given are:

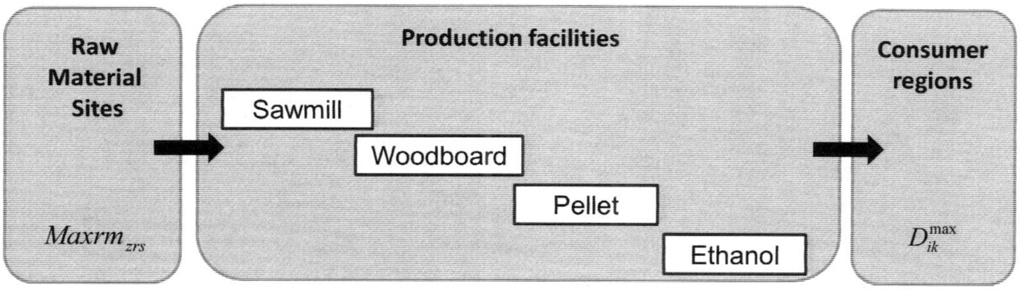

- Raw materials:
  - Types (z = 1, 2,..., Z) and
  - Qualities (r = 1, 2,..., R)
- Cost parameters
- Distances between SC nodes
- Production facilities (p = 1, 2,..., P)
  - Products (i = 1, 2,..., I)
  - By-products (b = 1, 2,..., B)
- Set of maximum plant capacities
- Process efficiency parameters

5

Problem Definition

Determine:

- CS design
  - Network configuration.
  - Production of each installed plant.
  - Flows between SC nodes:
    - raw materials,
    - residues,
    - products, and
    - by-products.

Objective function: **Profit Maximization**

6

Case Study

Forest industry

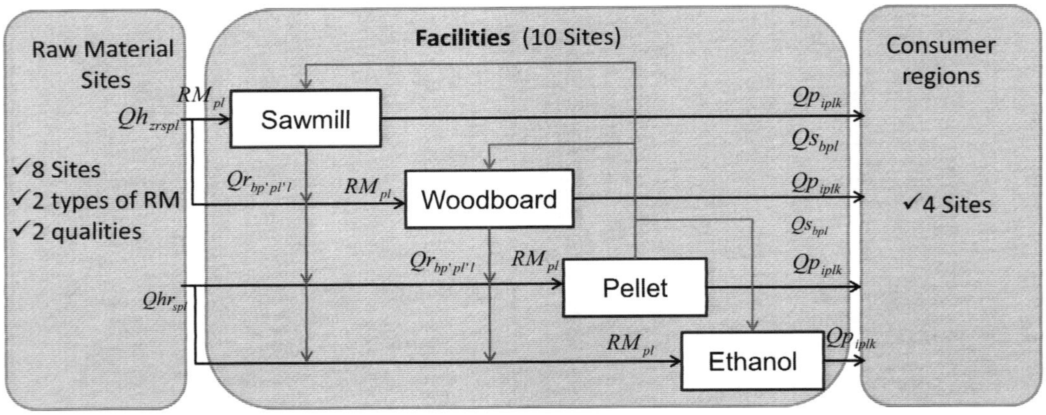

7

Case Study

Forest industry

		Sawmills	Woodboard	Pellets	Ethanol
Input	Raw Material	Log	Log	Residues	Residues
	Byproducts		Woodchips Firewood chips Sawdust	Woodchips Firewood chips Sawdust Bark	Woodchips Firewood chips Sawdust Bark
Output	Products	Lumber 1 Lumber 2	Board 1 Board 2	Pellets	Ethanol
	Byproducts	Woodchips Firewood chips Sawdust Bark	Bark		
energy source		Byproducts Pellets	Byproducts Pellets	Pellets	Pellets External energy

8

Case Study

Mathematical Model

Raw material sites

Maximum capacity

$$\sum_{p \in ZP_{zrp}} \sum_{l} Qh_{zrspl} \leq Maxrm_{zrs} \qquad \forall s, z, r$$

Residues production

$$\sum_{p \in RP} \sum_{l} Qhr_{spl} \leq \left(\sum_{p} \sum_{l} \sum_{z} Qh_{zrspl} \right) fhr \qquad \forall s$$

Production facilities

Input material

$$RM_{pl} = \sum_{\substack{r,z/ \\ p \in ZP_{zrp}}} \sum_{s} fl_z \, Qh_{zrspl} + \sum_{p'l'} \sum_{\substack{b/ \\ b \in BP_{bp} \\ p \in PB_{pb}}} fe_b \, Qr_{bp'pl'l} + \sum_{\substack{s/ \\ p \in RP}} fr \, Qhr_{spl} \qquad \forall p, l$$

Production

$$\sum_{\substack{t,i/ \\ i \in IP_{ip}}} Pm_{iptl} = f_{ip} \, RM_{pl} \qquad \forall p, l$$

Generation byproducts and use

$$Qrt_{bpl} = fc_b RM_{pl} \qquad \forall p, l, b \in BP_{bp}$$

$$Qrt_{bpl} = \sum_{\substack{p',l'/ \\ p' \in PB_{p'b}}} Qr_{bpp'll'} + Qb_{bpl} + Qs_{bpl} \qquad \forall b \in BP_{bp}, p, l$$

Production scale

$$\sum_{t} w_{ptl} = x_{pl} \qquad \forall p, l$$

$$\sum_{i \in IP_{ip}} Pm_{iptl} \leq w_{ptl} \; Pc_{tp}^{\max} \qquad \forall p, l, t$$

Energy balance

$$\sum_{ti} Pm_{iptl} \, ed = \sum_{b} Qb_{bpl} \, cr_b + Eext_{pl} \qquad \forall p, l$$

Demand constraints

Maximum demand

$$\sum_{p,l} Qp_{iplk} \leq D_{ik}^{\max} \qquad \forall p, i \in IP_{ip}, k$$

Product flow

$$\sum_{t} Pm_{iptl} \geq \sum_{k} Qp_{iplk} \qquad \forall p, i \in IP_{ip}, l$$

Objective function

$$G = I - RMc - Tc - Ic - Pc - Ec$$

Income

$$I = \sum_{\substack{i,p,l,k/ \\ i \in IP_{ip}}} S_{ip} \; Qp_{iplk} + \sum_{\substack{b,p,l/ \\ b \in BP_{bp}}} Sr_b \, Qrs_{bpl}$$

Raw material costs

$$RMc = Crm \sum_{\substack{z,r,p,l/ \\ p \in ZP_{zrp}}} Qh_{zrspl} + Chr \sum_{\substack{s,p,l/ \\ p \in RP}} Qhr_{spl}$$

Transportation costs

$$Tc = Ctrm \sum_{\substack{z,r,s,p,l/ \\ p \in ZP_{zrp}}} Dhp_{spl} \, Qh_{zrspl} + Cthr \sum_{\substack{s,p,l/ \\ p \in RP}} Dhp_{spl} \; Qhr_{spl} + \sum_{\substack{b,p,p',l,l'/ \\ b \in BP_{bp} \\ p' \in PB_{p'b}}} Ctr_b \; Dpp_{pp'll'} \; Qr_{bpp'll'} + \sum_{i,p,l,k} Ctp_{ip} \, Dpk_{plk} \; Qp_{iplk}$$

Installation costs

$$Ic = fam \sum_{p,t,l} \alpha_{pl} \left(PC_{tp}^{\max} \right)^{\beta_i} w_{ptl}$$

Production costs

$$Pc = \sum_{\substack{i,p,t,l \\ i \in IP_{ip}}} CP_{ipt} PM_{iptl}$$

Energy costs

$$Ec = \sum_{pl} Ce \;\; Eext_{pl}$$

9

Case Study

Cluster configuration

Proposal

- **Discounts on installation and production costs.**
- **Benefits depend on the number and relative size among plants.**

10

Case Study

Cluster configuration

Mathematical Model

Number of installed industries

$$\sum_{n=1}^{N_{plant}} n\, y_{ln} = \sum_{p,t} w_{ptl} \qquad \forall l \qquad\qquad \sum_{n=1}^{N_{plant}} y_{ln} \le 1 \qquad \forall l$$

Relation between factories

$$aux1_{ptl} \ge w_{ptl} + y_{l\,n1} - 1 \qquad \forall\, p,l,t \qquad\qquad aux1_{ptl} \le y_{l\,n1} \qquad \forall\, p,l,t$$

$$aux2_{p\,p'\,tt'l} \ge w_{ptl} + w_{p't'l} + y_{l\,n2} - 2 \qquad \forall\, p,p',l,t,t',p \ne p' \qquad\qquad aux2_{p\,p'\,tt'l} \le y_{l\,n2} \qquad \forall\, p,p',l,t,t',p \ne p'$$

$$\vdots \qquad\qquad\qquad \vdots$$

and so on   and so on

Installation cost

$$Ic = \sum_{l,n} ICC_{l\,n}$$

$$ICC_{l\,n1} = \sum_{p,t} ICPaux1_{ptl} \qquad \forall l$$

$$ICC_{l\,n2} = \sum_{\substack{p,p',t,t' \\ p \ne p'}} parI2_{tt'}\, ICPaux2_{pp'l} \qquad \forall l$$

$$ICPaux2_{pp'tt'l} \le ICP_{ptl} \qquad \forall p,p',l,t,t',p \ne p'$$

$$ICPaux2_{pp'tt'l} \le ICP^{UP}\ aux\,2_{pp'tt'l} \qquad \forall p,p',l,t,t',p \ne p'$$

$$ICPaux2_{pp'tt'l} \ge ICP_{plt} - ICP^{UP}\left(1 - aux2_{p\,p'\,tt'l}\right) \forall p,p',l,t,t',p \ne p'$$

$$ICP_{plt} = fam\ \ \alpha_{pl}\left(PC_{pt}^{\max}\right)^{\beta_l} w_{plt} \qquad \forall\, p,l,t$$

Production cost

$$Pc = \sum_{l,n} PCC_{l\,n}$$

$$PCC_{l\,n1} = \sum_{p,t} PCPaux1_{ptl} \qquad \forall l$$

$$PCC_{l\,n2} = \sum_{\substack{p,p',t,t' \\ p \ne p'}} parP2_{tt'}\, PCPaux2_{pp'l} \qquad \forall l$$

$$PCPaux2_{pp'tt'l} \le PCP_{ptl} \qquad \forall p,p',l,t,t',p \ne p'$$

$$PCPaux2_{pp'tt'l} \le PCP^{UP}\ aux\,2_{pp'tt'l} \qquad \forall p,p',l,t,t',p \ne p'$$

$$PCPaux2_{pp'tt'l} \ge PCP_{plt} - PCP^{UP}\left(1 - aux2_{p\,p'\,tt'l}\right) \forall p,p',l,t,t',p \ne p'$$

$$PCP_{ptl} = \sum_{i \in IP_p} Cp_{ptl}\, Pm_{iptl} \qquad \forall\, p,l,t$$

11

Case Study: Results

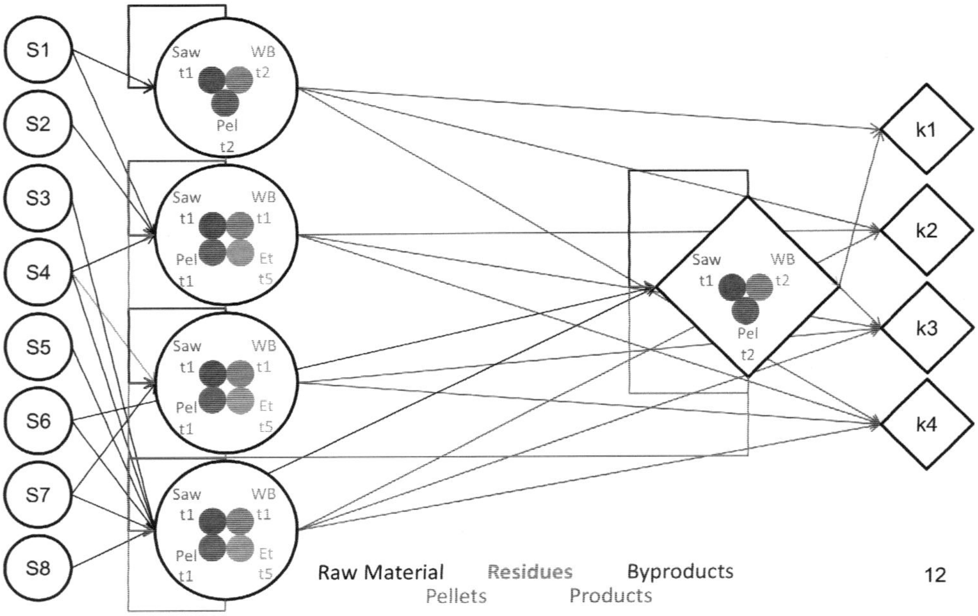

12

Case Study: Results

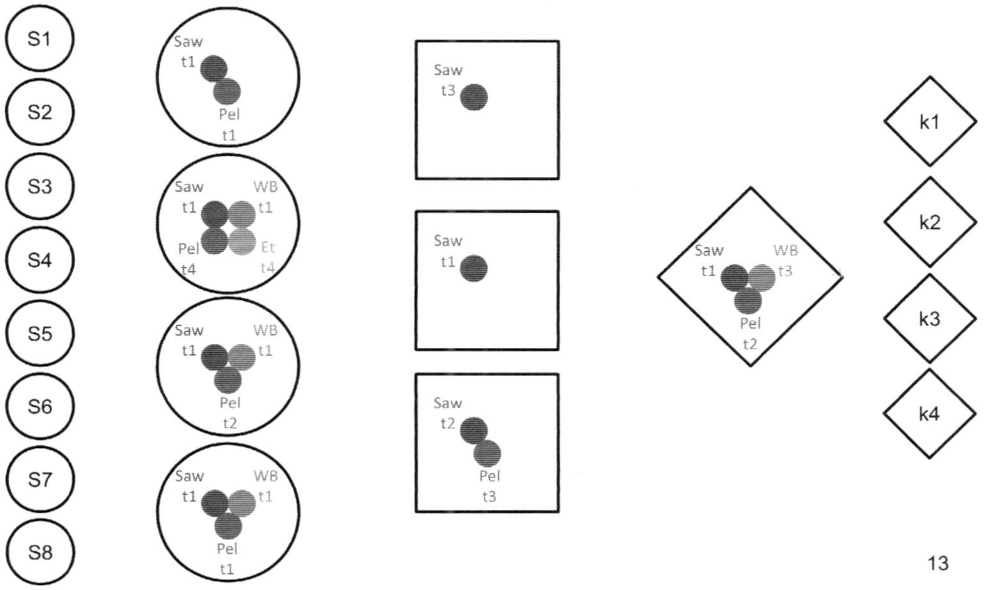

13

Case Study: Results

		Individual factories	Industrial clusters
Configuration	Sawmills	8	5
	Woodboards	4	5
	Pellets	6	5
	Ethanol	1	3
Location of plants	Harvested areas	12	15
	Intermediate point	4	-
	Consumer region	3	3
Formation of clusters	Four plants	1	3
	Three plants	3	2
	Two plants	2	-
	Individual plant	2	-
Use of raw materials	Sawmills	62.5 %	41.8 %
	Woodboards	37.5 %	58.2 %
Use of residues	Pellets	69.4 %	65.8 %
	Ethanol	30.6%	34.2 %

14

Case Study: Results

		Individual factories	Industrial clusters	
Production	Lumber	481685 m^3	322212 m^3	- 33.1 %
	Woodboards	975000 m^3	1200000 m^3	**23.1 %**
	Pellets	101348 t	100919 t	-0.4 %
	Ethanol	15000 m^3	16875 m^3	**12.5 %**
Use of installed capacity	Sawmills	99.1 %	89.5 %	
	Woodboards	100 %	100 %	
	Pellets	93.8 %	78.8 %	
	Ethanol	100 %	75.0 %	
Demand satisfaction	Sawmills	81.6 %	54.6 %	
	Woodboards	78.6 %	96.8 %	
	Pellets	0.9 %	0.0 %	
	Ethanol	1.4 %	1.6 %	

15

Case Study: Results

			Individual factories	Industrial clusters
Byproducts	Sawmills	Sold	-	-
		Energy	8.4 %	21.4 %
		Woodboards	91.6 %	73.9 %
		Pellets	-	4.3
		Ethanol	-	0.4
	Woodboards	Sold	-	-
		Energy	100 %	100 %
		Pellets	-	-
		Ethanol	-	-
Pellets	Sold		4.3 %	-
	Energy	Sawmills	13.7 %	0.6 %
		Woodboards	74.1 %	91.6 %
		Pellets	7.9 %	7.8 %
		Ethanol	-	-

16

Case Study: Results

Economic report *[M $/year]*

		Individual factories	Industrial clusters	
Sawmills	Income	153.80	107.29	
	Raw material costs	35.55	24.55	
	Installation cost	1.57	0.89	- 43.3 %
	Production cost	12.01	5.64	- 53.0 %
Woodboard	Income	317.94	385.50	
	Raw material costs	17.66	28.64	
	Installation cost	39.43	38.43	- 2.5 %
	Production cost	78.81	61.23	- 22.3 %
Pellet	Income	1.29	0.00	
	Raw material costs	8.44	8.01	
	Installation cost	0.75	0.61	- 18.7 %
	Production cost	4.09	2.61	- 36.2 %
Ethanol	Income	17.69	19.90	
	Raw material costs	0.62	0.69	
	Installation cost	4.91	4.61	- 6.1 %
	Production cost	2.78	1.21	- 56.5 %
Transportation and energy cost		36.27	39.22	
Net benefit		**253.82**	**303.01**	

17

Case Study: Results

Computational Results

Model Statistics

	Equations	Continuous variables	Discrete variables	CPU time (sec.)
Individual factories	2772	11164	280	2.2
Industrial clusters	10314253	377645	1684480	21466.0

18

Conclusions

✓ A MILP model for the optimal SC design is presented.

✓ The cluster installation in a SC is analyzed, considering cost discounts for facilities clusters.

✓ The different tradeoffs between clusters and individual facility configuration are evaluated.

✓ The effect of the production scale in the overall SC are highlighted through the proposed example.

✓The ethanol production is increased in cluster configuration.

✓ The presented approach represents a useful tool for analyzing production, resources and services integration in a SC involving different facilities in the diversification of an industry.

19

Thank you

nvanzetti@santafe-conicet.gov.ar

gcorsano@santafe-conicet.gov.ar

mmontagna@santafe-conicet.gov.ar

INGAR
CONICET
U T N

20

Techno-economic analysis and optimal design of integrated biorefineries under market and operational uncertainties

S. Salas, A. Geraili, J.A. Romagnoli

Department of Chemical Engineering

Louisiana State University

Baton Rouge, LA 70803

PSE @ LSU

LSU

PROCESS SYSTEMS ENGINEERING AT LOUISIANA STATE UNIVERSITY

Outline

- **Motivation and Project Objective**
- **Optimization-Based Support Framework**
  - **Framework formulation**
  - **Application case study**
- **Uncertainty Formulation**
  - **Uncertainty at strategic level**
  - **Uncertainty at the operational level**
- **Results**
- **Conclusions**

LSU

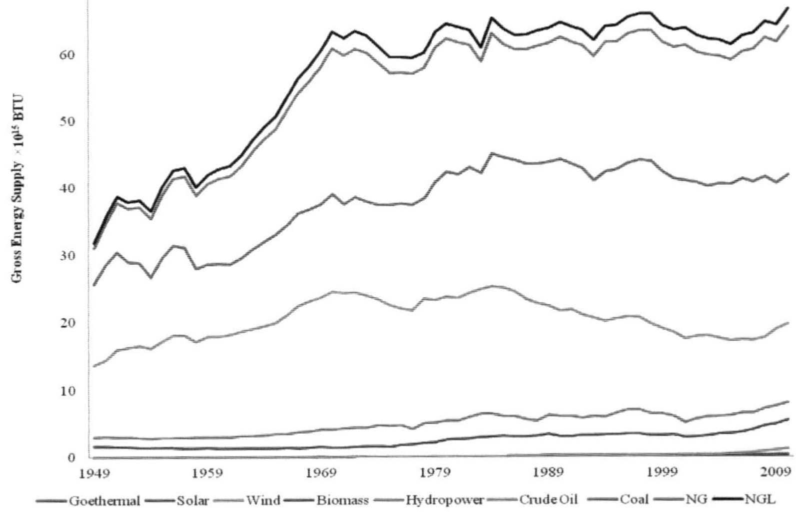

3

Motivation

Disadvantages of fossil fuels:

- Finite nature
- Volatility in prices
- Environmental hazards

Need to embrace a diversified energy strategy to supply future needs, especially domestically derived renewable energy

4

Motivation

- **Biomass** has become one of the most commonly used renewable sources of energy in the last two decades.
  - After a boom in the corn-based ethanol (first-generation biofuel) the interest has gradually shifted towards more viable renewable resources → **lignocellulosic feedstocks**

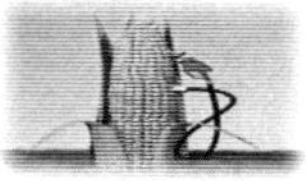

- Cellulosic ethanol is an example of such alternative fuel which is considered as a second-generation biofuel and is derived from cellulose instead of starch.

- ***The fuels and chemicals produced from lignocellulosic feedstocks*** are extremely attractive
  - Raw materials can be composed of "left-over" wastes of food crops and forest harvests that do not interfere with the human food chain.
  - Large variety of lignocellulosic materials: these types of produced fuels and chemicals can overcome the challenge of limited feedstock availability

5

Motivation

- **The commercial scale** production of biobased fuels and chemicals has been hampered due to several factors:
  - the lack of capital investment,
  - technological complexities
  - product market immaturities

  driven by the ***nascent nature of these industries***.

- Development of ***decision-making frameworks for renewable energy systems*** provides a suitable tool for analysis and investigation of these industries for potential improvements.
  - The use of decision support to aid in decision-making process will lead more sound actions being taken by stakeholders based on a more complete picture of what is actually happening around them.

6

Multi-layered framework: The concept

➢ Overall objective:
Developing a decision support system for sustainable design of renewable energy systems under uncertainty

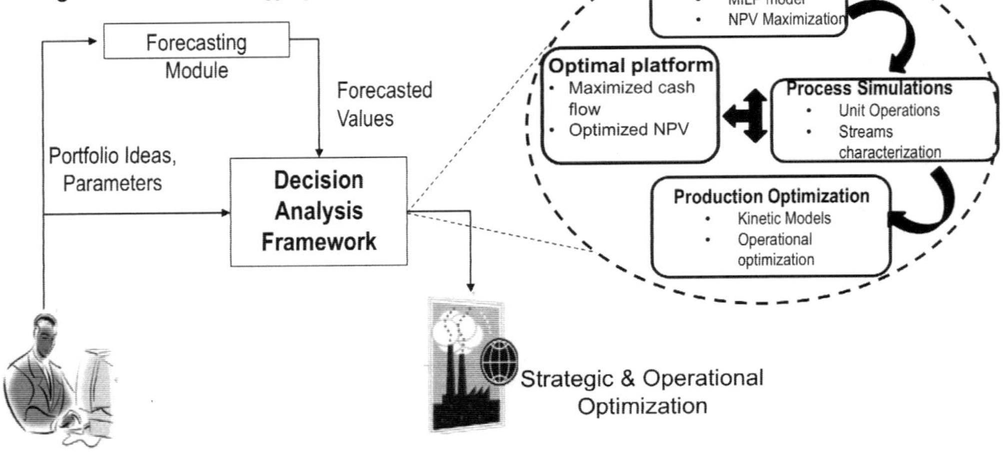

This strategy has the advantages of :

- introduce the nonlinear process characteristics
- optimization at the operational level, thus maximizing the cash flow.

7

Lignocellulosic biorefinery

The optimization strategy is applied to a biorefinery that utilizes lignocellulosic feedstock(s) to produce biobased fuels and chemicals.

Switchgrass serves as the selected feedstock for the biorefining process

Fermentation-based sugar conversion platform with 3 products: cellulosic ethanol, bio-succinic acid, and bioelectricity.

Assumed to be located in the ***southwest region of Louisiana***

Utilizes biomass crops from **land available within a 100 mile** radius of the plant.

The production chain comprises of ***6 major systems***:

- **feedstock pre-treatment**
- **sugar hydrolysis**
- **sugar fermentation**
- **product purification**
- **heat and power generation**
- **wastewater treatment**.

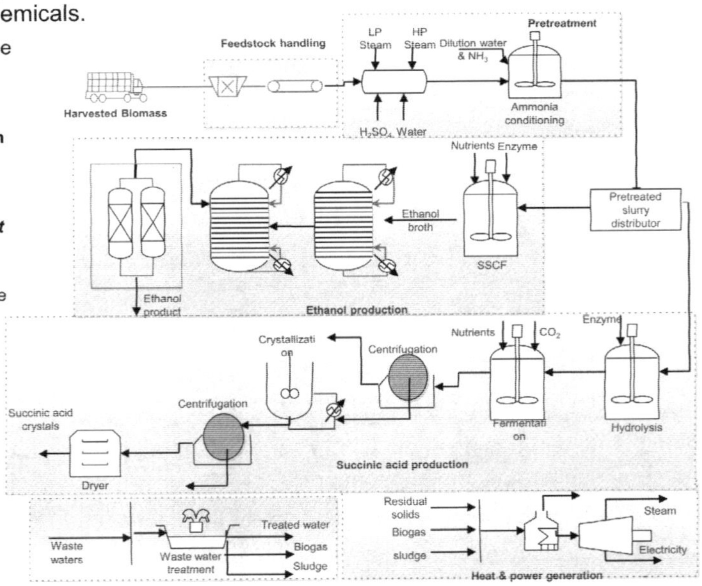

Uncertainty Formulation

Two types of uncertainty

- Endogenous Uncertainty
  - Biological platform is inherently uncertain → Variable process yields
- Exogenous uncertainty
  - Prices –illiquid markets leading to price swings
  - Demands – nascent markets →long-term demand trajectory unknown

How does uncertainty manifest itself → *RISK*

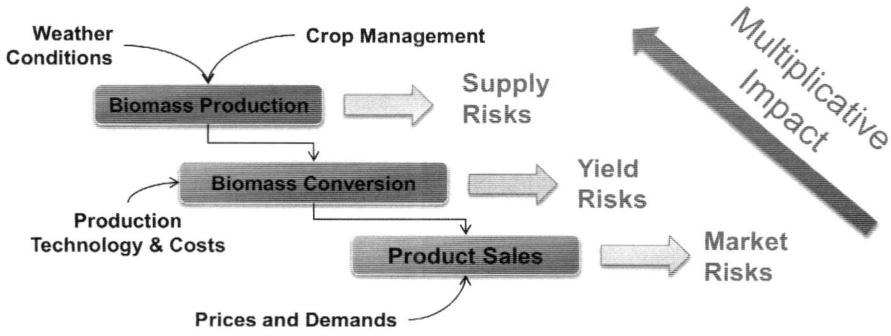

Strategic Model Formulation: Under Uncertainty

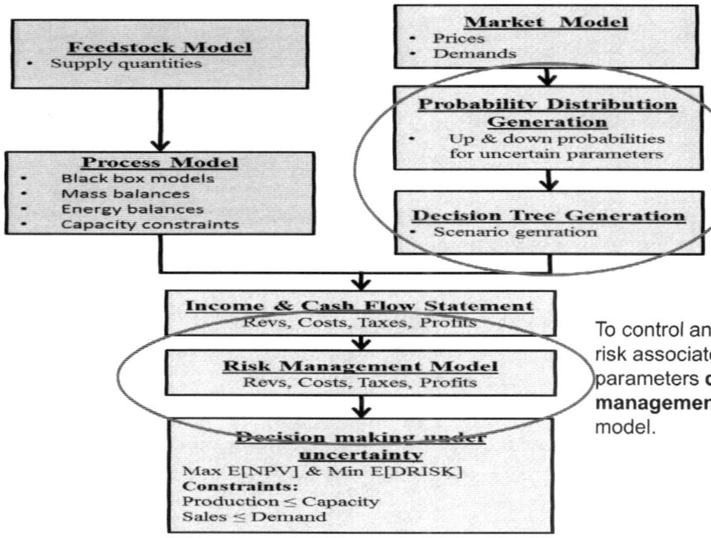

In the strategic layer, **different scenarios are developed** based on stochastic forecasts for uncertain market parameters including price and demand of bioproducts.

To control and manage the financial risk associated with uncertain market parameters **downside risk management** is introduced to the model.

Structure of the proposed strategic decision-making model under uncertainty

10

Decision tree analysis

- Prices and demands of bioproducts are impacted primarily by crude oil prices
- Bioproduct market can be represented as a decision tree

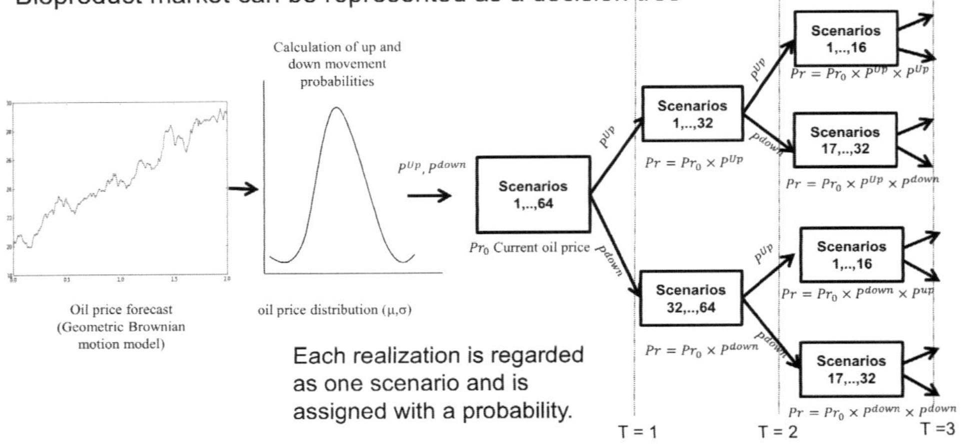

- Avoids high-dimensional numerical integration in the solution of the problem
- The expected net present value can be calculated as finite sums and each constraint can be duplicated for each scenario
- Address only a finite number of selected realizations of uncertainty in the optimization.

11

Risk management model

Downside risk approach: consistent measure of risk with good mathematical properties enabling efficient optimization by means linear programming techniques.

Downside Risk management: The risk associated with scenarios whose profits are not desirable is minimized.

Downside risk is calculated based on the area underneath the curve of probability function which has lower profitability than the defined threshold (π).

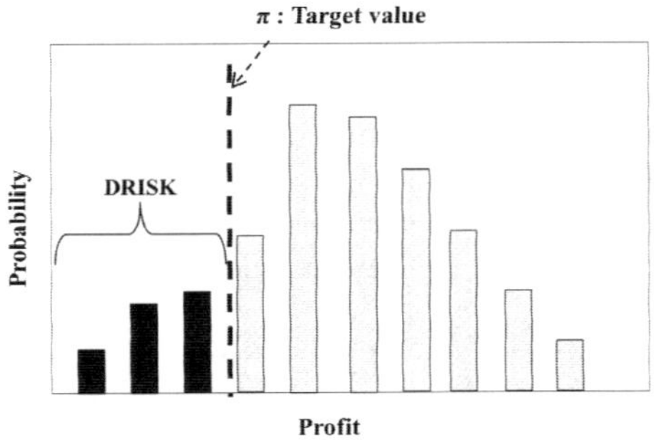

A positive variable δ_s deviation between the NPV of each scenario and the defined threshold is considered

$$\delta_s \geq \pi - NPV_s, \quad \delta_s \geq 0$$

Downside risk is calculated as the expected value of positive deviations from the defined target π

$$DRISK_\pi = \sum_s P_s \cdot \delta_s$$

Steps in Operational Model Formulation

Potential sources of uncertainties: Technical parameters (Complete set of kinetic parameters of biological reactions)

Step1: Identification of significant uncertainty sources

Sensitivity Analysis

Step2: Generating samples for parameters and variables

Sampling (Latin Hypercube method)

Step3: Extensive simulation based on samples

Monte Carlo Simulation

Step4: Finding optimal process conditions

Analysis of results

Sobol global Sensitivity analysis:

- ✓ Ranking uncertain individual model input parameters **(Sensitivity index)**
- ✓ Reducing the complexity of the stochastic model **(Selection of influential parameters)**

Latin hypercube sampling (**LHS**) statistical method for generating a sample of plausible collections of parameter values from a multidimensional distribution

13

Results-Strategic level

To illustrate the effectiveness of the proposed strategy, different case studies are considered including: ***Base case, stochastic model***, and ***stochastic model coupled with risk management model.***

higher profitability (NPV) is obtained by the stochastic model which is 44% higher than deterministic solution.

processing capacity of the plant also increases 40% in comparison to the deterministic model

more sugar is allocated to succinic acid production

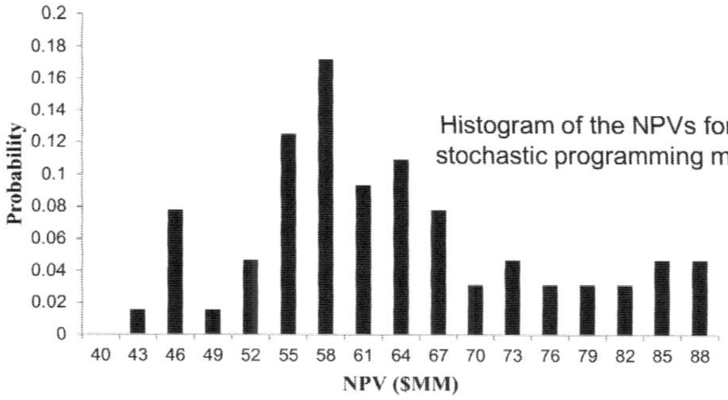

Histogram of the NPVs for the stochastic programming model

Table : Comparison of base case and stochastic model

	Feedstock capacity (1000 ton/yr)	Sugar allocation ratio (for ethanol production)	Ethanol (MM gal)	Succinic acid (1000 ton)	Net present value ($MM)
Base case	155.0	0.66	7.1	4.00	45.0
Stochastic case	218.0	0.62	11.1	6.00	62.8

Results-Strategic level

stochastic model coupled with risk management model.

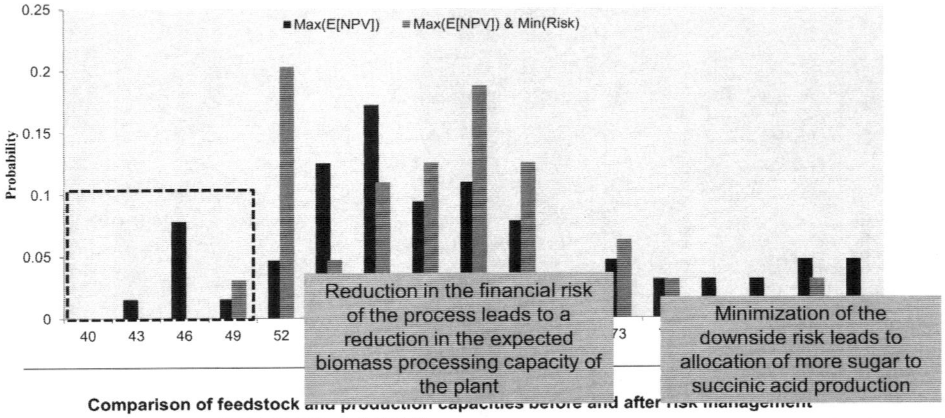

Comparison of feedstock and production capacities before and after risk management

Scenario	Expected NPV ($MM)	Downside risk	Feedstock Capacity (1000tons/yr)	Sugar allocation ratio (for ethanol production)
Stochastic case	62.8	10%	218	0.62
Risk managed case	60.0	3%	152	0.59

15

Although the capacity of the plant has decreased, **net present value of scenarios have much higher chances to be between the desired target**

LSU

Results- operational level optimization

Hydrolysis temperature and enzyme loading are selected as important operating variables to be optimzed

Monte Carlo simulation results for annual cash flow maximization

Scenario	Mean	95% Confidence interval	% saving (95% confidence interval)
Base case	8700	120	--
Sample 14	9570	83	30.8
Sample 47	9153	104.3	13.1

The most feasible designs:

High average cash flow and narrow uncertainty range – ranking the scenarios based on only the mean value does not reflect the variability associated with the uncertainty of parameters

Comparison of deterministic and stochastic cases

Scenario	Hydrolysis Temperature (°C)	Enzyme loading ratio (g enzyme/ Kg Cellulose)
Deterministic case	33.45	25
Optimal (stochastic) case	39.3	13.1

Optimal values of the decision variables obtained from Monte Carlo stochastic

Conclusions

- **A robust decision support framework for design of sustainable renewable energy production systems is developed:**
  - ✓ Mixed integer linear programming (MILP) models are suggested for the purpose of strategic planning.
  - ✓ Rigorous non-linear process modelling and process-level optimization are combined with strategic planning based on distributed strategy.

- **Scope of proposed framework is expanded by incorporating uncertainty analysis to the model.**
  - ✓ long-term uncertainties such as price and demand of the products are included in framework based on a hypothetical market model.
  - ✓ Scenario-based stochastic programming model is developed to transform the strategic optimization model into a deterministic approximation.
  - ✓ Financial Risk metric is added to the model to **control the variability** of scenarios performance.

17

Thank you!!

LSU

18

Economic, Environmental and Social Cost Optimization of Biomass Supply Chain for Electricity Generation in WI, USA

Sangpil Ko,
Pasi T. Lautala

Michigan Technological University

I. Introduction

Objectives and Literature Review

1.1. Introduction

- Building an efficient procurement system one of the main obstacles to extend bioenergy industry (25-50% of delivered cost)
- Transportation a key component of the system, affected greatly by interrelationship between feedstock sources and facility locations.

1.2. Objective

- To build a mathematical model that uses region specific data to minimize sustainable transportation costs from economic, environmental and societal factors.
- To calculate the upstream air emissions of forest residues as a result of residues collection at forest, transloading, and transportation by truck or multimodally.

1.3. Past Research vs. Current Project

1) Bioenergy plant location problem: two general methods

- Mathematical optimization programming *(e.g., Velazquez and Fernandez, 2010; Kim et al., 2011)*
- Using geographical information systems *(e.g., Panichelli and Gnansounou, 2008; Kurka et al., 2012)*

* **Limitation**: past research has relied on generalized values from the literature without detailed evaluation of transportation costs

→ Our model combines input data from the literature with detailed data specific to the region studied.

2) Trade-off evaluation with economic and external cost/benefit factor of bioenergy

- Economic cost and environmental cost *(e.g., Ayoub et al., 2009; Montross et al., 2013; Fan et al., 2013)*
- Economic cost and social benefits created by the bioenergy plant *(e.g., Perimenis et al. 2011; You et al., 2012; Roni et al., 2016)*

* **Limitation**: lack of integrated economic and external cost model on biomass transportation cost

→ We expand the trend by combining all three main components needed for the sustainable transportation cost evaluation.

II. Sustainable factor for biomass transportation

What are the factors for evaluation?

5

2.1. Sustainable Factors for Biomass Transportation

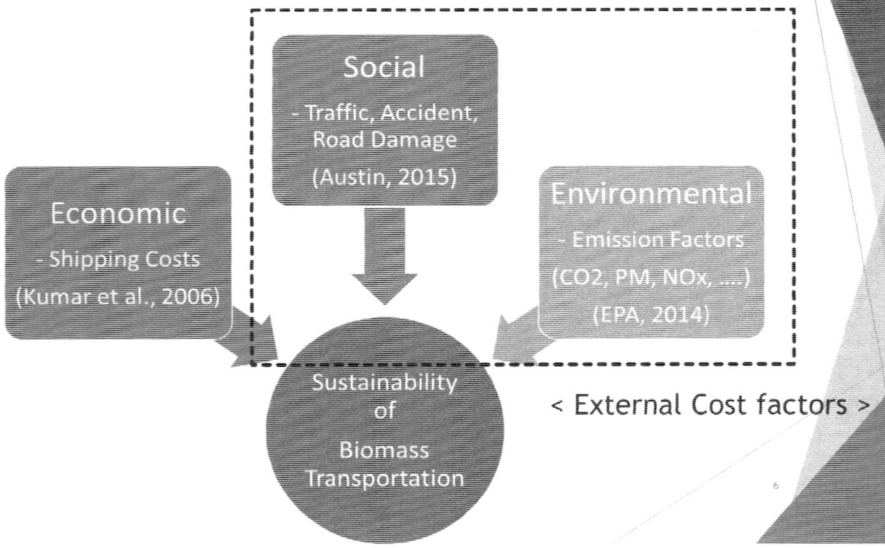

6

2.2. Economic cost

1) Economic cost factors for biomass transportation: Rates

- Transportation rates are typically formulated based on Distance Fixed Cost (DFC) and Distance Variable Cost (DVC).

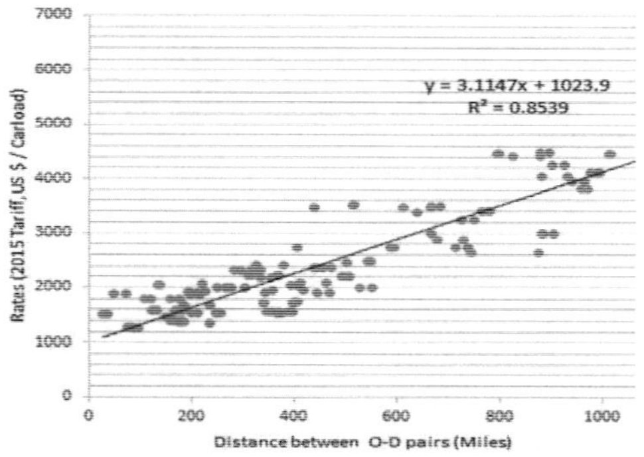

✓ Truck Costs **from 1,295 O/D pairs by local provider** for lumber and wood commodities in WI/MI (UP)

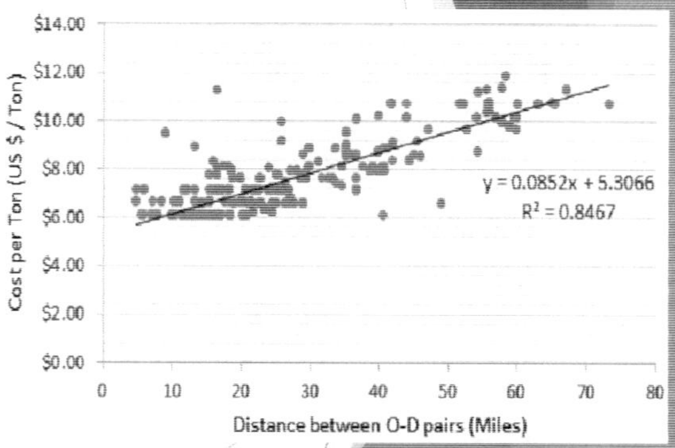

✓ Rail Cost **(tariffs) from 145 O/D pairs by CN** for lumber and wood commodities in WI/MI (UP)

2.2. Social cost

2) Social cost factors for biomass transportation

- The costs of freight shipments not passed on to consumers, and they categorized three social impacts from the surface freight transportation: pavement damage, accidents, and congestion (US GAO, 2015).

✓ Pavement Damage

: Pavement preservation costs (Government spending) per million ton-miles in 2010.

✓ Traffic Congestion

: Cost of delay to road users in 2000 (from 1997 Highway cost allocation study by FHWA).

✓ Accident Risk

: (The number of Fatalities per billion ton-mile, avg. 2003-2007) X (The latest value of human life used by DOT)

2.2. Environmental cost

3) Environmental cost factors for biomass transportation

- Three kinds of emissions factors were considered in this study to evaluate their effect on air quality: GHG emissions, Nitrogen Oxide (NOx), and Particulate Matter (PM).

✓ External costs of GHG (CO2)

: includes changes in net agricultural productivity, human health, property damages from increased flood risk, and the value of ecosystem services due to climate change.

✓ External costs of Air emissions (PM and NOx)

: typically includes the health effects on humans by determining the medical and lost work cost of premature death, asthma, and associated respiratory illnesses.

Unit Costs Estimates	CO2	PM (<10 μm)	NOx	Sources
2014 US Dollars per Green Ton	42.25	7255.63	4724.60	• CO2: IWGSCC (2013) • PM/NOx: Matthews et al. (2001)

* All costs are adjusted to Year 2014 USD using Consumer Price Index by the U.S. Department of Labor (2015)

III. Sustainable Biomass Transportation Model

Combinational Model for Trip Assignment and Mode Choice

10

3.1. Set and Decision Variables

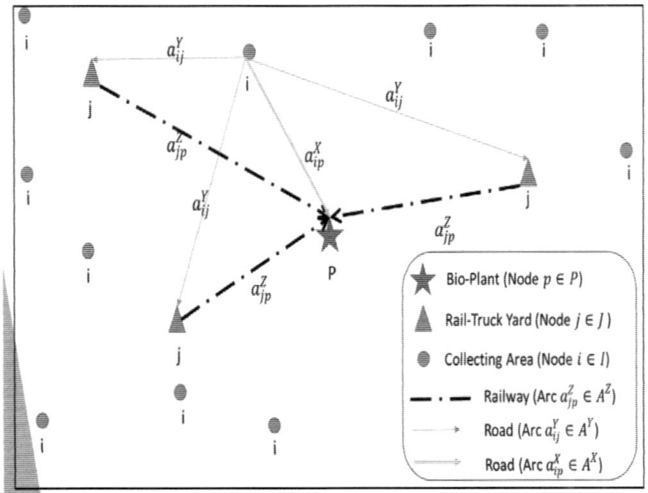

< Basic sets for nodes and arcs in mathematical model >

- Three decision variables for the mode of biomass transportation: X_{ip}, Y_{ij}, and Z_{jp}

✓ X_{ip}: number of trucks shipped along arc A^X (between supply area i to bioplant p).

✓ Y_{ij} : number of trucks shipped along arc A^Y (between supply area i to rail siding j).

✓ Z_{jp} : number of rail shipment along arc A^Z (between rail siding j and bioplant p).

11

3.2. Mathematical Model

Min

$$w_1\left\{\sum_{i=1}^{I}\sum_{p=1}^{P} C^X_{ip}X_{ip} + \sum_{i=1}^{I}\sum_{j=1}^{J} C^Y_{ij}Y_{ij} + \sum_{j=1}^{J}\sum_{p=1}^{P} C^Z_{jp}Z_{jp}\right\}$$ < Economic costs >

$$+ w_2\left\{\sum_{i=1}^{I}\sum_{p=1}^{P} S^X_{ip}X_{ip} + \sum_{i=1}^{I}\sum_{j=1}^{J} S^Y_{ij}Y_{ij} + \sum_{j=1}^{J}\sum_{p=1}^{P} S^Z_{jp}Z_{jp}\right\}$$ < Social costs >

$$+ w_3\left\{\sum_{i=1}^{I}\sum_{p=1}^{P} E^X_{ip}X_{ip} + \sum_{i=1}^{I}\sum_{j=1}^{J} E^Y_{ij}Y_{ij} + \sum_{j=1}^{J}\sum_{p=1}^{P} E^Z_{jp}Z_{jp}\right\}$$ < Environmental costs >

< Weighting factor >

S.t.

$$\sum_{p=1}^{P} t^X X_{ip} + \sum_{j=1}^{J} t^Y Y_{ij} \le Q_i \qquad \forall\, i \in I$$

$$\sum_{i=1}^{I} t^X X_{ip} + \sum_{j=1}^{J} t^Z Z_{jp} \ge O_p \qquad \forall\, p \in P$$

$$\sum_{i=1}^{I} t^Y Y_{ij} = \sum_{P=1}^{p} t^Z Z_{jp} \qquad \forall\, j \in J$$

< Constraints >

$$X_{ip}, Y_{ij}, Z_{jp} = Positive\ Integer \qquad \forall\, i \in I, j \in J, p \in P$$

where

$$C^X_{ip} = (D^X_{ip} \cdot \theta^X_v + \theta^X_f) \times t^X \qquad \forall\, i \in I, p \in P$$

$$C^Y_{ij} = (D^Y_{ij} \cdot \theta^Y_v + \theta^Y_f) \times t^Y \qquad \forall\, i \in I, j \in J$$

$$C^Z_{jp} = \left[(D^Z_{jp} \cdot \theta^Z_v + \theta^Z_f) + l_j + (r_j \times k_{jp}) + \left\{h_j/365 \times \left(\frac{1}{2}k_j + m_{jp}\right)\right\}\right] \times t^Z \qquad \forall\, j \in J, p \in P$$

$$S^X_{ip} = D^X_{ip} \times (\gamma^X_{ip} + \delta^X_{ip}) \times t^X \qquad \forall\, i \in I, p \in P$$

$$S^Y_{ij} = D^Y_{ij} \times (\gamma^Y_{ij} + \delta^Y_{ij}) \times t^Y \qquad \forall\, i \in I, j \in J$$

$$S^Z_{jp} = D^Z_{jp} \times (\gamma^Z_{jp} + \delta^Z_{jp}) \times t^Z \qquad \forall\, j \in J, p \in P$$

$$E^X_{ip} = D^X_{ip} \times (e^X_c + e^X_p + e^X_n) \times t^X \qquad \forall\, i \in I, p \in P$$

$$E^Y_{ij} = D^Y_{ij} \times (e^Y_c + e^Y_p + e^Y_n) \times t^Y \qquad \forall\, i \in I, j \in J$$

$$E^Z_{jp} = D^Z_{jp} \times (e^Z_c + e^Z_p + e^Z_n) \times t^Z \qquad \forall\, j \in J, p \in P$$

12

3.3. Holding costs at Rail yards

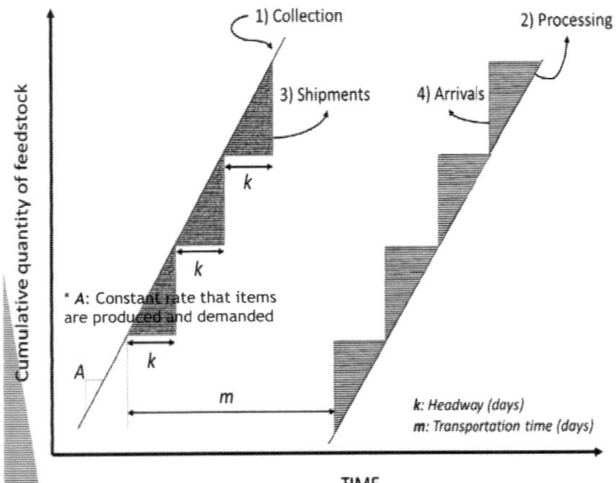

< Cumulative quantity of feedstock by rail between yard and plant >
(Source: Adapted from Daganzo, 2005)

- If let ***r*** be the coefficient of rent cost ($) per ton-year, rent cost per a ton of feedstock can be calculated with:

Rent cost per ton = ***{r × (maximum accumulation)} / A***
= r × k

- Meanwhile, if let ***h*** be the penalty paid for holding one ton of forest residue for one year, inventory cost in rail yard is identical with:

Inventory cost per ton = ***{h × (average wait per ton)} / A***
= h/365 × (½k + m)

- Thus, holding cost per one ton of forest residue is expressed as:

Holding cost (per ton) = Rent cost + Inventory cost (per ton)
= (r × k) + {h/365 × (½k + m)}

IV. Case Study: Bio-electricity plant in WI, USA

14

4.1. Case Description

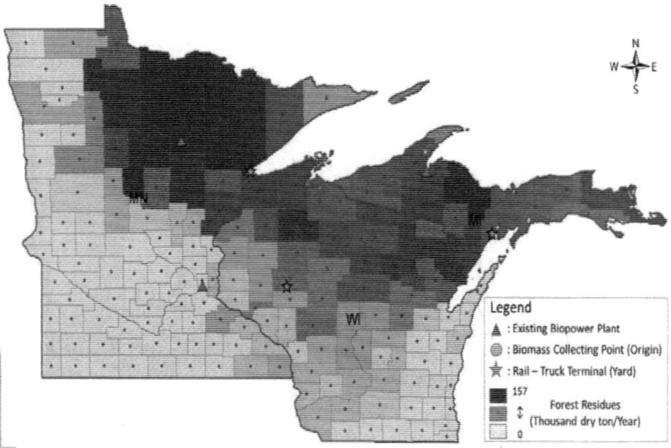

< Study area and availability of forest residue in each county >
(Source: USDA, 2012; NREL, 2015)

- **Objective:** To Compare Potential Bioenergy Plant Locations in Wisconsin, US using Sustainable Transportation Model
- **Procurement areas:** WI, MN, and UP MI
- **Three Rail-Truck yards** as potential satellite yards.
- **Basic unit of collecting areas** is county of each state: 173 collecting areas (counties).
- **Resource competition:** biomass feedstock in collection areas overlapped with that of existing plants may not be available for new plant

4.2. Assumptions

- A series of assumptions were also used to simplify the biomass transportation system, including:
  - All tonnage "green ton" basis (50% moisture content).
  - 10,000 dry ton per year of fuel for each 1 MW of installed capacity (Mayhead and Shelly, 2012).
  - 1 dry ton of woody biomass generates 1 MWh electricity energy (Ashton et al., 2008).
  - Forest residues from harvesting areas are the only feedstock considered.
  - Economic costs (rates) for truck and rail are for round trips.
  - Social costs of loaded and empty (return) miles are the same, while environmental costs are different.
  - Seasonality not considered in holding costs.
  - 3-day headways for rail shipments and 2-day rail hauling time to plant.

4.3. Preprocessing: Identifying potential locations for the plant

- Before transportation alternatives could be analyzed, a set of bioenergy plant candidate sites had to be established.

- The selection of candidate sites was conducted through a three-step process:

1) Finding sites where community interest
2) Screening for protected areas and
3) Confirming the site suitability to narrow down the candidate site

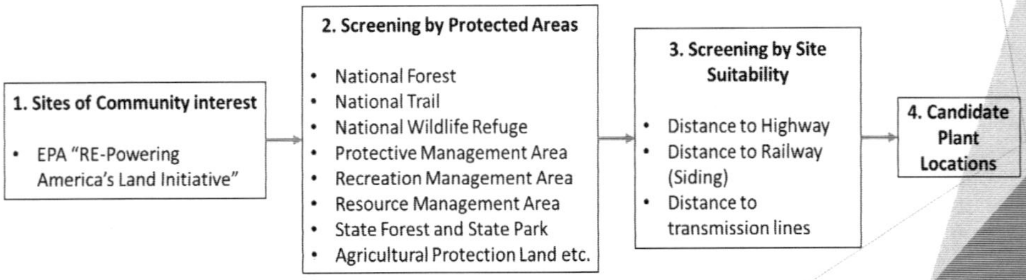

< Preprocessing steps to consider basic criteria >

4.3. Final Candidate sites

Plant	Site name	Distance (miles)		
		Transmission lines	Highways	Rail
Plant 1	MURPHY OIL USA	0	0	0
Plant 2	AMERICAN TOY AND FURNITURE	0	0	0
Plant 3	OUTAGAMIE COUNTY LF	0	0	0
Plant 4	FOX RIVER NRDA/PCB RELEASES	0	0	0
Plant 5	SC JOHNSON & SON INC	0	0	0
Plant 6	EPEC POLYMERS, INC.	0	0	0
Plant 7	UNIVERSITY OF WI MADISON ARLINGTON FARM	0	0	0
Plant 8	MOSS-AMERICAN CO., INC.	0	0	0
Plant 9	KOHLER CO	0	0	0

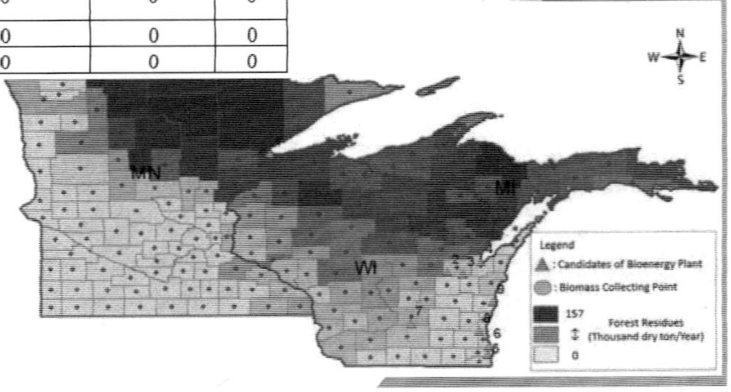

4.4. Data: Economic Costs

- Each **truck load has 28 tons capacity** for all plant (based on existing operations).
- Each **rail has 25 cars, and 90 tons** would be loaded for each rail car.
- All DFC and DVC were used based on actual Truck and Rail rate data obtained from a local forest products company in the study area (WI/MI (UP)).
- Miscellaneous economic cost factor concerning transportation operation are as follows:

Category	Cost type	Unit costs ($/ton)	Sources
Economic Factors	Rent cost	0.46 (Annual)	Morey et al. (2010)
	Inventory cost	254.77 (Annual)	Eksioglu et al. (2009)
	Trans-loading cost	4.32	Zhang et al. (2012)

* All costs are adjusted to Year 2014 USD using Consumer Price Index by U.S. Department of Labor (2015)

4.5. Social Costs

- Regionally applicable social costs.

Cost Factor	Data	Note
Pavement Damage	Budget for road maintenance of WisDOT	a
	All roads miles of WI	b
	Total number of crossings	c
	Pavement length railroad grade crossing	d
	Ratio of rail crossing out of all road miles of WI	e
	Annual ton-miles for truck in WI	f
	Annual ton-miles for rail in WI	g
Traffic Congestion	Congestion cost of rural highway	h
	National Rural VMT: Average 2009-2013	i
	WI Rural VMT: Average 2009-2013	j
	Distribution of truck for congestion cost	k
	Number of train crossings Rural Highway-Crossing per year in WI	l
	Daily traffic on crossings (Rural Highway-Crossing) for WI	m
	Average train speed	n
	Average train length	o
	Average hourly wage (value of time)	p
Accident Risk	Annual number of fatalities per VMT for Large Truck Fatalities: Average 2009-2013 for WI	q
	Annual number of fatalities per VMT for Train crash Fatalities: Average 2009-2013 for WI	r
	Latest value of human life by US DOT	s

Social Cost Factor	Values (Cent/Ton-Mile)		Equations	
	Truck	Rail	Truck	Rail
Pavement Damage	0.6317	0.0004	$(a*(1-e))/(f*10^6)$	$(a*e)/(g*10^6)$
Traffic Congestion	0.1500	0.0296	$((h*10^6)*(j/i)*k)/(f*10^6)$	$(l*((o/n)*60)*(m/24/60)*(p/60))/(g*10^6)$
Accident Risk	1.7476	0.1383	$(q*(s*10^6))/(f*10^6)$	(r*(s*10^6))/(g*10^6)

20

4.6. Environmental Costs

- **Monetary Costs** = {(**Emissions produced by mode** (g per ton-km)} **X** {**Unit external costs estimates** ($ per ton)}
- Environmental emissions were calculated by multiplying the fuel and chemical inputs by their emission inventories obtained from SimaPro 8.

	Fuel Type	Input for 1000 short ton/day (bone dry)
Residue collection and processing	Diesel	8,812
	Lubricating oil	80
	Hydraulic fluid	83
	Grease	220
	Gasoline	226
Yard equipment	Diesel	1376
	Lubricating oil	59
	Hydraulic fluid	62
	Grease	163

< Material and energy use (in *liter*) for residue acquisition >

Emission Category	Feedstock collection	Trans-loading	Emission Category	Feedstock transportation (Truck)	Feedstock transportation (Rail)
GHG (kg CO_2 eq/t)	31.9	4.99	GHG (g CO_2 eq/ton-km)	138.8	43.4
PM <2.5um (g/t)	16.3	0.287	PM <2.5um (mg/ton-km)	17.4	16.5
PM >10um (g/t)	1.58	0.3	PM >10um (mg/ton-km)	19.2	20.6
PM >2.5, <10um (g/t)	1.65	0.093	PM >2.5, <10um (mg/ton-km)	14.9	8.86
NOx (g/t)	517	2.814	NOx (mg/ton-km)	466	611

< Emission factors of the forest residues supply chain (one way loaded) >

4.7. Cost factors of empty return

- Emission factors for empty returns were adjusted to reflect difference in environmental impacts from loaded and empty miles.
- Assumption: emission savings of empty miles is equivalent to energy savings from weight reduction

Mode	Commercial Truck (40t)	Freight Train (Electric)	Freight Train (Diesel)
Relative Energy Savings (% / 10% weight reduction)	3.1%	5%	5%

(IFEU, 2002)

✓ Empty Truck: {28ton /(13ton + 28ton)} = 68.3% weight decrease -> **21% energy savings**

→ 79% emissions of full-loaded truck

✓ Empty Freight Rail: { (25cars*90 ton)/[193ton +{25cars * (30ton+90ton)}] = 70% weight decrease -> **35% energy savings**

→ 65% emissions of full-loaded freight rail

V. Result and Discussion: Bio-electricity plant in WI, USA

23

5.1. Evaluating transportation cost: Economic costs only

- Network Analyst of ArcGIS for Desktop 10.3.1, and CPLEX coded by Python language were used.
- 25, 50 and 100 MW plant capacity
- Plant 2 (P2) offers the least estimated cost alternative for all capacity scenarios.
- Average transportation cost ranged from $29 to $48 per MWh, and it increased with increasing plant capacity.

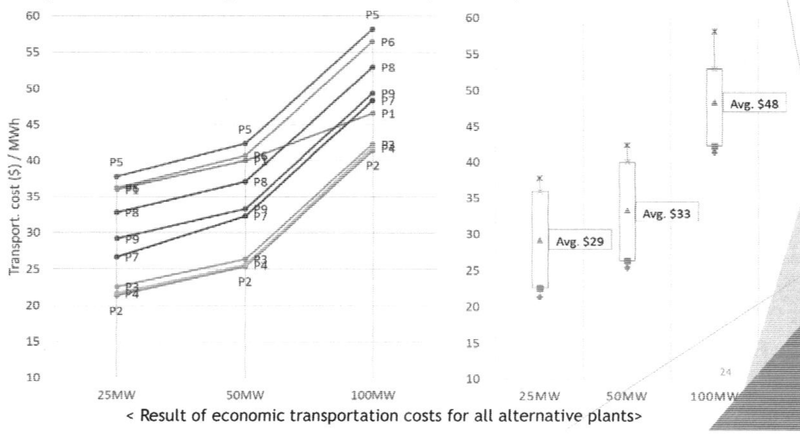

< Result of economic transportation costs for all alternative plants>

24

5.1. Evaluating transportation cost: Economic and External costs

- When external costs are fully considered (i.e., weighting factor $w2 = w3 = 1$), they account for considerable portion of total transportation cost.
- The external cost of transportation for all bioenergy plants ranged between \$12~\$31 (25MW), \$17~\$36 (50MW), and \$31~\$43 (100MW) per MWh.

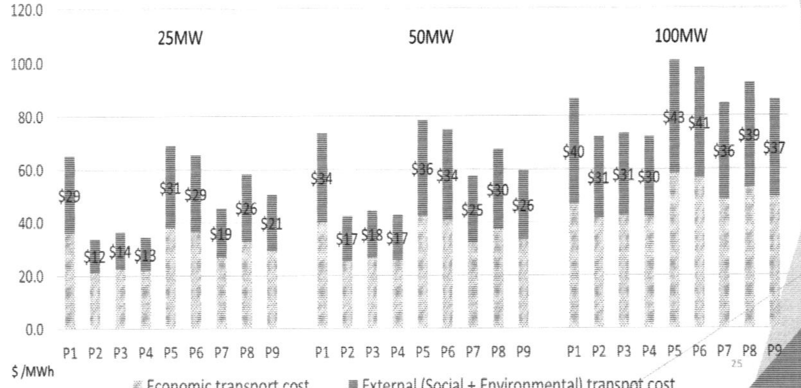

< Breakdown of economic and external transportation costs for all alternative plants >

5.2. Impact of external costs on modal share

- Truck haulage dominates all shipment scenarios when only economic cost is considered.
- When external costs are included, however, rail shipments increased for 50 and 100 MW annual capacity scenarios, respectively.
- Delivery with rail was not selected in small capacity scenario (25MW) even when external costs included

	Economic Cost Only			Economic + External Cost		
Plant	25MW	50MW	100MW	25MW	50MW	100MW
P1	0%	0%	0%	0%	5%	13%
P2	0%	0%	0%	0%	0%	21%
P3	0%	0%	0%	0%	0%	25%
P4	0%	0%	0%	0%	0%	28%
P5	0%	0%	12%	0%	7%	68%
P6	0%	0%	12%	0%	7%	68%
P7	0%	0%	0%	0%	0%	52%
P8	0%	0%	0%	0%	5%	60%
P9	0%	0%	0%	0%	0%	59%
Average	0%	0%	3%	0%	3%	44%

< The volume percentage of multimodal shipments (using Rail) for all alternative plants and capacities >

5.3. Result on Upstream emissions of forest residues

- Transloading has small effect on emissions. The remainder dependent on the transportation modes and fairly consistent across all plant locations.
- Understandably, emissions are higher, if more residues are transferred between truck and rail system.
- For 25 MW capacity scenario, the GHG emission profile of EXT and ECN are identical (curves overlay with each other in two Figures) because all residues are transported by trucks for the same distance.
- Overall, plant options 2, 3, and 4 have the lowest GHG emissions per dry ton for all capacity scenarios, mainly as a result of shorter transportation distances from collection site to the bioenergy plant.

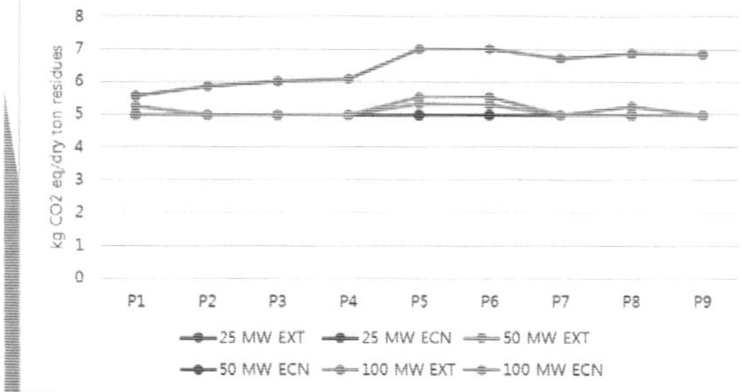

< GHG emissions of residue transloading for all plant locations >

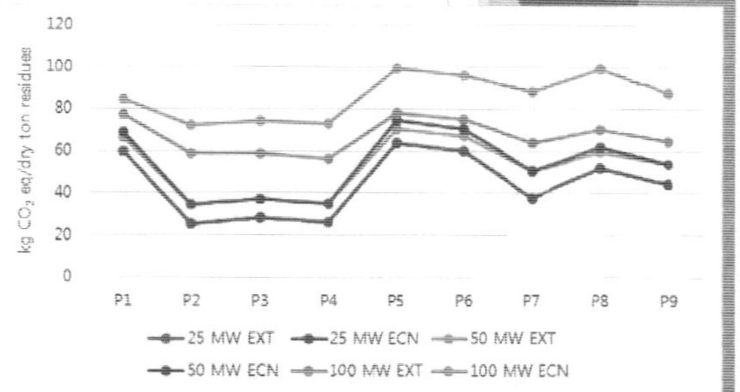

< GHG emissions of residue transportation for all plant locations >

5.4. Sensitivity Analysis: Effects of Weighting factor

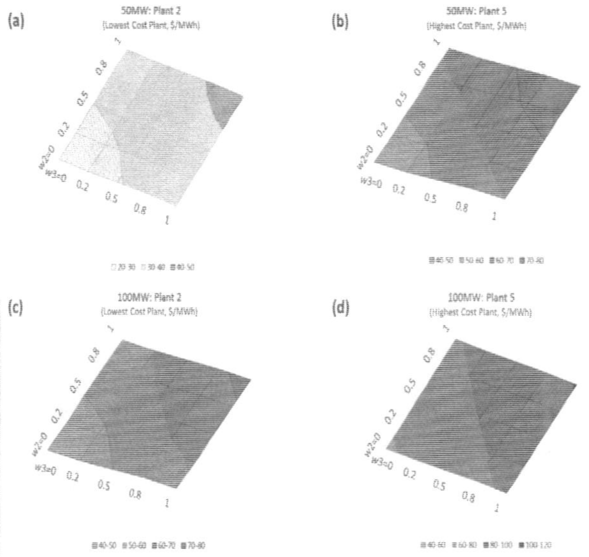

- The risk perception of external factors may differ between agencies, states, or countries.
- In case of 50MW, the external cost of transportation for Plant 5 could increase significantly while the impact on Plant 2 would be much less significant.
- In contrast, for 100MW case the cost variation was more significant for Plant 2.

< Result of external cost weighting factor sensitivity analysis for plant 2 and 5 (plants had lowest and highest transportation cost per electricity produced, respectively) >

5.4. Sensitivity Analysis: Effects of Rail capacity per train

- The impact on the quantity of feedstock delivered by rail for 50MW and 100MW capacity was investigated, when rail capacity is either decreased or increased.
- The shipment size (given that shipping rate per ton remains the same) does not affect significantly the amount of biomass moving multimodally, whether only economic, or both economic and external factors are considered.

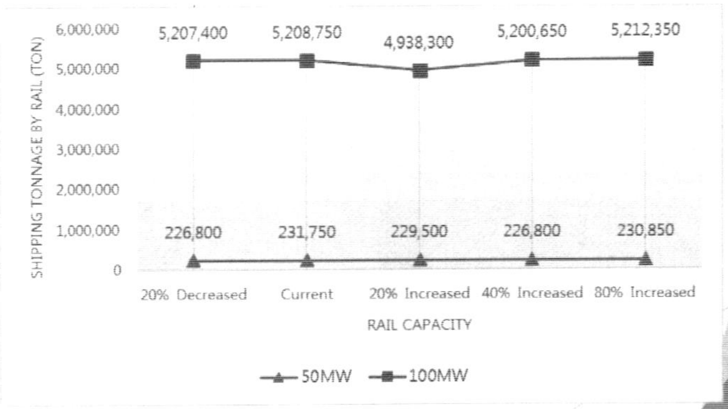

< Impact of rail shipment size on tons shipped >

5.5. Conclusion

1) Plants surrounded with densified forest and with limited competition have lower transportation costs.
2) Transportation cost increased steeply when the capacity changed from 50MW to 100MW. Size matters!!
3) When external costs are fully considered, they account for considerable portion of total transportation cost (approximately 60~80% of the economic costs).
4) With external costs, the share of multimodal (using rail) shipments increase as the plant capacity increases to minimize total transportation costs.
5) Total GHG and air emissions increase as the plant capacity increases, due to longer transportation distance, but emissions per dry ton are actually lower for higher capacity plants, if external costs are optimized together with economic cost.
6) Changes in weighting factors have different impact on plants of varying capacity which suggests that policies on inclusion/exclusion of external costs may affect the desirability of plant location.

30

Q & A
Thank you!
31

Use of Remote Sensing and Geographic Information Systems for Calculating Volume of Production, Geographical Location and Transport of Raw Plant Material

Utilizacion De La teledetección Y Sistemas De Informacion Geografica PARA Calculo De Volumen De Produccion, Ubicación GeográFica Y Transporte De Materia PRIMA Vegetal**Scandaliaris P.; Soria F.; Fandos C. and J. Carreras I Baldrés**
Estación Experimental Agroindustrial "Obispo Colombres", Av. William Cross 3150, (T4101XAC), Las Talitas. Provincia de Tucumán, Argentina

Keywords: Sugar cane - Agricultural Crop Residue (RAC) - Cogeneration - Remote Sensing - GIS

The cultivation of sugar cane, is one of the most important productive activities in the province of Tucuman, with an approximate area of 270,000 ha for the harvest 2015. Historically the sugar cane was strongly related to the fire, because harvest was done using the same burning to reduce trash and improve eficientica the harvesting process. From the decree of the Provincial Law 6,253 and its amendment, the 7,459, which prohibit the burning of sugar cane as an alternative method of harvest (except for controlled burning) it began to avoid the use of fire and increase progressively green harvesting system with subsequent maintenance of the maloja on the surface; which in turn is an effective tool to improve production, because many positive effects that this practice has on production.

Additionally the crop residue represents a large biomass potential for generating electricity. This potential is not being exploited properly, as only two sugar factories in the country are prepared to add electricity to the national grid.

From the deficient energy situation in Argentina, the idea of use of this resource for power generation emerges as a viable alternative to help cover part of the growing need of the country economic and environmental alternative.

The aim of this study was to show the ability of remote sensing, together with Geographic Information Systems (GIS) to speed up the search, collection and transport of agricultural crop residues (RAC) for power generation.

For the study used remote sensing techniques, which provide information on the location and RAC tonnage of sugar cane, and geographic information systems (GIS). The integration of these two techniques results obtained power by combining various layers of information.

The study was conducted under an agreement between the EEAOC and a recognized national energy company, which is interested in the use of this raw material for generating electricity. In this conference it highlighted the methodology used to develop project.

Aspectos políticos y socioeconómicos después de la COP 21
Argentina frente al cambio climático

Dr. Alieto Aldo Guadagni.
Academia Argentina de Ciencias del Ambiente

2ª. Conferencia Panamericana sobre sustentabilidad en biocombustibles y bioenergía

Septiembre del 2016.

1

IMPACTO DEL CAMBIO CLIMATICO GLOBAL

EL ORIGEN DEL PROBLEMA AMBIENTAL.

Como nos advertía Discépolo en Cambalache:

**"SIGLO XX,
PROBLEMATICO Y FEBRIL"**

2

SIGLO XX: PROBLEMÁTICO Y FEBRIL

CRECIMIENTO EXPONENCIAL DE LA POBLACION

Población mundial	Mill. de habitantes	Aumento anual
✓ Época de Nerón (Siglo I)	250	----
✓ Año 1800	1000	420.000
✓ Fin 2da.Guerra Mundial (1945)	2300	9.000.000
✓ Año 2014	7200	70.000.000
✓ Hacia el año 2050	9100	53.000.000

CRECIMIENTO MAS ACELERADO DE LA PRODUCCION

- En el Siglo XX el PBI mundial se multiplicó 19 veces
- La producción de bienes y servicios durante el siglo XX es mayor a toda la producción acumulada desde Adán y Eva hasta fines del Siglo XIX.
- En los cuatro siglos anteriores al Siglo XX, la producción mundial apenas se había multiplicado 7 veces.

3

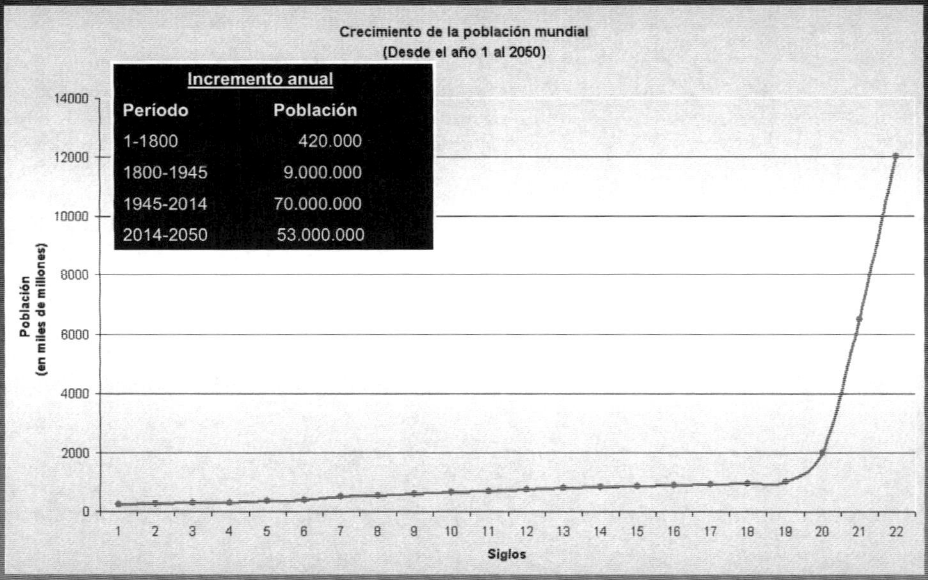

4

EN CUANTO HAY QUE REDUCIR LAS EMISIONES DE CO2?

- ✓ El nivel actual de gases invernadero en la atmósfera es de alrededor de 400 ppm de CO2.
- ✓ Antes de la revolución industrial el nivel era de apenas 280 ppm.
- ✓ Reducir las emisiones exigirá un gran esfuerzo porque el PBI mundial seguirá creciendo
- ✓ Como el PBI en el 2050 será por lo menos el triple del actual esto exigirá reducir en un 75% las actuales emisiones por unidad del PBI, si se desea abatir las emisiones totales en, por ejemplo un 25 por ciento.
- ✓ Esto exigiría una verdadera revolución tecnológica "verde".
- ✓ Relación costo-beneficio de este esfuerzo: 5% x 1% en términos del PBI mundial.
- ✓ El problema: externalidades, bienes públicos globales y *"free riders"*.

El gran conflicto: naciones contaminantes de ayer y de mañana.

5

CAMBIO CLIMÁTICO Y CRECIMIENTO ECONÓMICO.

- ❖ El impacto del cambio climático ya ha comenzado a sentirse en el planeta (tormentas, inundaciones y sequías).
- ❖ A este ritmo de incremento de las emisiones, la temperatura global podría incrementarse en 5° C, hacia fines de este siglo.
- ❖ Los países en desarrollo serán los mas vulnerables, ya que dependen mas de los recursos naturales y muchos están ubicados en zonas tropicales.
- ❖ **Es falso un dilema entre crecimiento económico y mas emisiones o menos crecimiento económico y menos emisiones.**
- ❖ Es posible - mediante innovaciones tecnológicas y reformas institucionales – aspirar a un planeta con crecimiento económico (esencial para abatir la pobreza) y menos emisiones contaminantes.

6

COMPLEJIDAD DE LAS NEGOCIACIONES

- Después del fenecido Protocolo de Kioto:
- El tradicional primer contaminador (EEUU, hoy 15,9 por ciento del total) no estaba obligado.
- El nuevo líder contaminador (China, hoy 28 por ciento del total) tampoco estaba obligado.
- El mundo en desarrollo (más del 50 por ciento del total de las emisiones por uso de combustibles) no estaba obligado.
- Grandes diferencias en las emisiones por habitante.
- Grandes diferencias entre el PASADO y el FUTURO

7

LAS SIETE NACIONES QUE AHORA LIDERAN LAS EMISIONES ENERGETICAS DE CO_2 (2013)

	(%) del total
1. China	28,0
2. USA	15,9
3. India	5,8
4. Rusia	4,8
5. Japón	3,8
6. Alemania	2,3
7. Corea	1,8
Subtotal	62,4
28. Argentina	0,6

Fuente: IEA, 2015.

8

LAS DOS GRANDES DIFERENCIAS CONFLICTIVAS

1. **El pasado (1850-2005)** Participación en las emisiones históricamente acumuladas desde 1850 por los países:

Ingresos altos	64 por ciento	16 por ciento de la población
Ingresos medios	34 por ciento	65 por ciento de la población
Ingresos bajos	2 por ciento	19 por ciento de la población

Los 6 países históricamente responsables: EEUU (28%), Alemania (10%), China (8%), Rusia (8%), UK (6%) y Japón (4%). Estos seis países representan el 64% del total históricamente acumulado.

2. **El futuro (hasta el 2030).** Las emisiones totales aumentaran alrededor del 40 por ciento.

   Más del 90 por ciento del futuro incremento proyectado de las emisiones, si no se toman medidas ya, corresponderá a los países en desarrollo.

9

LAS NEGOCIACIONES INTERNACIONALES AVANZAN…PERO MUY DESPACIO:

2009: Copenhague
2010: Cancún
2011: Durban
2012: Doha
2013: Varsovia
2014: Lima
2015: Paris

AHORA: RUMBO A MARRAKESH. Diciembre 2016
EN BUSCA DEL TIEMPO PERDIDO

10

LOS RECURSOS FOSILES NO SE ESTAN AGOTANDO

	PETROLEO			GAS		
	RESERVAS Barriles (miles de millones)	PRODUCCION Millones de barriles día	R/P	RESERVAS (Billones de M³)	PRUDUCCION (Miles de millones de M³)	R/P
1980	683,4	62959	29,7	71,6	1431,9	50,0
2000	1258,1	74983	46,0	139,2	2410,3	54,4
2014	1700,1	88673	52,5	187,1	3460,6	54,1
Incremento (%) 1980 - 2014	148,8%	40,8%	76,8%	161,3%	141,7%	8,2%

Fuente: BP. Statistical Review of World Energy – June 2015

LA AMENAZA DEL CALENTAMIENTO CLIMATICO NO SE RESOLVERA POR EL AGOTAMIENTO DE LOS RECURSOS ENERGETICOS FOSILES

11

LOS GRANDES PRODUCTORES DE PETROLEO, GAS Y CARBON (2014)

PETROLEO %	GAS %	CARBON %
Arabia Saudita 12,9	USA 21,4	China 46,9
Rusia 12,7	Rusia 16,7	USA 12,9
USA 12,3	Irán 5,0	India 6,2
China 5,0	Qatar 5,1	Australia 7,1
Canadá 5,0	Canadá 4,7	Indonesia 7,2
Subtotal 47,9	52,9	80,3
Otros países 52,1	47,1	19,7

Del consumo energético mundial: petróleo 33%, carbón 30% y gas 24%. TOTAL: 87%
Fuente: B.P.

12

LAS ULTIMAS ESTIMACIONES (2014) DE LAS EMISIONES DE CO_2

- Situación actual (año 2012): 31,5 Gt.
- Proyección hacia el año 2035: 37,2 Gt.
- Nivel máximo tolerado por el escenario 450: 21,7 Gt.

Fuente: International Energy Agency, Special Report. "World Energy Investment Outlook", June 2014.

13

(9) ARGENTINA: SEGUNDA MITAD DEL SIGLO XX HASTA HOY

- Aumento generalizado de las precipitaciones al este de los Andes (zona subtropical). Mayor frecuencia de precipitaciones intensas.
- Cambios hidrológicos en la Cuenca del Plata. Mayores caudales, mas energía pero más inundaciones.
- En la cordillera (Cuyo y Patagonia Norte) menos precipitaciones y mas temperatura. Menos caudales, menos energía. Recesión de glaciares.
- Patagonia (aumento de temperatura). Recesión de los glaciares.

(Fuente: Vicente Barros, "Aspectos científicos del cambio climático", 15 de noviembre del 2012)

14

(14) CAMBIO CLIMÁTICO: Evidencias y causas

Informe de la ROYAL SOCIETY y la U.S. National Academy of Sciences
Febrero 2014

Interrogantes y Respuestas

- ¿Está aumentando la temperatura en el planeta?

Si. Gran parte del aumento ocurrió desde 1970.

- ¿Cómo saben los científicos que el cambio climático está siendo causado por la humanidad?

Por la aplicación de principios básicos de física.

- ¿Por qué son importantes las emisiones de CO_2 causadas por la humanidad?

Por la masiva utilización de fósiles y la rápida deforestación.

- ¿Cuál es el rol del Sol en el cambio climático de las últimas décadas?

Muy reducido.

- El clima siempre cambió. ¿Por qué hoy son importantes los cambios climáticos?

La velocidad del actual cambio climático es muy superior a la del pasado, esto hace muy difícil la adaptación del hombre y la naturaleza.

- La actual concentración de CO_2 en la atmósfera tiene algún precedente histórico.

Es la más alta del último MILLON DE AÑOS.

Hubo niveles superiores en el muy distante pasado (varios millones de años).

15

- ¿Es posible que después de cierto nivel de acumulación de CO_2, nuevas emisiones no aumenten más la temperatura?

No.

- ¿Es posible que la reciente disminución en el ritmo de crecimiento del calentamiento global signifique que no aumentará la temperatura en el futuro?

No.

- ¿El cambio climático aumenta la intensidad y frecuencia de las inundaciones, sequías, huracanes y tornados?

Si.

- ¿Está aumentando el nivel del mar?

Si. En los últimos 20 años aumentó 3,2 milímetros por año. Desde 1901 aumentó 20 cm.

- ¿Cuáles son las previsiones acerca del futuro aumento de la temperatura?

A este ritmo de aumento en las emisiones de CO_2 hacia el año 2100 la temperatura habrá aumentado entre 2,6 y 4,8 C°, sobre el nivel actual.

- ¿El aumento de la temperatura preocupa a los científicos?

Si. Porque tendrá "un serio impacto sobre la humanidad y la naturaleza".

- Si ahora se eliminaran todas las emisiones de CO_2, ¿el clima podría no volver a las condiciones de hace 2 siglos?

No.

Fuente: National Academy of Sciences – The Royal Society
"Climate change: Evidences and Causes" Febrero 2014.

16

(20) MAS ADVERTENCIAS CIENTIFICAS EN EL 2014

1) Cambios observados y sus causas:

*"La influencia humana en el clima es clara, las emisiones recientes de gases invernadero son las mas altas de la historia"

*"Los recientes cambios climáticos tienen gran impacto en los sistemas naturales".

2) Cambios recientes en el clima:

*"Los cambios registrados desde 1950 no tienen precedentes"

*" La atmosfera y el océano se ha calentado"

*"Disminuye la cantidad de hielo y nieve y sube el nivel del mar".

17

(20) MAS ADVERTENCIAS CIENTIFICAS EN EL 2014

3) Causas del cambio climático:

*"La concentración en la atmosfera de CO_2, metano y otros gases no tiene precedente en por lo menos los últimos 800.000 años".

*"Esta concentración ha sido la causa principal del aumento de la temperatura desde 1950"

4) Los riesgos de los cambios futuros:

*"Si continúan las emisiones de gases seguirá aumentando la temperatura"

* "PROBABILIDAD CRECIENTE DE CAMBIOS CLIMATICOS IRREVERSIBLES CON SEVEROS IMPACTOS EN LA POBLACION Y LOS ECOSISTEMAS". SE REQUIERE UNA SUSTANCIAL REDUCCION DE LAS EMISIONES.

.UN-IPCC Climate Change 2014, Approved Summary for Policy Makers, November 2014

18

(27) EL SOSTENIDO AUMENTO DE LA TEMPERATURA 1880-2015 (mar y tierra)

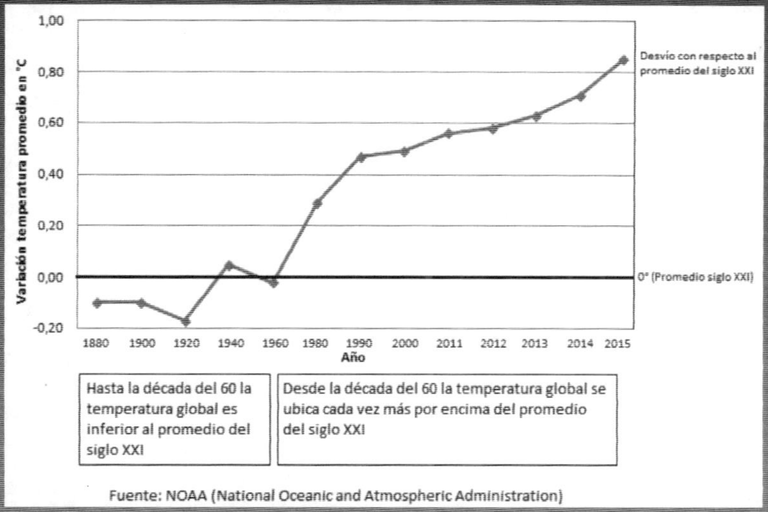

19

(30) ANTARTIDA: UNA CATASTROFE SILENCIOSA POR LA ALTERACION DEL ECOSISTEMA.

- Una colonia de pingüinos quedo devastada por el cambio climático.
- Los pingüinos Adelaida están sufriendo una catástrofe silenciosa. Desde el 2010 murieron 150.000 en la Antártida, su hábitat natural (quedan apenas 10.000)
- El responsable de la matanza es un villano que cada vez golpea con mas fuerza: el cambio climático.
- La población de pingüinos de Magallanes que llega todos los años a Punta Tombo, en Chubut, se redujo un 36% en las dos ultimas décadas.

Fuente: Clarín, 16 de febrero del 2016.

20

(31) EL CALENTAMIENTO DEL ARTICO

- De acuerdo a los datos de la National Snow and Ice Data Center (NSIDC), la extensión del hielo del Ártico se esta reduciendo
- Esta reducción indica en la actualidad niveles no sólo inferiores al promedio sino también inferiores a los niveles del año 2012, que había sido el año con el menor registro de hielo en el Ártico

Fuente: Miami Herald, March 6th, 2016.

21

(34) ADVIERTEN QUE ESTE AÑO 2016 SE MANTENDRA LA TENDENCIA DE DESASTRES CLIMATICOS EN TODO EL MUNDO.

- Nuevos records de temperaturas máximas en enero y febrero del 2016, según la Organización Meteorológica Mundial (OMM).
- La extensión de los hielos marinos en el Ártico es hoy la mas reducida registrada por satélite (NASA- NOAA)

"LAS TEMPERATURAS SORPRENDENTEMENTE ELEVADAS QUE SE HAN REGISTRADO HASTA AHORA EN EL 2016 ESTAN CAUSANDO CONMOCION EN LA COMUNIDAD CIENTIFICA DEL CLIMA"

Fuente: David Carlson, Director del Programa Mundial de Investigaciones Climáticas (OMM). Clarín, 21-3-16

22

(38) MEDICIONES DEL CO2 EN EL OBSERVATORIO DE MAUNA LOA (HAWAI).

- Según este Observatorio de la US National Oceanic and Atmospheric Administration, instalado en 1956, se registraron estos niveles de CO2 (ppm):

1959	316
1980	339
2000	370
2015	401

2016 (enero) 2,5 mas que en enero del 2015

Fuente: FS-UNEP Colaboration Center, "Global trends in renewable energy investment 2016"

(42) JUNIO DEL 2016: 14 MESES CONSECUTIVOS DE TEMPERATURAS RECORD EN EL PLANETA

- Junio 2016 fue 1,62 grados F mayor al promedio del Siglo XX.

- La temperatura promedio del primer semestre del 2016 fue 1, 89°F mayor al promedio del Siglo XX.

Fuente: NOAA, July 19, 2016

EL FMI Y EL PRECIO DEL CO_2 HACIA EL 2010

- El precio "eficiente" del CO_2 es de 57,5 U$D por tonelada (7,7 U$D por barril de petróleo).
- Este precio reducirá los emisiones en los 20 países estudiados en un 13,5% , lo cual equivale a un 10,8% de reducción global.

Fuente: IMF, Working Paper 14/174. "How Much Carbon Pricing is in Countries' Own Interests?" Ian Perry, Chandara Veung y Dirk Heine. Septiembre 2014.

25

CON ESTAS PROPUESTAS RESPETAREMOS EL LIMITE DE LOS 2°C?

- Analizadas todas las propuestas presentadas para la reunión de París, se estima que la temperatura global subiría 2,7°C (según expresó Christiana Figueres)
- Si todos los países hubiesen presentado una oferta similar a la Argentina, la temperatura mundial subiría 4°C (según afirma Climate policy team of Climate Analysis)

26

COMO SE ALINEAN LOS PAISES LATINOAMERICANOS EN LAS NEGOCIACIONES EN NNUU SOBRE CAMBIO CLIMATICO (Grupos de Negociación)

- AILAC: Paraguay, Chile. Colombia, Costa Rica, Guatemala, Panamá y Perú.
- BASIC: Brasil, India, China y Sudáfrica.
- INTEGRIDAD AMBIENTAL: México, Suiza, Corea y Mónaco.
- LMDC: Irán, Irak, Kuwait, Qatar, Arabia Saudita, Libia, Siria, Venezuela, Ecuador, Cuba y Argentina.

27

LOS PETROLEROS EUROPEOS PIDEN EN PARIS UN IMPUESTO AL CO2.

- "We need governments across the world to provide us with clear, stable, long term ambitious policy frameworks. This would reduce uncertainty and help estimulate investments in the right low- carbon technologies:"
- "We believe that the price on carbon should be a key element of these frameworks... this discourages high carbon options and encourages the most efficient ways of reducing emissions."
- "We call on governments, including at the UNFCCC negotiations in Paris and beyond –to introduce carbon pricing systems where they do not yet exist at the national or regional levels."
- "Pricing carbon obviously adds a cost to our production, but carbon pricing policy will contribute to provide our business and their many stakeholders with a clear roadmap for future investment."

Nota dirigida a Mr Laurent Fabius, Presidente de COP 21 y a la Secretaria Ejecutiva de la UNFCCC. Firmada por: BG Group, BP, ENI, Shell, State Oil y Total (29-5-2015)

28

NACIONES UNIDAS ESTA PROPICIANDO EL IMPUESTO AL CO2

- "El último informe del IPPC evidencia la importancia de imponer un precio al CO2".
- En este año alrededor de 40 naciones y 20 jurisdicciones subnacionales están propiciando esquemas de cupos o impuestos a las emisiones.
- Estas jurisdicciones ya representan el 22 por ciento de la emisión global.
- Según la ONU (Statement "Putting a Price on Carbon", June 2014) "Imponer un precio al CO2 es inevitable en un esquema efectivo de reducción de las emisiones".

UN – Secretary General´s Special Envoy on Climate Change, 27 June 2014.

29

EL ACUERDO DE PARIS NO ALCANZA PARA PRESERVAR EL PLANETA

- La denominada COP es la conferencia de las 195 naciones firmantes de la "Convención Marco de las Naciones Unidas sobre el Cambio Climático" acordada en 1992. En diciembre tuvo lugar en Paris la COP21, han transcurrido ya 21 COP en los últimos 23 años y se han registrado avances, pero aún estamos lejos de asegurar la preservación del clima en el planeta. Esta amenaza global, requiere una solución global con compromisos de todas las naciones.
- Preservar el planeta de los daños del calentamiento, que ya comenzaron a sentirse en nuestro país afectando, por ejemplo la disponibilidad de agua por el retroceso de los glaciares andinos al mismo tiempo que aumentan las inundaciones en el Noroeste, exigirá un compromiso de toda la humanidad. **La meta es exigente, ya que requiere que dentro de 20 años deberíamos estar emitiendo un 30 por ciento menos de CO2 que hoy, esto exigirá la modificación del patrón energético y reforestación**. Como dentro de 20 años la población mundial crecerá en 1400 millones de habitantes y además el PBI mundial será el doble del actual, se trata de reducir las emisiones contaminantes por unidad de PBI más de un 65 por ciento.

30

EL ACUERDO DE LA COP 21 EN PARIS: ES SUFICIENTE?

- En la COP 21 celebrada en Paris en diciembre del 2015 las 195 naciones participantes acordaron "mantener el aumento de la temperatura media mundial muy por debajo de 2C* con respecto a los niveles preindustriales, y proseguir los esfuerzos para limitar ese aumento de la temperatura a 1,5 C* con respecto a los niveles preindustriales, reconociendo que ello reduciría considerablemente los riesgos y los efectos del cambio climático" (Art.2-a).
- Según el acuerdo logrado en esta COP 21 la reducción de las emisiones de CO2 será fijada por cada país, según estas propuestas INDC. Estas metas voluntarias serán fijadas periódicamente por cada nación, pero no serán estrictamente obligatorias, ya que se trata de una oferta voluntaria que presenta cada país.
- **El texto del acuerdo logrado en la COP 21 no es riguroso, ya que no se determinan acciones obligatorias capaces de reducir de una manera efectiva las futuras emisiones contaminantes**

31

PERO, LAS EMISIONES ENERGETICAS CONTINUARAN AUMENTANDO EN EL FUTURO SEGÚN BP

	Crecimiento %	Porcentaje en el total	
Fósiles		**2014**	**2035**
Petróleo	20	32	29
Gas	44	24	26
Carbón	10	30	25
Subtotal:		86	80
No contaminantes			
Nuclear	50	4	5
Hidráulicas	45	7	7
Renovables	285	3	8
		14	20
TOTAL	**34**	**100**	**100**

Aumento de población: 21%. Aumento de PBI 107%. Intensidad energética: -35 %

AUMENTO DE EMISIONES CO2: 20%

Fuente: BP "Energy Outlook 2035", Febrero 2016.

32

DECLARACIONES DEL MINISTRO DE PETROLEO DE ARABIA SAUDITA

- "Los combustibles fósiles son buenos. Y además son necesarios. Son un abundante recurso natural. Durante los últimos 250 años han transformado nuestras economías y nuestras sociedades. Su producción en masa ha resultado en consumo masivo. Han creado y sostenido el crecimiento económico y la prosperidad".

- "Los combustibles fósiles continuaran jugando un rol vital en el panorama energético. Nos guste o no nos guste."

Fuente: His Excellency Ali Al -Naimi. Minister of Petroleum and Mineral Resources. Houston, Texas, February 23rd, 2016 . CERAWEEK 2016

33

(40) TRUMP ATACA EL ACUERDO CLIMATICO DE PARIS.

- El candidato Trump afirma que no es "a big fan" del acuerdo climático.

- También ha afirmado que el cambio climático es un concepto que fue inventado por China para perjudicar la competitividad de los negocios de los EEUU.

- Laurent Fabius quien ha sido Ministro de Relaciones Exteriores de Francia resalta la importancia de las elecciones presidenciales de los EEUU, ya que sostiene que "si un negador del cambio climático fuese electo esto amenazaría dramáticamente la acción colectiva global."

Fuente: Buenos Aires Herald, May 20th, 2016

34

CONSUMO POR TIPO DE ENERGIA HACIA EL 2040.(Según US-DOE)

	2012	**%**	**2040**	**%**	**Δ%**
Petróleo	183,6	33,4	246,0	30,2	34
Gas	124,2	22,6	211,4	25,9	70
Carbón	153,3	27,9	180,2	22,1	17,5
Fósiles	461,1	83,9	637,6	78,2	38,3
Nuclear	24,5	4,5	46,0	5,6	88
Otros	63,7	11,6	131,4	16,2	106
TOTAL	549,3	100	815	100	48,3

La amenaza: El consumo de energía fósil aumentara un 38 por ciento

Fuente US-Department of Energy. International Energy Outlook 2016, May 2016

35

ULTIMAS ESTIMACIONES DEL GOBIERNO DE LOS ESTADOS UNIDOS

(Emisiones - Gt de CO2)

	2012	2040	Δ %
Países OCDE	12,8	13,8	8
Otros países	19,5	29,4	51
Total Países	32,3	43,2	33,7

El nivel máximo de emisiones toleradas en el escenario 450 ppm es de apenas 21,7 Gt para el 2035. Esta estimación del gobierno de los EEUU fue realizada después de la COP21 (Paris), el día 11 de mayo del 2016.

Fuente: USA, Energy International Agency. US Energy Outlook 2016- May 2016.

36

EMISIONES DE CO2 POR COMBUSTIBLE

(Gt de CO2)

	2012	2040	Δ %
Petróleo	11,7	15,5	32
Gas	6,6	11,2	70
Carbón	14,0	16,5	18
TOTAL	32,3	43,2	33,7

Fuente: USA, Energy International Agency. US Energy Outlook 2016- May 2016.

37

HACIA UNA NUEVA ERA ENERGETICA

- **El actual modelo económico, basado en los combustibles fósiles, deberá ser sustituido por otro con nuevas energías limpias, y mayor eficiencia energética. No ha concluido aún la era de los combustibles fósiles, pero hemos comenzado a transitar hacia el fin de esta era, nacida a fines del siglo XVIII.**
- Los flujos de inversiones comenzaran a orientarse hacia las nuevas tecnologías y la mayor eficiencia en el uso de la energía, esto requerirá más inversiones en investigación y desarrollo. Las nuevas energías renovables impulsaran nuevos sectores productivos, capaces de absorber crecientes flujos de inversiones. Los costos de las renovables han disminuido, mientras que estas inversiones se han multiplicado más de seis veces, Los dos últimos años han sido testigos de una gran expansión de estas nuevas energías en todo el mundo

38

EL GRAN QUEBRANTO FINANCIERO DEL PETROLEO

- La amenaza climática no se solucionara por el agotamiento de los fósiles, ya que nunca hubo tanto petróleo y gas como hoy; en 1980 las reservas petroleras mundiales cubrían apenas 30 años de consumo, mientras que hoy cubren 53 años. La utilización plena de estos recursos fósiles, ya contabilizados en los balances empresarios, no sería compatible con la meta de no cruzar la barrera de un aumento de 2 grados centígrados. BP acaba de informar (13-X-2015) que **"si las reservas existentes de fósiles fueran utilizadas totalmente en los próximos años se emitirían más de 2,8 trillones de toneladas de CO2, bien por encima del límite de un trillón consistente con la meta de no superar el límite de 2 C* de aumento de la temperatura global".**
- **EL QUEBRANTO FINANCIERO DE RESPETAR ESTE LIMITE EQUIVALE A 2/3 DEL PBI MUNDIAL**
- **Las carencias en los acuerdos logrados en la COP 21 son muchas, por esta razón la Agencia Internacional de Energía pudo afirmar que "El rumbo del viaje está cambiando, pero el destino todavía no son los 2 grados."**

39

"EL CLIMA COMO BIEN COMUN"

El Papa Francisco expresa:

- " el clima es un bien común, de todos y para todos. A nivel global es un sistema complejo relacionado con muchas condiciones esenciales para la vida humana".
- "Hay un consenso científico muy consistente que indica que nos encontramos ante un importante calentamiento del sistema climático".
- "La humanidad esta llamada a tomar conciencia de la necesidad de realizar cambios de estilos de vida, de producción y de consumo, para combatir este calentamiento".
- "Los peores impactos probablemente recaerán en las próximas décadas sobre los países en desarrollo".

40

LA ENCÍCLICA DEL PAPA FRANCISCO SOBRE ECOLOGÍA

En la reciente encíclica Laudato SI, el Papa nos recuerda la vigencia del mandato bíblico:" Labrar y cuidar el jardín del mundo" (libro del Génesis 2,15), se trata de "cuidar", no de degradar nuestro planeta. Es evidente que enfrentar eficazmente la amenaza climática exige una solución global.

También es evidente que el creciente riesgo causado por más emisiones globales plantea la urgente necesidad de una autoridad global, ya que está comprometido un importante bien común global. Por esta razón es necesario que las negociaciones internacionales apunten, como propone Stefano Zamagni, a la creación de una Organización Mundial del Ambiente (OMA).

Para garantizar la salvaguardia del ambiente en nuestro planeta, es urgente que los países acuerden la creación de una autoridad global, que tenga un poder efectivo y cuya legitimidad sea reconocida por todas las naciones

41

QUE NOS DICE LA ENCICLICA CARITAS IN VERITATE? (Cont.)

- PARA GARANTIZAR LA SALVAGUARDIA DEL AMBIENTE URGE LA PRESENCIA DE UNA VERDADERA *AUTORIDAD POLITICA MUNDIAL*... DICHA AUTORIDAD DEBERA ESTAR RECONOCIDA POR TODOS, GOZAR DE PODER EFECTIVO...

42

EL RETROCESO ENERGETICO ARGENTINO (2003-2015)

- Reservas de gas (-) 50%
- Reservas de petróleo (-) 16%
- Producción de gas (-) 16%
- Producción de petróleo (-) 25%
- Crecimiento de la demanda de energía eléctrica (MW) (+) 77,6%
- Crecimiento de la capacidad instalada de generación (MW) (+) 37,8%

Fuente: Ministerio de Energía

43

ARGENTINA: MERCADO ELECTRICO MAYORISTA (2002/2015)

Generación (GWH)	2002	%	2015	%	Incremento%
Fósil	33629	43	91853	65	173
Limpias	44499	57	50057	35	12
TOTAL	78128	100	141.910	100	59

DE CADA 100 KWH DE AUMENTO DE GENERACION ENTRE 2002 Y 2015 NADA MENOS QUE 91 FUERON CONTAMINANTES

CONSUMO DE COMBUSTIBLES PARA GENERACION (2002-2015):

- Los combustibles fósiles han cubierto casi totalmente el incremento en la generación de energía eléctrica. El consumo de carbón se multiplicó 15 veces, fueloil 79 veces, y gasoil 172 veces.
- Oferta hidroeléctrica en fuerte retroceso relativo, ya que se reduce de casi la mitad de la producción en el 2002 a apenas el 29 por ciento en el año 2015.
- Fuente: CONEA

44

UN HECHO PREOCUPANTE:

En Argentina aumentan las emisiones contaminantes (CO_2) de la generacion electrica.

Incremento de la generación eléctrica (MWH) 2002-15

Energía contaminante	173%
❖ Energía no contaminante	12,5 %

Aumento de las emisiones de CO_2 por generación de energía eléctrica

❖ Año 2002	emisiones (millones Tn)	12,5
❖ Año 2015	emisiones (millones Tn)	46,5
❖ Incremento en las emisiones		272 %

Fuente: CONEA (2016))

45

La nueva agenda energética debería contemplar las siguientes iniciativas:

❖ Inversiones en obras hidroeléctricas y centrales nucleares
❖ Inversiones en nuevas energías (eólica, solar y mareomotriz). Biocombustibles
❖ Modernización tecnológica del transporte publico en los grandes núcleos urbanos: Gran Buenos Aires,. Rosario, Córdoba, Mendoza y Tucumán. Aprovechar la valiosa experiencia de Curitiba, Bogotá (Transmilenio), Santiago de Chile (Transantiago) y Méjico, DF.
❖ Rehabilitación y modernización del ferrocarril de cargas y del transporte urbano e interurbano.
❖ Nuevas normas técnicas en Mercosur para mejorar la eficiencia en la utilización de combustibles en la industria automotriz.
❖ Procesos en la industria manufacturera que sean energéticamente eficientes.
❖ Nuevos códigos de edificación que aseguren la conservación energética.
❖ Artefactos eléctricos que sean energéticamente eficientes.

46

NUESTRO APORTE A LA PRESERVACION DEL PLANETA:

- Nuestra propuesta presentada en Paris por el anterior gobierno es insuficiente, ya que si todas las naciones hubiesen presentado una propuesta similar a la nuestra la temperatura global subiría 4 grados centígrados, sobrepasando ampliamente la meta de no cruzar la barrera critica de los 2 grados.

- La tarea del gobierno será elaborar ahora una nueva planificación ambiental que apunte a respetar el medio ambiente, y debería incluir un programa concreto y financiable de iniciativas de eficiencia y conservación energética.

- Si nosotros acordamos actuar a partir de ahora con más sensatez ambiental deberíamos, por ejemplo, acordar una propuesta nacional que propicie la expansión de las nuevas energías limpias, y también de las tradicionales como la hidroelectricidad (con cuidado del ambiente local). Nuestro gran potencial hidroeléctrico nos está ofreciendo oportunidades de aprovechamiento futuro, pero sin "capitalismo de amigos". Es hora de revertir el proceso de "fosilización" de la energía que se impulsó durante la última década.

- **Un hecho es ya evidente: respetar el límite de los 2C* exige una acción global de las naciones, pero también un aporte de nuestro país que contribuya eficazmente al esfuerzo global.**

47

"EL OPTIMISMO ES UNA ESTRATEGA PARA HACER UN FUTURO MEJOR, PORQUE SI UNO NO CREYERA QUE EL FUTURO PODRIA SER MEJOR,
NO ESTARIA PROBABLEMENTE DISPUESTO A ASUMIR LA DECISION Y LA RESPONSABILIDAD DE QUERER HACERLO"

(Noam Chomsky)

48

Los biocombustibles en el MERCOSUR y su relación con las políticas de la Unión Europea

CARMEN VICIÉN, LUCIA LONGO, DANIEL TOMASINI, MARÍA MARTA DI PAOLA, GERARDO PETRI, RITA MARRA Y DANIELA GUARÁS

DEPARTAMENTO DE ECONOMÍA, DESARROLLO Y PLANEAMIENTO AGRÍCOLA DE LA FACULTAD DE AGRONOMÍA DE LA UBA

FAUBA

Estado de situación

- Políticas en biocombustibles en el ámbito internacional: complejidad de la situación.
- Biocombustibles de primera generación empleados comercialmente en la actualidad: biodiesel a partir de aceite de soja y palma, y bioetanol de caña de azúcar y maíz.

Las políticas públicas

- **Debate en Foros Internacionales**: precios de los alimentos y eliminación de las políticas distorsivas.
- **G20**: preocupación acerca de la seguridad alimentaria.
- Desacuerdos **productores o importadores netos de alimentos**
- **Políticas domésticas** de producción de biocombustibles (incentivos)
- **Incertidumbre científica** sobre el impacto en la seguridad alimentaria y el ambiente.

Difíciles decisiones de política pública.

Consumidores y productores

- **Unión Europea y Estados Unidos:** mayores consumidores y productores de biodiesel y etanol.
- **MERCOSUR** :
  - recursos naturales (superficie, suelo y clima) para producción primaria e industria energética alternativa.
  - Desarrollo económico y exportación de energía.
  - Situación heterogénea: políticas públicas de apoyo y desarrollo de la industria.

La Unión Europea

- Estrategia: asegurar el abastecimiento energético y reducir la emisión de gases de efecto invernadero.
- Directiva sobre Energías Renovables (junio de 2009)
- Importación de biocombustibles para cumplimiento de metas: MERCOSUR
- Enmiendas Directiva sobre Energías Renovables (2015).

Directiva sobre Energías Renovables (UE)

- Marco común para fomentar el uso de energías renovables, definiendo criterios de sostenibilidad.
- Medidas sobre materias primas cultivadas dentro o fuera del territorio de la Unión Europea
- Biocombustibles sean considerados "sostenibles"

Directiva sobre Energías Renovables (UE): criterios

- Fijación de porcentaje mínimo de reducción de emisiones de GEI en producción y/o uso de biocombustibles.
- Cambios en el uso del suelo en la producción de la materia prima por terceros países.
- Necesidad de ratificación de ciertos acuerdos internacionales ambientales y sociales.
- Términos sin definiciones consensuadas: "rica biodiversidad".

Directiva sobre Energías Renovables (UE): criterios

- Pautas uso de biocombustibles:
  - reducir las emisiones de gases efecto invernadero como mínimo en un 35% respecto a los combustibles fósiles a 2013 (dicho porcentaje se incrementará a un mínimo del 50% a partir del 1 de enero de 2018)
  - Garantizar que la materia prima utilizada en la producción no provenga de sitios de elevada importancia por su biodiversidad.

Directiva sobre Energías Renovables (UE): criterios

- Cambios indirectos en el uso del suelo:
  - expansión de otros cultivos en tierras no cultivadas
  - Incorporación de áreas no cultivadas por terceros ante reducción de exportaciones.
- Posiciones encontradas sobre uso indirecto del suelo: productores , consumidores, sociedad civil (riesgos sociales, seguridad alimentaria y/o ambientales)

La generación de incertidumbre

- Directiva sobre Energías Renovables y otras medidas en frontera afectan:
  - Mercado interno
  - Productores externos: impacto económico e impacto ambiental.
- **Incertidumbre en los productores del MERCOSUR (y otros) por la necesidad de certificar estas condiciones para poder acceder al mercado europeo.**

Preferencias según los países

- Países clave en producción y consumo han desarrollado **diferentes preferencias:** biocombustibles y alimentos.
- En Brasil:
  - **flexibilidad** de la industria azucar-alcohol
  - Condiciones de los **mercados**.
  - **Reducción** de exportaciones (desde 2008)

Derechos antidumping: situación

- UE: (11/2013) impone derechos antidumping a la Argentina
- OMC: (04/2014) establece Panel para resolver el reclamo argentino
- OSD - Órgano de Solución de Diferencias:
  - UE (20 mayo de 2016) apela cuestiones de derecho e interpretación jurídica del informe del Grupo Especial.
  - Argentina (25 de mayo de 2016) apelación cruzada.

- Detalles en: https://www.wto.org/spanish/tratop_s/dispu_s/cases_s/ds473_s.htm

Derechos antidumping: impactos

- Impacto en la industria: UE primer destino de exportación.
- Producción de aceite crudo de soja en Argentina (2012) :
  - 6,35 millones de toneladas
  - 40% se destinó a la producción de biodiesel.
- Aranceles de la UE (antidumping): sobreoferta de aceite de soja.
- Exportaciones a Estados Unidos (cambio en las reglas):
  - Compensación.
  - Aumento de 80% (2016) respecto al año anterior (1,4M tons).
- Perú, segundo mercado después de Estados Unidos, evalúa la aplicación de derechos antidumping.

Proyecto

- Estudio de la vinculación entre las políticas de apoyo a la producción de biocombustibles en el MERCOSUR y la demanda derivada de las medidas de la Unión Europea.

- Cuestiones que hacen a
  - cambios en los precios
  - regulaciones ambientales y
  - aspectos comerciales (derechos antidumping)
- que pueden afectar
  - demanda materias primas,
  - cambios en uso del suelo
  - estabilidad de resultados en la empresa.

Modelo sectorial

- Construcción de **modelos sectoriales** de agregación sucesiva para considerar los impactos de las políticas públicas, tanto a nivel de empresas como de territorios.
- El análisis combina aspectos microeconómicos y agregados permitiendo la revisión del **impacto de las políticas públicas** vinculadas a la producción y comercialización de materias primas y biocombustibles.

Las interacciones en el modelo

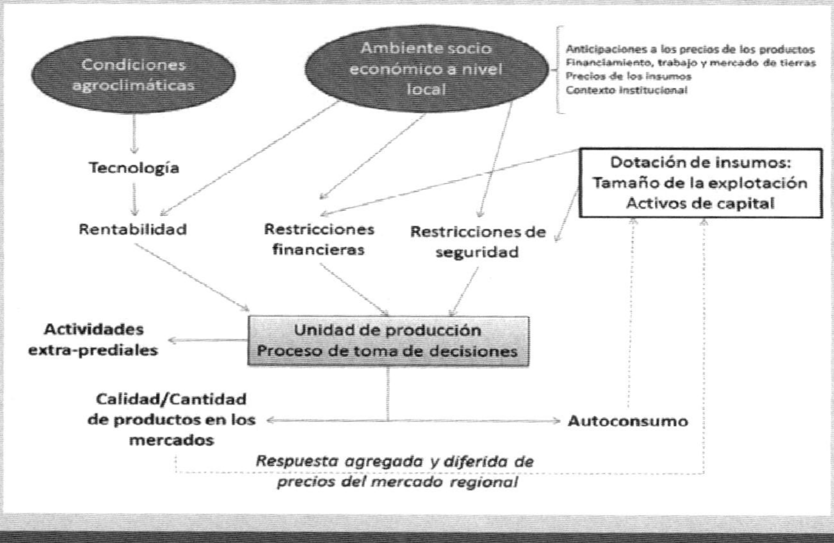

El modelo agregado

El modelo incluye:

- Módulo de oferta primaria (agregado de empresas agropecuarias)
- Módulo de producción de biocombustibles.

Análisis de impactos:

- Instrumentos de políticas (internas y externas)
- Precios y resultados
- Asignación de tierras
- Adaptación a mercados

Los casos considerados

- Se consideran en forma prioritaria los casos de **bioetanol** (azúcar y maíz) y **biodiesel** (soja) en al menos dos zonas de **Argentina y Brasil** teniendo como base relevamientos realizados en el marco de proyectos anteriores y de trabajos de integrantes del equipo.
- También se realizan estudios de caso que tienen en cuenta la situación en todos los países del **MERCOSUR** y la influencia de las políticas de la **Unión Europea**.

CAN SMALL BE SUSTAINABLE?

Reaching Smallholders with Sustainability Initiatives in Argentina, Brazil, and the U.S.

Jesse Abrams, Diana Cordoba, Theresa Selfa, Erik Nielsen, and Cassandra Moseley

BACKGROUND

- Bioenergy production increasingly expected to demonstrate commitment to sustainability (social, ecological)
- Pressure from NGOs, consumers, governments
- Proliferation of third-party certification schemes (RSPO, RTRS) and metastandards (UK Renewable Heat Incentive Timber Standard)

BACKGROUND

- Both social movements and certification schemes focus on large producers
- Are smallholders left out of global markets?
- Is sustainability on smallholder production units neglected?
- *We compare policies in Argentina, Brazil, and the U.S. that attempt to specifically bring smallholders into sustainable production*

FOCAL PUBLIC POLICIES

- Argentina: Ley 25.080 (1999 Investments for Cultivated Forests law)
- Brazil: 2004 National Program of Biodiesel Production (PNPB) and Social Fuel Seal
- State of Wisconsin, U.S.: 1985 Managed Forest Law and Group Certification

	Principal policy mechanisms	Linkage to environmental sustainability criteria	Dependence on implementation by non-state actors	Requires specific appropriations?	Vulnerability to changes in executive
AR: 25.080	Direct subsidies, tax breaks	Medium	Low	Y	Medium
BR: SFS	Market access, tax breaks	Weak	High	N	High
US: MFL	Tax breaks	Strong	Medium	N	Medium

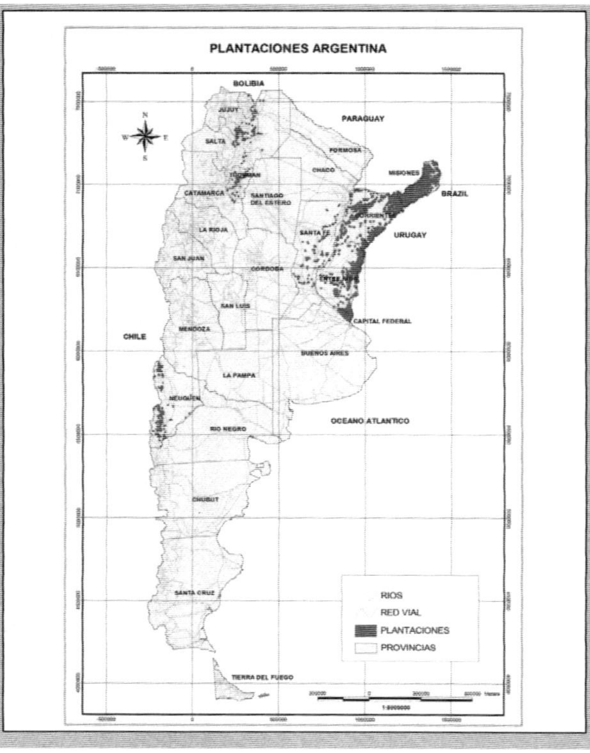

Source: Niklitschek 2012

ARGENTINA: 25.080

- 1999 law designed to stimulate domestic production of high-quality wood fiber
- Two principal policy mechanisms:
  - Direct planting, improvement, and maintenance subsidies: 80% of costs on projects up to 300 ha., 20% on projects 300-500 ha.
  - "Fiscal stability:" tax exemptions, IVA refunds, no new taxes or tax increases for 30 years, accelerated amortization on forestry-related activities

ARGENTINA: 25.080

- Planted species principally conifers (64%), eucalyptus (23%), willows (9%)
- Wood residues not currently a major source of bioenergy, but pellet plants exist in NE provinces with capacity for much more (Uasuf 2010)

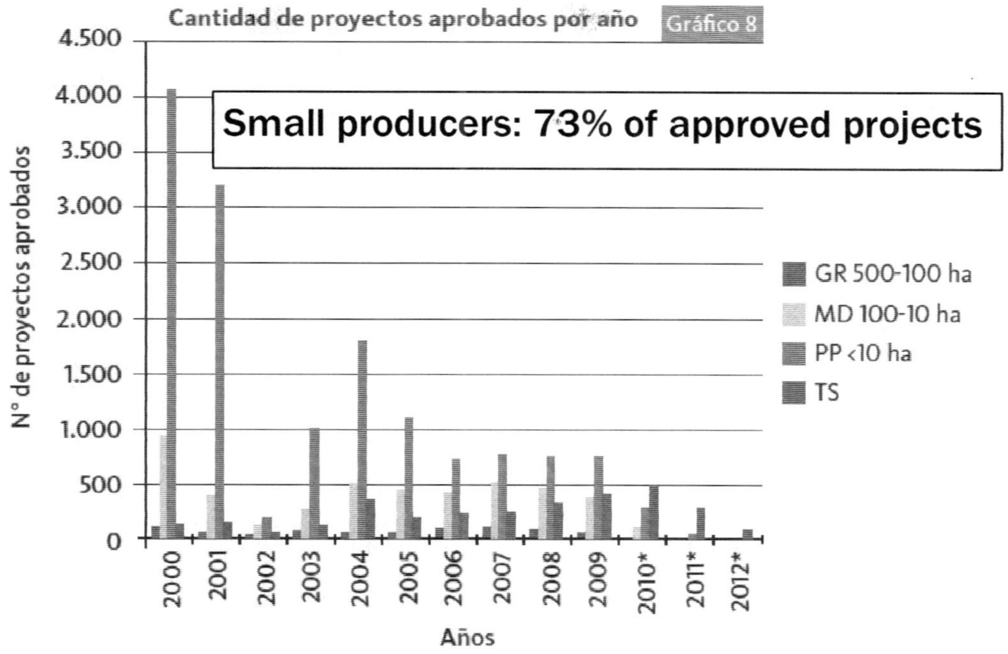

GR: grandes productores, MD: medianos productores, PP: pequeños productores, TS: tareas silviculturales

Fuente: Ministerio de Agricultura, Ganadería y Pesca. Subsecretaría de Agricultura. Dirección de Producción Forestal. Área Promoción.
* Ver aclaración en página 68

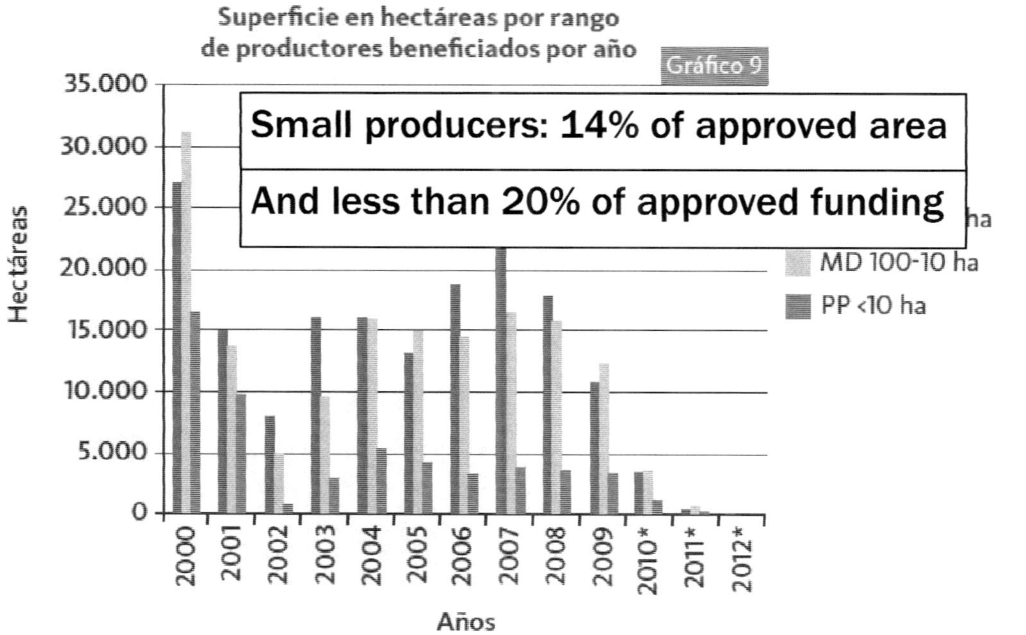

GR: grandes productores, MD: medianos productores, PP: pequeños productores

Fuente: Ministerio de Agricultura, Ganadería y Pesca. Subsecretaría de Agricultura. Dirección de Producción Forestal. Área Promoción.
* Ver aclaración en página 68

ARGENTINA: 25.080

- Ecological sustainability linkages:
  - Provincial native forest plan (under 26.331 "Forest Law") required to receive subsidies
- National / provincial governments and INTA promotion of certification:
  - FSC
  - CerFoAr
- Tax benefits self-sufficient, but subsidies require annual appropriations
- Reach of law to smallholders is limited by lack of full appropriation

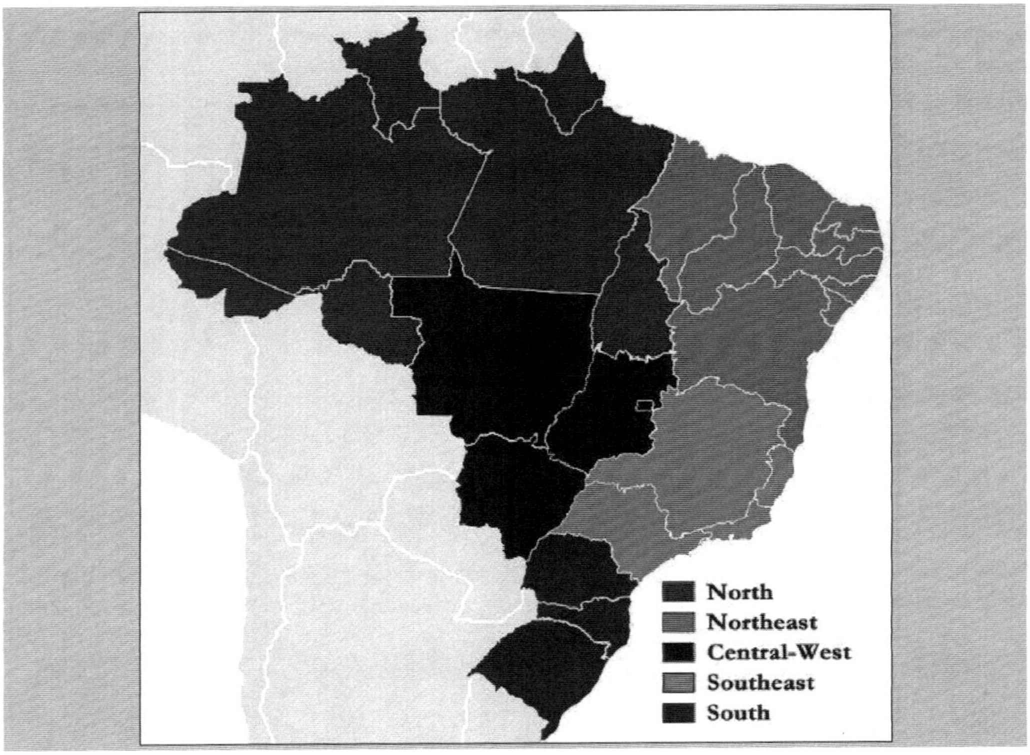

BRAZIL: PNPB AND SOCIAL FUEL SEAL

- Initiated in 2004 by President Silva and then-Energy Minister Rouseff
- Overseen by National Petroleum, Natural Gas, and Biofuel Agency (ANP)
- Currently: mandate of 7% biodiesel at the pump, up to 10% by 2019
- Two policy mechanisms: tax abatements and "Social Fuel Seal" program (required to access internal markets)

BRAZIL: PNPB AND SOCIAL FUEL SEAL

- Social Fuel Certification: Presidential Decree No. 5.297 (December 6, 2004) plus later modifications
- Certification of compliance with minimum sourcing from small farmers (15-40% depending on region)
- Administered by Ministry of Rural Development (MDA)
- Participating smallholders receive preferential pricing and are guaranteed technical assistance
- Combines bioenergy promotion with regional development interests (emphasizing N, NE and semi-arid)
- Biodiesel feedstock currently dominated by soybean oil (78%)

SFS AS OF 2015

- 42 of the 58 biodiesel producers in Brazil have SFS
- Representing 91% of the installed capacity
- And 99% of production
- 99% of smallholder-acquired oils come from soy
- >72,000 participating families in 2014
- Representing >$13M USD
- And 78 cooperatives

Data: Brazilian Secretary of Family Agriculture

BRAZIL: PNPB AND SOCIAL FUEL SEAL

- No explicit inclusion of ecological sustainability criteria
- Rural Workers' Unions certify smallholders and mediates contracts between them and the biofuel processors
- Does not rely on annual appropriations...
- ...but does rely on political support from the executive branch

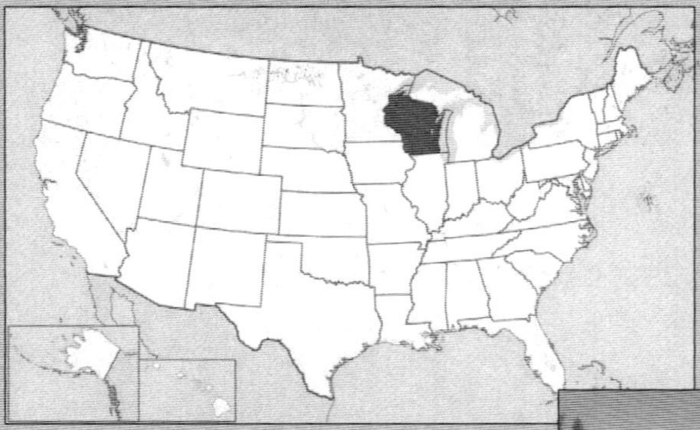

WISCONSIN: MANAGED FOREST LAW

- 1985 law that provides tax benefits to smallholders:
  - Must adopt a forest management plan
  - Must regularly harvest timber
  - Must receive authorization before cutting and pay taxes on sales
  - “Open” and “closed” lands options
- State DNR automatically certified MFL lands until July 2015
  - ATFS starting in 2005
  - FSC starting in 2008
- MFL participants can now “opt in” to certification

WISCONSIN: MANAGED FOREST LAW

- Several large wood-fired biomass energy facilities in Wisconsin, including 50MW cogen in Rothschild (2013)
- As of 2016: 2.6M MFL acres (~1M hectares) certified ATFS/FSC (~28% of smallholder lands)
- Certification implies adherence to biomass harvesting guidelines
- Implies expanded role for state DNR

WISCONSIN: MANAGED FOREST LAW

- Current state government strongly anti-regulatory
- Removed authority from DNR, including some oversight of MFL activities
- Concern that DNR's role as intermediary could be in jeopardy

POLICY COMPARISON: STRENGTHS

- AR 25.080:
  - Successful at promoting forest plantings, including among smallholders
- BR PNPB:
  - Successful at integrating / assisting large number of smallholders
  - Integrates multiple interests (unions, businesses, gov't agencies)
  - Does not rely on annual appropriations
- US MFL:
  - Successful at certifying large proportion of forest smallholders
  - Does not rely on annual appropriations

POLICY COMPARISON: WEAKNESSES

- AR: 25.080
  - Inadequate appropriations has led to lack of full potential
  - Integration with ecological sustainability criteria is largely voluntary
- BR: PNPB
  - Weak integration with ecological sustainability criteria
  - Vulnerable to changes at the executive level
- US: MFL
  - Requires a strong and competent regulatory agency
  - Vulnerable to anti-regulatory political winds

REFLECTIONS ON POLICY INSTRUMENTS

Policy design and performance reflective of institutions at national scale:

- Role and strength of government
  - Bureaucratic presence / competence
  - Ability to consistently provide funding
- Role and strength of non-state actors
  - Policy implementation
  - Ongoing political support
- Vulnerability of policies to political changes
  - Stability of policies / agencies
  - Legal foundation of policies

ACKNOWLEDGEMENTS

National Science Foundation OISE-PIRE, "Sustainability, Ecosystem Services, and Bioenergy Development across the Americas," grant #1243444

RCN

Collaborators in the U.S. and Argentina

THE RELATIONSHIP BETWEEN SENSE OF PLACE AND PERCEPTIONS OF ENVIRONMENTAL, SOCIAL AND ECONOMIC IMPACTS OF THE EXPANSION OF MONOCULTURE PLANTATIONS

A CASE STUDY OF UBAJAY, ENTRE RIOS PROVINCE, ARGENTINA

by Victoria Marini
Advisor: Theresa Selfa
Steering committee:
Jesse Abrahms & Elizabeth Vidon

Outline

1. Problem statement
2. Purpose of research
3. Study area
4. Literature review
5. Research method and Design
6. Findings
7. Conclusions and Implications

The problem statement

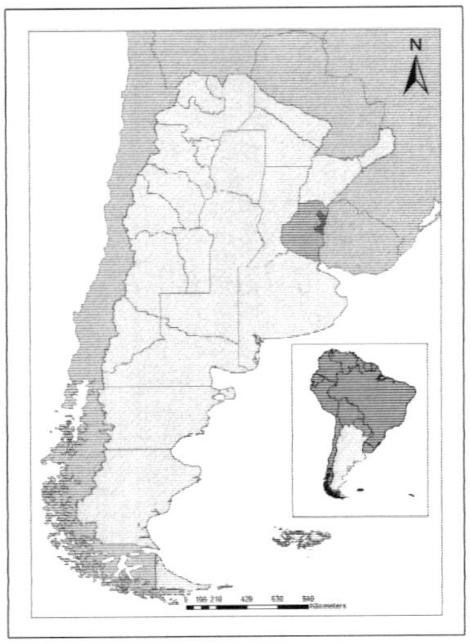

In Argentina, in the northeast coast of the Uruguay River of Entre Rios Province, land use change has resulted due to the expansion of Eucalyptus and Pine plantations.

These monoculture plantations expanded from ca. 30,000 ha in 1996 to ca. 100,000 ha in 2001

Law 25080

The problem statement

Environmental
Impacts: affect water resources, losses of soil organic carbon, acidification of soil, modify climate, biodiversity decline

Economic
Impacts: jobs

Social
Impacts: decrease in quality of employment, decline in rural population, poor working conditions, land ownership and land prices

(Jobbágy et al., 2003; Andersson et al., 2016; Bardomás, 2009)

Purpose of research

- Understand environmental and socio-economic impacts associated with the forestry industry expansion
- Framework for understanding the perceptions that local residents have of these impacts based on their sense of place

Study area

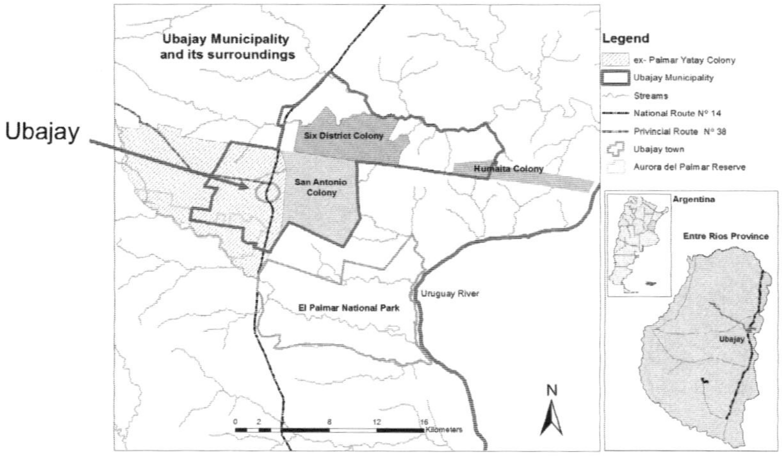

Total population of 3,507 inhabitants

Largest number of hectares planted (13,840.5 ha)

One of the largest concentrations of sawmills of the Province

Next to the highways and harbor

Eucalyptus and pine nursery in the area

Soil and climatic conditions enhance the growing of eucalyptus plantations

Study area

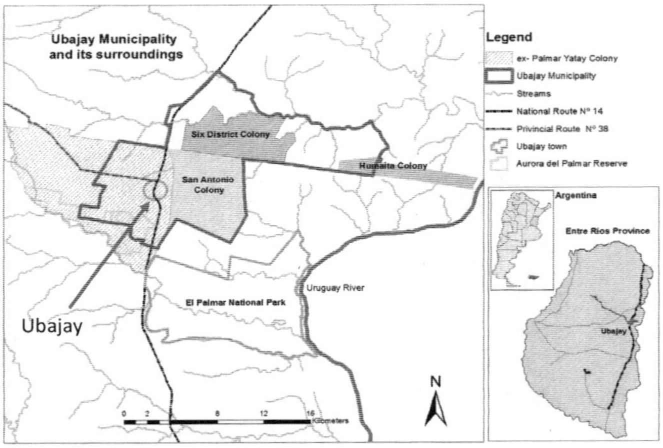

Literature review
Sense of Place theory

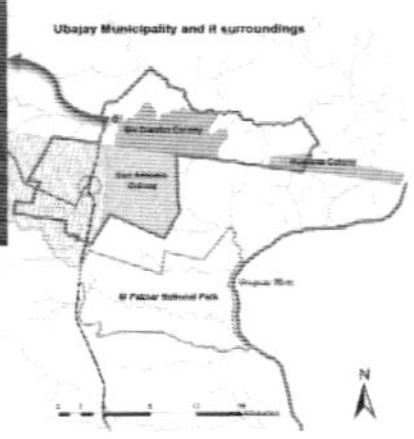

"The meaning and importance of a setting held by an individual or group, based on an individual's and group's experience with the setting"

(Tuan, 1977; Relph, 1976; Stedman, 2003; Williams & Roggenbuck, 1989)

Literature review
Place theory

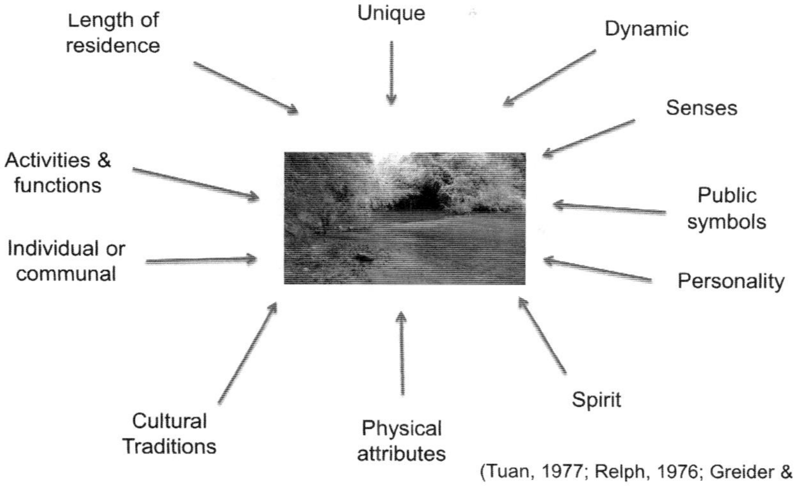

(Tuan, 1977; Relph, 1976; Greider & Garkovitch, 1994)

Research method and design

Grounded theory

- Sampling
  - 32 adults over 35 years old (14 women and 18 men)
  - Combination of sampling methods
    - Stratified random sampling (age and gender)
      Sampling frame: Ubajay voters' list of the Presidential elections of 2011
    - Snowball sampling (key informants)

- Data collection (Validity of results)
  - Interviews
    - Face-to-Face, open-ended, in-depth questions
    - Administered during summer 2015 (Spanish, one hour)
    - Interviews asked about the reasons why places where considered important, perceptions of plantations and some demographic questions

Research method and design

- Data collection
  - Map data collection
    - paper map

Research method and design

- Data analysis
  - Transcribed and coded using Nvivo software

- Map analysis
  - Vector analysis
  - Raster analysis
    - Satellite images

ID	Important places	Latitude	Longitude	Frequency mentioned
#1	Aurora del Palmar	-31.81511380	-58.32175570	4
#2	Ubajay	-31.79499960	-58.31442770	32
#3	El Palmar	-31.86499710	-58.25986110	21

Results

1. Map with these important places
2. Reasons why places are considered important
3. Present the association between their sense of place and perceptions of this landscape transformation.

Places most frequently mentioned around Ubajay Municipality

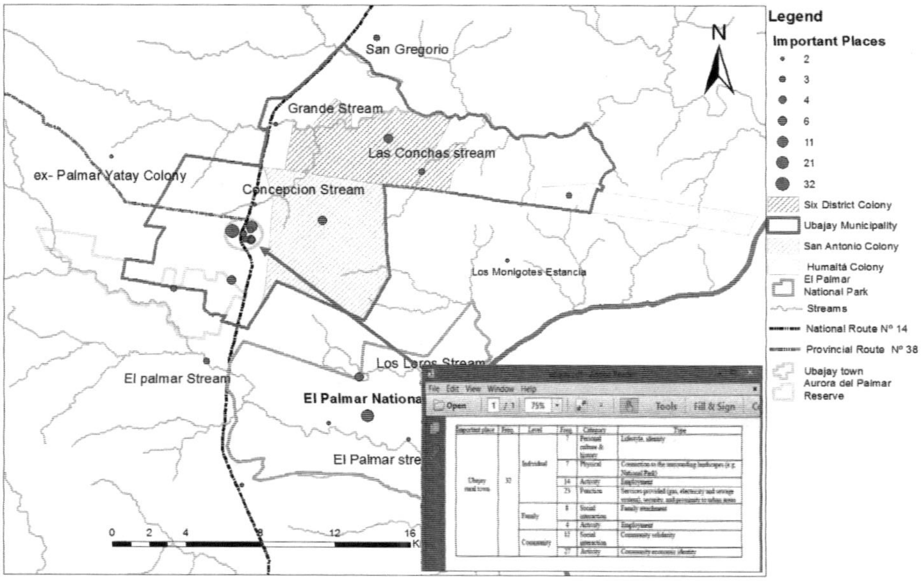

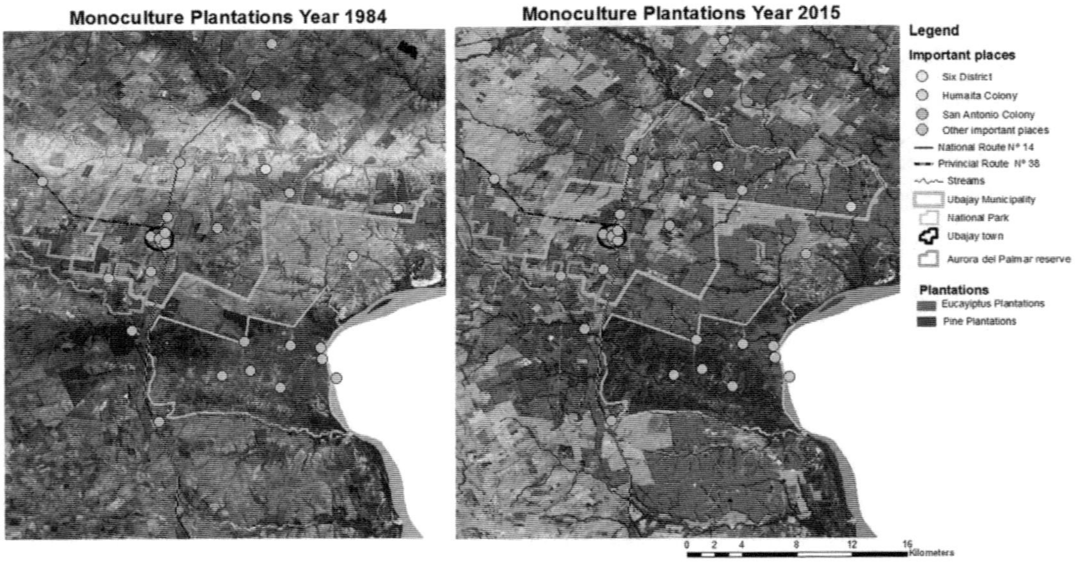

This image shows the overlap between the location of some of these places and the eucalyptus plantations.

Reasons of Important places

Ubajay- Activity

Interviewee I said, “the sawmill feeds many people. Gave life to the town. I don’t know if Ubajay would still exist if it weren’t for the sawmills”.

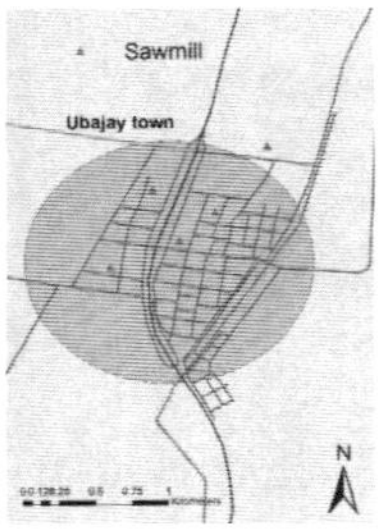

Collective identity

‘Capital of the Wood’

Reasons of Important places

Ubajay- Function

Another important aspect that came out of the interviews is how this town provided them with a good quality of life since it was a peaceful town, it had many services (gas, electricity, sewage system, and bus transportation) and it was very near to urban areas.

Reasons of Important places

Ubajay- Community

Interviewee VI explained, “This is a very peaceful town. People are very nice. People are very caring. For instance, when something happens to a neighbor, someone makes a call to the radio and immediately, everybody is there trying to help.”

Community solidarity rooted in mutual needs

Reasons of Important places

Uruguay River and Streams- Identity

Interviewee XVI explained, "I always look for places with water. You know, we, people from Entre Rios, are always looking for streams and lakes. I love the area of the countryside where I can go and sit on the shore of the river and just stay there."

Reasons of Important places

National Park "El Palmar"- Aesthetics & Wilderness

Interviewee VI explained, "Sunsets at the Park are special. I love taking pictures. It is what we always do because it is the nearest wild area that we have. The Park belongs to us."

This Park was also unique for respondents because it is one of the few places in the region that still preserves the natural landscape.

Reasons of Important places

National Park “El Palmar”- Biodiversity

Interviewee XIV described, “while we are on the trail, you can see all sort of animals: capybaras, foxes, rheas. And we always take pictures as if it was the first time we saw one of those. It is a special moment.”

Reasons of Important places

Ubajay- Collective Identity

Interviewee IX said, “I identify with my town because it is where I grew up, and it is where I work. Whenever I visit another place, I always say: I come from the capital city of the wood and the heart of El Palmar.”

UBAJAY
corazóndelpalmar

Reasons of Important places

The colonies- Personal culture and social interactions

Interviewee III explained to me how, with her grandmother, she used to make cheese and dry fruits that were used as dessert. She also narrated to me how her brothers had to help her father to work the land.

Reasons of Important places

The colonies- Aesthetics and biodiversity

Reasons of Important places

The colonies- Collective identity of natives to the colonies

Interviewee XX said, "I totally identify myself with the smell of the surrounding citrus fields that existed when I was little. I remember the peach, mandarin, and orange trees that no longer exist in the area […] The whole idea was to eat fruit by the tree. I still love eating mandarin next to the tree. No, there is no way I can forget this feeling."

Sense of place & Perceptions

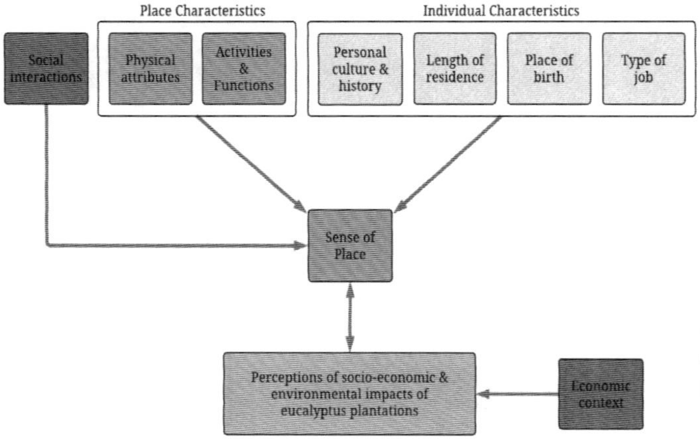

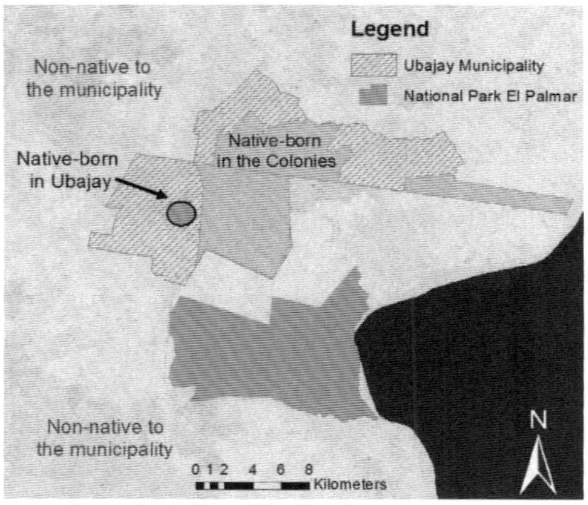

Non-native to Entre Rios Province

Sense of Place & Perception of the impacts of eucalyptus plantations

	Positive	Negative
Natives to the colonies		• Tradition & customs • Rural depopulation • Change in local climate • Water scarcity • Biodiversity loss • Fires
Native to Ubajay	• Jobs from plantations • Air-quality provision	• Fires
Sawmill workers	• Air-quality provision	• Bad labor conditions of plantations
All interviews		• Visual impact

Sense of Place & Perception of the impacts of sawmills

	Positive	Negative
Native to the colonies	• Jobs from the sawmills	• Health impacts • Migration
All other interviewees	• Both jobs from the plantations and the sawmills • Migration	• Health impacts

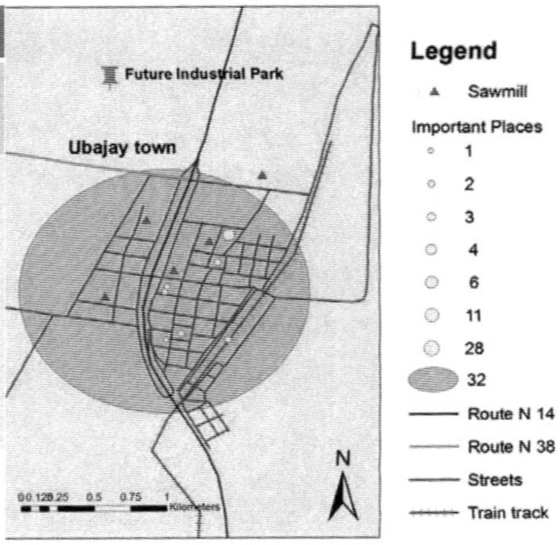

Colonies' attachment & Rural depopulation

Interviewee VIII pointed out, "I miss all my friends that used to live next to me. This no longer exists. Now everything is what we call 'tapera', an abandoned house."

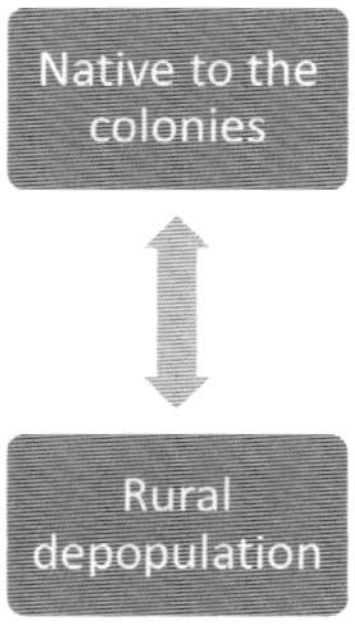

Colonies' attachment & Loss of traditions and customs

Interviewee XI said, "When I think about San Antonio colony, it reminds me of my childhood experiences and how I used to enjoy those natural landscapes. However, this landscape has changed so much that I can hardly recognize it. It is over. All that is left are my memories. For this reason whenever I want to find nature, I go to the National Park. The National Park is what I mostly like."

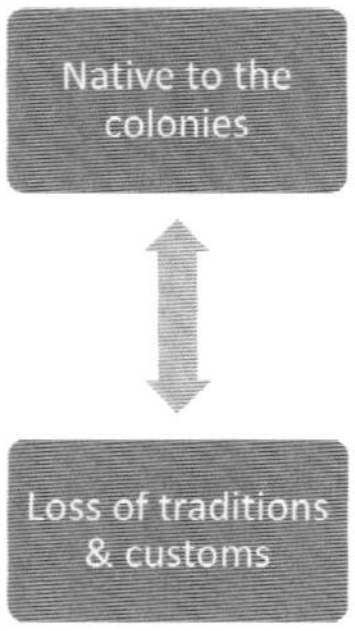

Colonies' attachment & Water scarcity

Interviewee XXXI said,
Plantations took all the water. They absorbed all the upland groundwater. In my land, we have a stream called 'Las Conchas' stream. That stream used to have a constant and fluent flow as well as crystalline water. The change is very obvious. When we were younger, we used to swim and play inside the stream, and you cannot longer do that (she cried here). You cannot tell your grandsons about this because we no longer have that amount of water there. Those were great times.

Colonies' attachment & Biodiversity loss

Interviewee IX mentioned, "Have you ever been to a deforested site? That is so sad. It is worse than a battlefield. It makes me angry. All animals are left without habitat. I do not get it. A ñandubay tree is more than 300 years old. [...] There is not even one bird nesting there nor grass in the soil, only weeds. There is no food for animals there".

Physical attachment & Visual impact

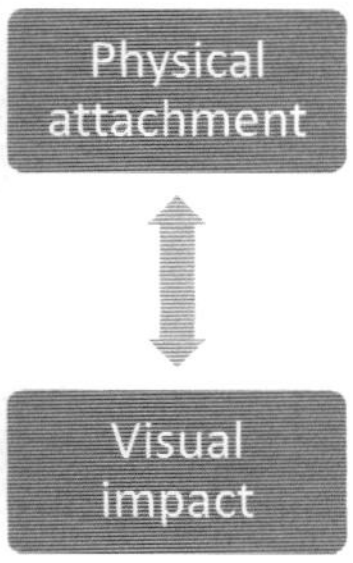

Ubajay' s community identity & Economic benefits

Interviewee V pointed out, "the eucalyptus plantations are part of the landscape. They are natural capital. They generate all the economic movement of Ubajay because one person provides the seeds, another plants the trees, another harvests the trees, another transports the wood to the sawmills, other people work for the sawmill, then you need the mechanics, and so on. It is amazing how this industry works."

Ubajay' s community identity & Economic benefits

Perceptions were also influenced by the agro-export model that Argentina has followed since the 1990s. A model based on a technological package technology that allowed for the expansion of soy.

Soy plantations

- Less jobs → depopulation
- Agrochemicals → pollute water and biodiversity loss
- Agrochemicals → cause cancer
- No wind protection
- Soy not consumed in Argentina

Eucalyptus plantations

- More jobs → population
- Plantations→ water scarcity and biodiversity loss
- Sawmills → health issues
- Wind protection
- Forest products consumed in Argentina

Conclusions & Implications

- <u>Sense of place is not a static phenomenon but can be affected by landscape change</u>.
- Perceptions vary according to the different senses of place
  - Important tool for policy-makers, land use management and local planning
- Perceptions are influenced by the economic context of Argentina.
- This research also contributes to:
  - Literature that focuses on mapping important places as well as place values.
  - The body of knowledge regarding the impacts of monoculture plantations on local rural areas

Important place	Freq.	Level	Freq.	Category	Type
Ubajay rural town	32	Individual	7	Personal culture & history	Lifestyle, identity
			7	Physical	Connection to the surrounding landscapes (e.g. National Park)
			14	Activity	Employment
			23	Function	Services provided (gas, electricity and sewage system), security, and proximity to urban areas
		Family	8	Social interaction	Family attachment
			4	Activity	Employment
		Community	12	Social interaction	Community solidarity
			27	Activity	Community economic identity
Six District, Humaitá and San Antonio Colonies	10	Individual	10	Personal culture & history	Lifestyle, identity
			10	Physical	Serenity, freedom, secure, nature aesthetic, connection and appreciation of the countryside landscape
		Family	10	Social interaction	Lifestyle and family interaction
			10	Activity	Livelihood and subsistence
		Community	10	Social interaction	Community attachment, lifestyle, friendship
Uruguay River	11	Individual	11	Personal culture & history	Identity
			8	Physical	Nature aesthetics, connection and appreciation of water landscape, relaxation, serenity, recreational activities (fishing)
		Family	5	Physical	Recreational activities (e.g. enjoy the beach, and spend the day)
El Palmar National Park	21	Individual	5	Personal culture & history	Identity
			21	Physical	Nature's aesthetic, connection and appreciation of wilderness, relaxation, serenity, spiritual well-being, exploration, activities (exercise, photography)
		Family	11	Physical	Recreational activities (e.g. camping and making barbecue, enjoying the beach)
		Community	6	Activities	Jobs provided in the Park and source of income from tourists
		Cultural heritage	4		History of the Jesuits mission as well as history of the Estancia La Niña and the quarries and the jobs that these two provided
		Religious	2		Mass made in the times of the Estancia and religious icons (e.g.sculpture of virgin)
Streams	11	Individual	4	Personal culture & history	Identity
			3	Physical	Connection and appreciation of water landscape, relaxation, serenity, recreational activities (Fishing)
		Family	4	Physical	Recreational activities (Bathing)
Aurora del Palmar Private Reserve	4	Individual	1	Personal culture & history	Identity
			4	Activities	Employment
			4	Physical	Aesthetics
		Family	3	Social interaction	Family connection

ANÁLISIS DEL MARCO INSTITUCIONAL PARA LA BIOENERGÍA EN ARGENTINA

Ing. Agr. Lucila Grassi
Septiembre de 2016

CONTEXTO ACTUAL

✓ **La promoción de las EE.RR. es un tema clave dentro de la política energética:**

Contribuye con:

- Seguridad de abastecimiento.
- Disminución del impacto del sector energético sobre el cambio climático.
- Modernización de la infraestructura energética en generación y distribución.

Ley 27.191:

- Meta: llegar a 2025 con una cobertura del 20% del consumo de energía eléctrica a partir de fuentes renovables.
- Dar seguridad a la inversión mediante la definición de una política de largo plazo.
- Generar un mercado competitivo con precios “a la baja”.

CONTEXTO ACTUAL

✓ **La revisión de las NDC son una prioridad del Gobierno Nacional**

CoP 21 – propuesta presentada:

- Objetivo de reducción incondicional de emisiones del 15% a 2030 respecto al BAU.
- Objetivo condicional de aumentar esa reducción en un 15% adicional, sujeto al financiamiento externo.

Ejes para la revisión:

- Limitación a la deforestación.
- Impulso a las energías renovables.

Sectores clave:

- Agroindustrial – Energías renovables

CONTEXTO ACTUAL

✓ **Actualización de tarifas de la energía eléctrica y el gas para usuarios finales.**

– Favorece la viabilidad económica de las fuentes renovables frente a las convencionales.

Resultado de las medidas → grandes expectativas en el sector de las energías renovables

Éxito en la convocatoria a la primera ronda del **Plan Renovar:**

Superó en más de 6x el cupo licitado

6.366 MW ofertados vs.

1.000 MW licitados

MARCO INSTITUCIONAL

✓ **Características generales:**

Estudios sectoriales ya evidenciaban alrededor de 2010:

- Existencia de numerosas iniciativas públicas y privadas vinculadas al desarrollo del sector, con superposición de objetivos y acciones.
- Insuficiente articulación y coordinación institucional.
- Dificultades en la generación y difusión de información sectorial.
- Importante potencial para el desarrollo de conocimientos, capacidades técnicas y tecnologías locales.

MARCO INSTITUCIONAL

✓ **En 2012 el Gobierno Nacional promueve la creación del Proyecto Probiomasa.**

- Promotores: Ministerio de Agricultura y Secretaría de Energía.
- Cuenta con apoyo técnico y administrativo de FAO.
- Objetivo principal: incrementar la producción de energía derivada de biomasa.
- Estrategia de intervención: implementación de una plataforma interinstitucional que articule las tres dimensiones principales de la bioenergía:

ENERGÍA – AGRICULTURA - AMBIENTE

MARCO INSTITUCIONAL ACTUAL

- ✓ **Intervención positiva de Programas como:**
  - Probiomasa (MINAGRI + MINEM)
  - Programa Nacional de Agroindustrias y Agregado de Valor (INTA)
  - Programa de Energías Renovables (INTI)
  - Argentina Innovadora (MINCyT)

Se avanzó en muchas cuestiones, sin embargo, al día de hoy persisten barreras:

Institucionales
Regulatorias
Tecnológicas
Económicas

MARCO INSTITUCIONAL ACTUAL

- ✓ **Estudio en 2016: Descripción y análisis del marco institucional actual para bioenergía en Argentina:**
  - Identificación de actores clave.
  - Identificación de programas e iniciativas.
  - Definición de roles, funciones y relaciones entre actores clave.
  - Definición de barreras institucionales, tecnológicas y regulatorias.
  - Análisis FODA.
  - Elaboración de recomendaciones.

IDENTIFICACIÓN DE ACTORES CLAVE

Categoría	Nómina de Actores
Administración Pública Nacional – Organismos Centralizados	Ministerio de Agroindustria (MINAGRI) - Ministerio de Energía y Minería (MINEM) Ministerio de Ambiente y Desarrollo Sustentable (MINAyDS) - Ministerio de Ciencia, Tecnología e Innovación Productiva (MINCyT) - Ministerio de Relaciones Exteriores y Culto (MRECIC) - Ministerio de Producción (MINPROD)
Administración Pública Nacional – Organismos descentralizados	Instituto Nacional de Tecnología Agropecuaria (INTA) - Instituto Nacional de Tecnología Industrial (INTI) - Consejo Nacional de Investigaciones Científicas y Técnicas (CONICET)
Organizaciones de la sociedad civil	Unión Industrial Argentina (UIA) - Asociación Forestal Argentina (AFoA) - Federación Argentina Industria de la Madera y Afines (FAIMA) - Fundación FADA - Fundación Proteger Cámara Argentina de Energías Renovables (CADER)
Universidades	Universidad de Buenos Aires(Facultad de Agronomía y CEARE) - Universidad Tecnológica Nacional (UTN) - Universidad Nacional de Luján (UNLu) - Universidad Nacional de Chilecito (UNdeC) - Universidad Nacional de Cuyo (UNCu) - Instituto Tecnológico de Buenos Aires (ITBA) - Universidad Nacional de San Martín (UNSAM) - Universidad Nacional del Centro (UNICEN)
Centros Tecnológicos	Estación Experimental Obispo Colombres (EEAOC) - Instituto de Investigaciones en Energías no Convencionales (INENCO) - Parque Tecnológico Misiones (PTMi) - Instituto de Desarrollo y Diseño (INGAR)
Sector privado (consultoras y proveedoras de tecnología)	Grupo IFES - EG Ingeniería - Biometanos del Sur – Tecnored – IBS - PFI Energía & Ecology Gonella S.A. - Fontanet

PROGRAMAS DESARROLLADOS

Programa	Alcance	Promotores
PROBIOMASA	NACIONAL	MINAGRI, MINEM
ARGENTINA INNOVADORA 2020	NACIONAL	MINCyT
PLATAFORMA BIOECONOMIA ARGENTINA	NACIONAL	MINCyT
BANCO DE PROYECTOS DE INVERSIÓN PRODUCTIVA	NACIONAL	MRECIC
Fondo Nacional para el Desarrollo de la Micro, Pequeña y Mediana Empresa (FONAPYME)	NACIONAL	MINPROD
Programa Nacional de Agroindustrias y Agregado de Valor	NACIONAL	INTA
Programa de Energía Renovable (PER)	NACIONAL	INTI
Programa "Centros de Investigación y Transferencia	NACIONAL	CONICET
Programa de tecnología Socialmente Apropiada	REGIONAL	FUNDACIÓN PROTEGER
IncUBAgro	NACIONAL	UBA (Facultad de Agronomía)
Programa de Bioenergía	PROVINCIAL	Universidad Nacional de Cuyo
Programa de Residuos	PROVINCIAL	Universidad Nacional de Cuyo
Proyecto biogás	PROVINCIAL	Universidad Nacional de Cuyo
Observatorio de Energía y Producción	PROVINCIAL	Universidad Nacional de Cuyo
Proyecto de gasificación de biomasa	REGIONAL	EEAOC
Mesa de bioenergía/Sub-mesa de Dendroenergía	NACIONAL	MINAGRI

INICIATIVAS EN CURSO

✓ **Se detectaron numerosas iniciativas vinculadas a:**

– Actividades de **formación y capacitación**: CEARE, ITBA, PROBIOMASA.

– Actividades de **difusión y sensibilización**: Fundación Proteger, FADA, PROBIOMASA.

– Creación de **centros de investigación** en Universidades: UNSAM, UNLu.

ROLES, FUNCIONES Y RELACIONES ENTRE ACTORES

✓ **Indicadores:**

– Cantidad de vínculos existentes con otras instituciones y programas.
– Calidad de los vínculos existentes:
  • permanentes, temporales; formales o informales.
– Alcance de acciones y programas desarrollados:
  • Nacional, regional, local.
– Impacto de los programas desarrollados.
– Capacidades institucionales y tecnológicas.

MAPA RELACIONAL DE ACTORES CLAVE

Línea roja: vinculación institucional formal.
Línea azul: vinculación a través de participación en iniciativas
Línea verde: vinculación mediante acuerdos específicos (proyectos de investigación, contratos de servicios, etc.)

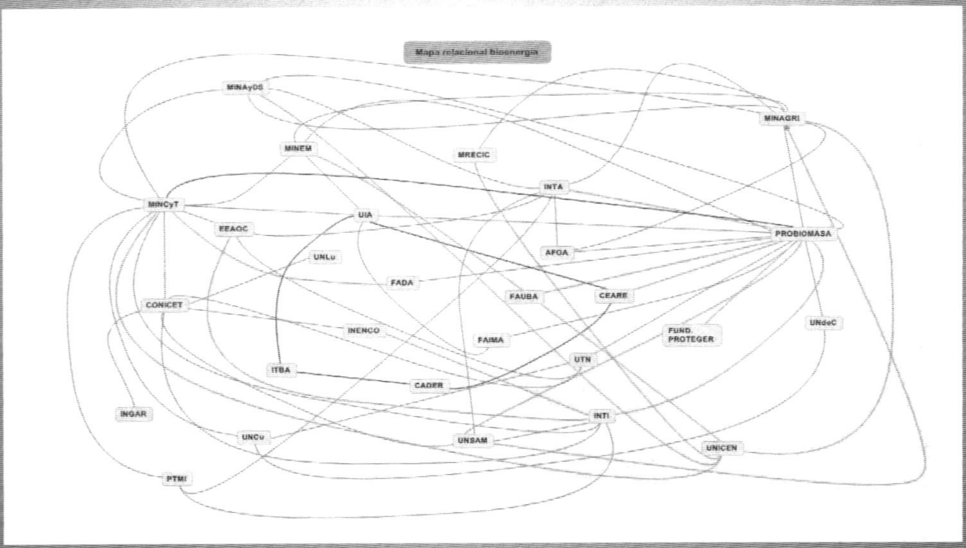

CAMBIOS ESTRUCTURALES EN LA APN

Dependencia anterior	Dependencia actual	Jerarquización institucional	Ampliación de competencias	Comentarios
Ministerio de Agricultura	**Ministerio de Agroindustria**	-	X	Creación de la **Secretaría de Agregado de Valor**
Secretaría de Energía	**Ministerio de Energía y Minería**	X	X	Creación de la **Subsecretaría de Energías Renovables**
Secretaría de Ambiente y Desarrollo Sustentable	**Ministerio de Ambiente y Desarrollo Sustentable**	X	X	Creación de la **Secretaría de Política Ambiental, Cambio Climático, Ambiente y Desarrollo Sustentable**
Secretaría de Industria	**Ministerio de Producción**	X	X	Creación de la **Secretaría de Industria y Servicios**

Objetivos:

- ✓ Favorecer la cooperación interinstitucional.
- ✓ Promover la coordinación de acciones.
- ✓ Dar transparencia a la generación, gestión y publicación de la información.
- ✓ Promover una mayor participación público-privada.

ACTORES CLAVE MÁS INFLUYENTES

- ✓ **Ministerio de Agroindustria:**

  Desde 2012: rol destacado en el fortalecimiento institucional del sector bioenergético.

  Nuevas competencias: asegurar el agregado de valor en cadenas productivas y una mayor articulación y cooperación institucional a fin de promover un desarrollo armónico de la bioenergía, desde el sector primario hasta sectores industriales asociados.

- ✓ **Ministerio de Energía y Minería:**

  Es uno de los actores centrales para el desarrollo de la bioenergía.

  Cambios recientes: marcado sesgo hacia la generación de energía eléctrica a partir de fuentes renovables, tanto desde lo institucional (la Subsecretaría de EE.RR. depende de la Secretaría de Energía Eléctrica) como desde lo regulatorio (la Ley 27.191 regula la generación de energía eléctrica a partir de fuentes renovables) → ¿hueco institucional en relación a la generación de energía térmica a partir de biomasa y otras fuentes renovables?

ACTORES CLAVE MÁS INFLUYENTES

- ✓ **Ministerio de Ambiente y Desarrollo Sustentable:**

  Nuevo mandato: impone un rol mucho más activo mediante el diseño de políticas e incorporación de nuevas tecnologías e instrumentos que defiendan al medio ambiente y contribuyan a combatir el cambio climático.

- ✓ **Ministerio de Ciencia, Tecnología e Innovación Productiva:**

  Viene desarrollando acciones de alto impacto:

  Creación de espacios de debate y de lazos entre los sectores privado y público, como los científicos y productivos, a fin de promover la innovación productiva y el crecimiento tecnológico en sectores estratégicos.

  Financiamiento para proyectos productivos innovadores que incluyen la integración de los sectores público y privado.

ACTORES CLAVE MÁS INFLUYENTES

✓ **INTA:**

Larga trayectoria de trabajo en bioenergía (desde 1980) + fuerte presencia territorial en las provincias de potencial bioenergético = adecuado abordaje de las características sitio-específicas de esta actividad.

Se destacan sus sistemas de extensión, información y comunicación, como herramientas para poner sus capacidades al servicio del aprovechamiento y transformación de biomasa en diferentes vectores aprovechables energéticamente y el consecuente agregado de valor en origen.

✓ **INTI:**

Trabaja en forma sistémica desde 2008 en el desarrollo de EE.RR.

Generador y proveedor de servicios tecnológicos en áreas de I + D; asistencia técnica; ensayos, análisis y calibraciones; certificación y transferencia del conocimiento.

PER: enfoque redárquico para al fortalecimiento del entramado productivo, la promoción de la innovación, y el aumento de la competitividad en procesos industriales.

COMENTARIOS SOBRE OTROS ACTORES CLAVE

- ✓ **Organizaciones de la sociedad civil:** representan a distintas cadenas de valor en diversas regiones del país. Tienen el potencial de desarrollar acciones vinculadas a la promoción de las energías renovables y la eficiencia energética.
- ✓ **Universidades:** se detecta algún grado de vinculación con sectores productivos pero se destaca el trabajo con CONICET en la creación de centros de investigación enfocados al desarrollo de la bioenergía con un enfoque regional. También es importante considerar la oferta de carreras de grado y postgrado relacionadas a la temática.
- ✓ **Centros tecnológicos:** El trabajo de estos centros es en general de enfoque regional. Pareciera necesario afianzar el trabajo conjunto entre éstos y el sector privado (productores, empresarios, técnicos) para promover el desarrollo de innovaciones en bioenergía, lo que permitiría sacar provecho de los recursos humanos tecnológicos y productivos de alta calidad disponibles en el país.
- ✓ **Sector privado:** Cuenta con capacidad técnica suficiente y vocación para la adopción y/o desarrollo de innovaciones tecnológicas adecuadas a las condiciones locales. Sin embargo, cuestiones como el escaso desarrollo del mercado bioenergético, la existencia de un marco regulatorio poco específico para la bioenergía y un contexto económico-financiero hasta hace poco desfavorable para la inversión en nuevas tecnologías, fueron en contra del desarrollo de los recursos técnicos locales, al no contar con suficiente experiencia en la implementación de proyectos.

IDENTIFICACIÓN DE BARRERAS

✓ **Institucionales:**

- Insuficiente integración y coordinación de políticas nacionales y provinciales de largo plazo para el desarrollo de la generación de energía a partir de biomasa.
- Insuficiente coordinación entre programas sectoriales.
- Escasa asignación de recursos (financieros) para el desarrollo de capacidades en algunas instituciones.
- Escasa disponibilidad de infraestructura para generar capacidades en algunas instituciones públicas y privadas.
- Amplia dispersión de esfuerzos y superposición de I+D.
- Insuficientes programas de sensibilización pública y de divulgación.
- Información sectorial incompleta, descentralizada y poco accesible.
- Existencia de vacíos de información (p.ej.: en relación al mercado de biomasa, calidad de la materia prima, disponibilidad de biomasa en determinadas regiones, etc.).
- Falta de internalización de los beneficios ambientales y sociales para la promoción de la generación de energía a partir de biomasa.

IDENTIFICACIÓN DE BARRERAS

✓ **Tecnológicas:**

- Insuficientes recursos humanos calificados: falta de profesionales y técnicos debidamente calificados para el diseño y formulación de proyectos, sobre todo en PyMEs.
- Reducida difusión de tecnologías apropiadas y baja experiencia en proyectos para la generación de energía derivada de biomasa.
- Inexistencia de normas de calidad específicas para la biomasa como materia prima y como producto energético.
- Debilidades técnicas específicas (ver sección temas clave).
- Dificultad de acceso a información específica sobre tecnologías de conversión de la biomasa en energía (para usuarios y consumidores).

IDENTIFICACIÓN DE BARRERAS

✓ **Regulatorias:**

- Falta de claridad en los objetivos de las políticas específicas de estímulo a la generación de energía a partir de residuos biomásicos.
- Inexistencia de un marco legal que promueva los usos térmicos de la biomasa.
- Complejo proceso de ingreso al MEM, especialmente para proyectos medianos y pequeños.
- Diversidad de normativas de ingreso al mercado eléctrico en general.
- Ausencia de regulación a nivel nacional para la generación distribuida de pequeña escala. Sin embargo, existen proyectos de ley que están siendo tratados a nivel legislativo. A nivel provincial, Santa Fe, Salta y Mendoza ya cuentan con marcos legales que les permiten inyectar energía a la red.
- Falta de un procedimiento estandarizado en la asignación de tarifas eléctricas.
- Complejas regulaciones provinciales de distribución de electricidad.
- Insuficiente control ambiental sobre los residuos biomásicos (sólidos y líquidos).
- Escaso control ambiental a nivel provincial y/o municipal del destino final de los residuos de biomasa (acumulación, quema a cielo abierto, etc.).

MATRIZ FODA

Situación institucional para la bioenergía

FORTALEZAS

Vocación del sector agropecuario y agroindustrial por la innovación y adopción de nuevas tecnologías.

Disponibilidad de recursos humanos científicos, tecnológicos y productivos de alta calidad.

Fuerte presencia territorial de actores clave.

Antecedentes de vinculación entre los sectores productivos locales y el sector científico.

OPORTUNIDADES

Condiciones agroecológicas favorables para la producción de biomasa.

La promoción de las EE.RR. es una prioridad para el actual gobierno.

Condiciones macroeconómicas favorables para el desarrollo de inversiones.

El desarrollo sustentable y el cambio climático son temas de alta prioridad en la agenda de gobierno.

Implementación de políticas que tienden a optimizar y eficientizar el diseño y aplicación de políticas públicas.

Existe la necesidad de incorporar potencia a partir de fuentes renovables para la generación de energía eléctrica.

DEBILIDADES

Escasa articulación institucional.

Insuficiente articulación Nación Provincias.

Insuficiente coordinación entre los sectores público y privado.

Falta de experiencia concreta con distintas tecnologías.

Inexistencia de un marco regulatorio específico para la bioenergía.

Aspectos regulatorios ambientales insuficientes y escaso control.

Insuficiente infraestructura institucional para la generación de capacidades técnicas.

Existencia de vacíos de información de base y difícil acceso a la información disponible.

AMENAZAS

Insuficiente coordinación de políticas sectoriales.

El productor agropecuario considera a la generación de energía a partir de biomasa como un tema ajeno a su actividad.

Mercado de la biomasa muy poco desarrollado.

Dificultades en el acceso al financiamiento para la incorporación de nuevas tecnologías.

Necesidad de internalizar beneficios ambientales y sociales de la bioenergía.

CONCLUSIONES

- ✓ El sector agropecuario y agroindustrial puede aprovechar el contexto actual para incorporar innovaciones tecnológicas que le permitan aprovechar residuos biomásicos para generación de energía, creando nuevas cadenas de valor, complementarias a las actuales.
- ✓ Deben superarse las barreras institucionales, consideradas el factor más limitante para el desarrollo sectorial.

CONCLUSIONES

- ✓ **Políticas recientemente asumidas ➔ resolución de muchas de las amenazas identificadas.**

...pero quedaría por abordar:

- ¿Qué actor/es de la/s cadena/s de valor debe/n asumir la internalización de los beneficios ambientales y sociales de la bioenergía?
- ¿Cómo lograr que los productores agropecuarios y agroindustriales asuman a la bioenergía como un tema propio a su actividad productiva?

CONCLUSIONES

✓ **Necesidad: generar instrumentos y políticas de desarrollo sectoriales de largo plazo y sostenibles**

- Diseñar estrategias que permitan:
  - Profundizar vínculos interinstitucionales.
  - Integrar iniciativas en base a una visión de desarrollo común y acorde a las prioridades de Gobierno, alineando con objetivos y necesidades territoriales.
  - Adecuar la asignación de competencias en función de los mandatos y recursos de los distintos actores.

¡MUCHAS GRACIAS!

lucilagrassi@gmail.com

An overview of biogeochemical issues related to bioenergy production

Conference: RCN Conference on Pan American Biofuel and Bioenergy Sustainability

Year: 2016

Proceeding: 2nd RCN Conference on Pan American Biofuels & Bioenergy Sustainability

Group: Abstracts

Session: Session 6: Large Scale Production Case Studies - Biogeochemical Aspects/ Casos De Estudio En La producción En Gran Escala - Aspectos BiogeoquíMicos)

Time: Thursday, September 15, 2016 - 2:30pm-2:50pm

Authors:

Titus, B., Canadian Forest Service
Gollany, H., ARS
Wassner, D., Universidad de Buenos Aires

Biomass production for bioenergy can affect biogeochemical processes by increasing the demand on soil nutrients needed to sustain site productivity because of incrementally greater removals of organic matter than would occur for traditional crop management, which in turn affects soil organic matter and associated carbon and nutrient storage and cycling. Management for feedstocks can affect soil physical, chemical and biological properties, but impacts can be mitigated through appropriate management.

Biogeochemical cycles related to feedstock production are also affected by over-arching issues such as increased atmospheric CO_2 concentrations, climate change, atmospheric deposition of nutrients, possible future scarcity of minerals used in fertilizers, and land use and land cover changes associated with establishing crops that supply feedstocks for bioenergy; the effects of impacts associated with these issues can also be cumulative.

Feedstock production can also have off-site impacts that include GHG emissions, eutrophication, and disposal of by-products at the end of the life cycle. Off-site impacts are not all negative: ash can be used to replenish some soil nutrients, biochar can improve soil quality, and some bioenergy processes (e.g., anaerobic digestion of manures) can be used to reduce GHG emissions and nutrient loading of the environment.

We will review the various biogeochemical issues that are relevant to bioenergy feedstock production, and provide some thoughts on future directions.

PAN AMERICAN BIOFUELS AND BIOENERGY SUSTAINABILITY
AN NSF RESEARCH COORDINATION NETWORK

INTA

Dendroenergía. Consideraciones sobre los bosques cultivados de Argentina.

Ana Maria Lupi

Instituto de suelos

Centro Investigaciones en Recursos Naturales -INTA

Lupi.ana@inta.gob.ar

Roberto Fernández

EEA Montecarlo (INTA) - Facultad de Ciencias Forestales (UNaM)

Buenos Aires, 15 de septiembre de 2016

Estimaciones y consideraciones para forestaciones potencialmente aplicables

- Consideraciones sobre aspectos ambientales
- No incluyen los residuos de procesos industriales.
- No incluye aspectos sociales ni económicos.

INTA · Ministerio de Agroindustria Presidencia de la Nación · CONICET · FCF

13 unidades de INTA, MAGyP, CONICET y UNaM

Acuerdo INTA- FAO en apoyo a PROBIOMASA-ARG/UTF/020ARG

Frente al escenario actual de CC , el consumo de la energía proveniente de fuentes no renovables es insostenible.

Es necesario trabajar en procesos que racionalicen y disminuyan la demanda de energía y en el desarrollo de fuentes de alternativas.

BIOENERGÍA

Forma de energía renovable que se genera a partir de materiales derivados de fuentes biológicas.

(biomasa de productos y residuos de las actividades agrícolas y forestales, de sus procesos de industrialización; de desechos humanos y animales, etc.)

PLANTACIONES DENDROENERGÉTICAS.

PLANTACIONES DENDROENERGÉTICAS

Objetivo: generar cantidades elevadas de biomasa, sin requerimientos especiales de calidad, en rotaciones de corta duración.

Características:

Especies de rápido crecimiento,

Altas densidades,

Elevado potencial de productividad en biomasa y,

Capacidad de rebrotar.

Productos de bajo precio, requieren minimizar costos de producción y de transporte -cuencas de abastecimiento-.

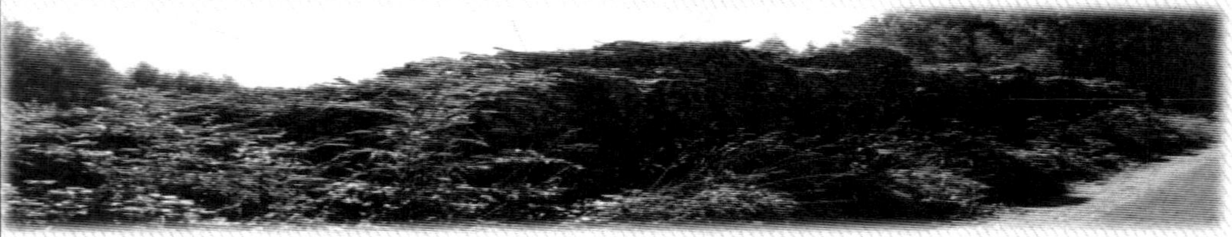

INFORMACIÓN DE BASE

- Plantaciones forestales de uso y difusión actual a nivel nacional.
- La información corresponde a especies cuyas propiedades tecnológicas y objetivo de producción se asocia a la obtención de madera de calidad o bien a producción de fibra.
- Las estimaciones realizadas provienen de simulaciones en base a antecedentes o modelos ajustados para objetivos de producción mencionados en el punto anterior.

ÁREAS ELEGIBLES-RESGUARDOS

• Resguardo por normas legales.

De nivel nacional y provincial, referidas a la conservación de los recursos naturales y al mantenimiento de los servicios ambientales del ecosistema.

Ley de Presupuestos Mínimos de Protección Ambiental -26.331- Áreas que deben ser conservadas. 10 mill. ha
Leyes provinciales de Ordenamiento Territorial.
Sitios RAMSAR,
Áreas con especies protegidas;
Áreas con vegetación protectora en márgenes de arroyos, bosques en galería, humedales; etc.

- Resguardo Climático-Hídrico

Precipitación media

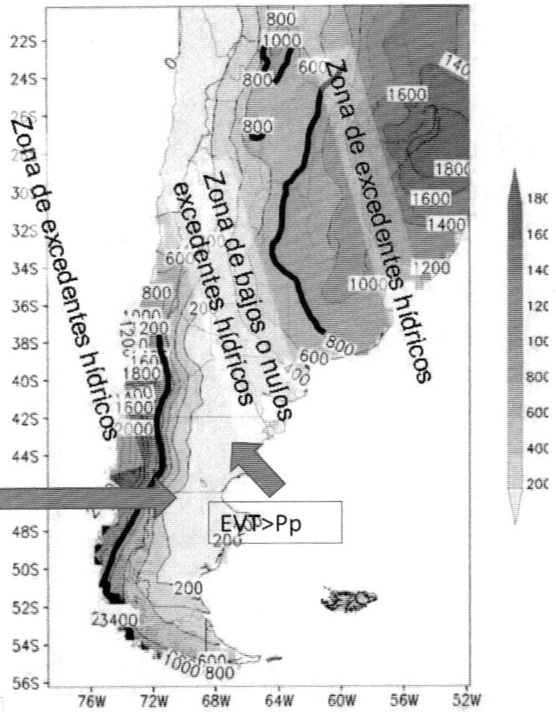

El agua podría limitar la productividad

-Uso de agua profunda
-Riego

Salinización secundaria

Plantaciones a altas densidades pueden generar conflictos por el consumo de agua y procesos de salinización secundaria.

> 800 mm anuales, con balance hídrico positivo

24% del país

Áreas Posibles

< 800 mm anuales, con balance hídrico neutro o negativo

76% del país

Áreas No recomendadas

Gyenge, Keller, Alvarez, Fernandez

• Resguardo por capacidad de uso de los suelos

Tierras Clase VIII

deben ser excluidas para establecer cultivos con fines energéticos

"tierras con limitaciones muy fuertes a extremadamente severas para uso y manejo de los suelos y no cuentan con las condiciones mínimas para uso agropecuario; por lo tanto, se deben dedicar a la conservación de los recursos naturales. La mayoría de estas tierras son importantes para la protección de los recursos hídricos, para refugio de la fauna, conservación de la flora, turismo ecológico y recuperación de los suelos"

18 ecorregiones

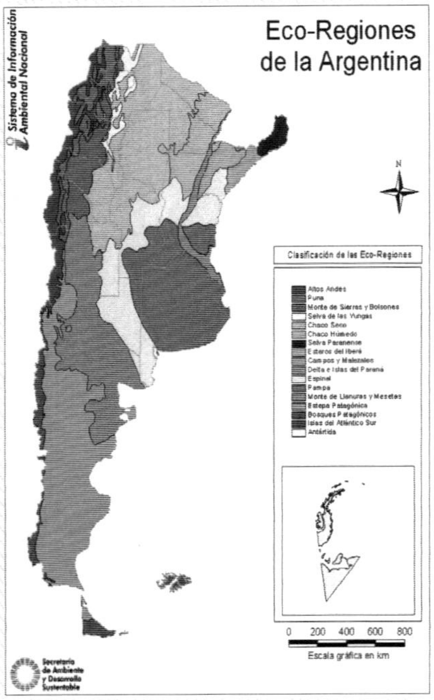

Bosques cultivados 1.120.411 ha

83 % en Mesopotamia y Delta.

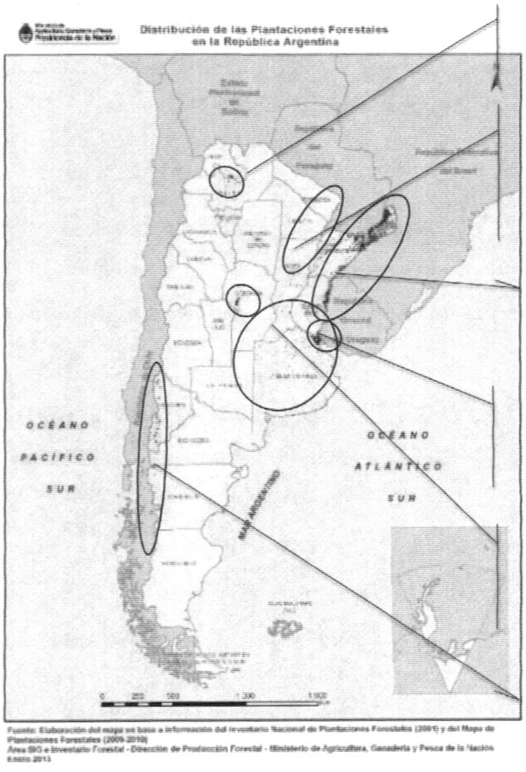

NOA:
Jujuy, Salta y Tucumán (parciales)

Central:
CE Chaco y Formosa. N Santa Fe. Valles de la región serrana y áreas S-SO Córdoba.

Mesopotamia:
Misiones, Corrientes y Entre Ríos

Delta:
islas del NE de Buenos Aires y S Entre Ríos.

Pampeana:
Buenos Aires, CS Santa Fe y SE Córdoba.

Patagonia Andina:
Neuquén, Río Negro, Chubut, y Santa Cruz.

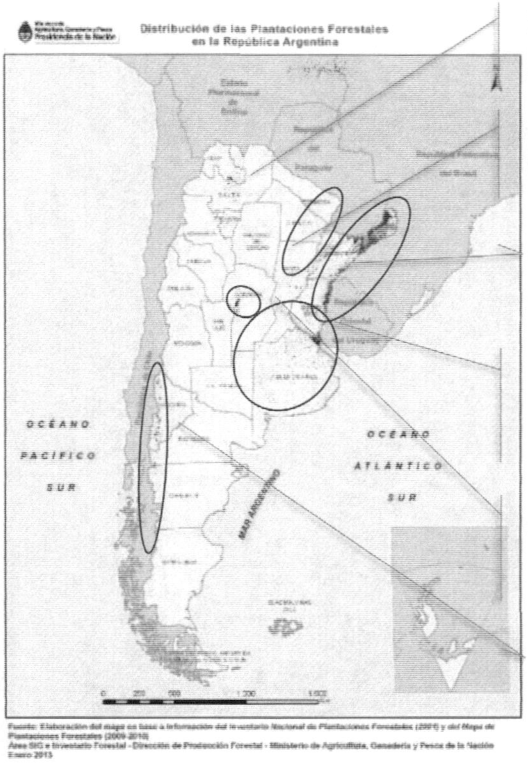

NOA.
Eucalyptus grandis , E. camaldulensis

Central *E. grandis, E. camaldulensis, E. viminalis y E. tereticornis; Populus*

Mesopotamia: *Pinus taeda, Eucalyptus grandis, e híbridos x tereticornis y x E. camaldulensis*

Delta: *P. deltoides y P. deltoides x canadensis*
Salix babylonica, S. matsudana x S. alba,
S. nigra

Pampeana: E. camaldulensis (NE Bs. As. y SO y N de Bs. As, Sur de Santa Fe y SE de Córdoba-). E. viminalis (-SO y N de Bs. As)

Patagonia: Pino Ponderosa (Neuquén, Río Negro y Chubut). Nothofagus antartica- Ñire- (Santa Cruz).

REGION MESOPOTAMICA.

	Densidad inicial (pl/ha)	Edad de corta (años)	biomasa (Mg/ha)
P. taeda. Suelos bien drenados y con drenaje moderado de Misiones y Corrientes.	2500	9-11	100-260
E. grandis. Suelos rojos arcillosos de Misiones y noreste de Corrientes y pedregosos de Misiones	2500	5-6	150-230
E. grandis. Lomadas y planicies arenosas del centro-oste de Corrientes y terrazas del Río Uruguay	2000 - 3000	4-7	95-200
E. grandis. Suelos mestizos y arcillosos de Entre Ríos y malezales de Corrientes.	2500	6-7	100

Fassola, Crechi, Keller, Caniza, Mastrandrea, Aparicio, Lopez, Gauchat, Von Wallis, Phar, Fernandez, Oberschelp, Harrand, Garcia, Licata, Tesón, De la Peña, Barth, Bedendo, Flores., Wink, Verdoljak.

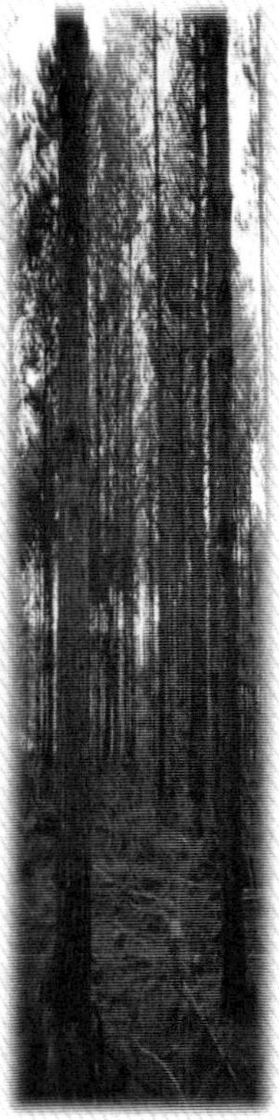

REGION PATAGONIA ANDINA.

	Densidad Inicial (pl./ha)	Edad de corta (años)	Biomasa (Mg/ha)
P. Ponderosa en sitios de media a alta calidad	7.500	23-28	260
ñire, *Nothofagus antártica en sitios de alta y baja calidad*	4.500	32-45	65-55

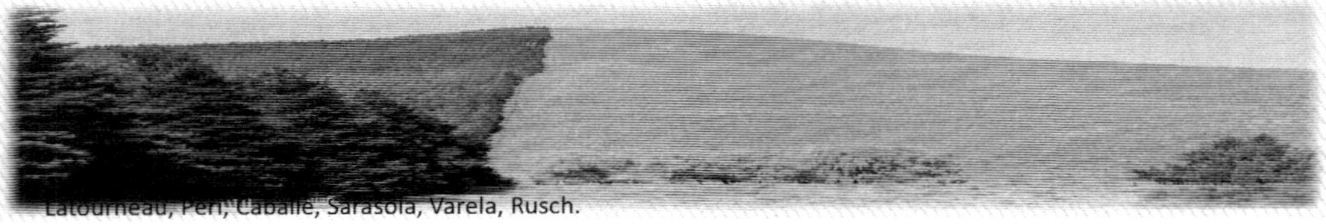

Latourneau, Peri, Caballe, Sarasola, Varela, Rusch.

REGION DELTA DEL PARANÁ.

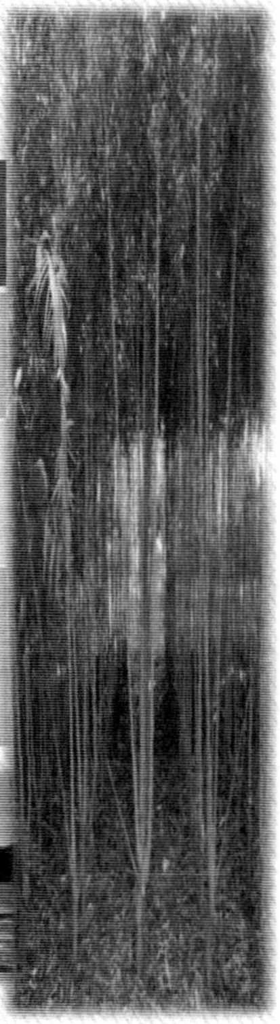

Distanciamiento (m)	Densidad (plantas/ha)	Biomasa (Mg/ha)
Estaqueros a 0,6 x 0,6 de *Populus deltoides.* 3 años	27.777	139
Estaqueros a 1,0 x 1,0 de *Populus deltoides.* 3 años	10.000	131
clones experimentales de *Salix* spp, estimado en estaquero. 3 años		64 – 165

Cerrillo, Cortizo, Alvarez, Garcia Cortez, Fracassi. Fernandez, Monteverde.

Menor desarrollo de la actividad forestal.

REGION NOROESTE: Información disponible de Producción de biomasa por clase de sitio para *E. grandis y parcial para E. camaldulensis.*

Fornes, Trápani, Speranza, Balduci

REGION PAMPEANA.

E. viminalis: Información parcial de producción de biomasa para en la pampa arenosa

E.camaldulensis: No se dispone estimación de biomasa. Escasa información sobre crecimiento

Salix y Populus en Pampa deprimida. No se dispone de información suficiente

Ferrere, Pathauer, Lupi, Fernandez, Gyenge, Dominguez Daguer, Signorelli

REGION CENTRAL.

E. viminalis y Populus, E. grandis y *E. camaldulensis* y *E. tereticornis*
No se dispone de información

Gomez, Zupan, Kees, Atanasio, Pernochi

SANIDAD FORESTAL. Posibles escenarios

Problemas intrínsecos

- Características de el sistema productivo (Monoespecífico, altas densidades, sin manejo)
- vulnerables a las plagas y enfermedades exóticas invasoras que existen para pinos, eucaliptos y salicáceas

Problemas extrínsecos

- Por condiciones microambientales actúan como fuente-sumidero de plagas y enfermedades.
- incertidumbre y complejidad a los sistemas de manejo sanitario vigentes, uno de cuyos objetivos es minimizar la dispersión de las plagas-enfermedades.

Botto, Eskivisky, Fernandez, Ramos, Landi, Mema, Braccini.

VIABILIDAD AMBIENTAL
Relaciones armónicas entre la productividad forestal y el entorno ambiental
escala de paisaje: mant. de la biodiversidad, ciclo hidrológico, calidad del agua y calidad del suelo
escala de pedio: mantenimiento de la capacidad productiva del suelo
Planificación del uso del suelo.
Áreas de preservación, de amortiguación, sitios aptos para la producción (potencial productivo, sensibilidad y requerimientos de uso y manejo).

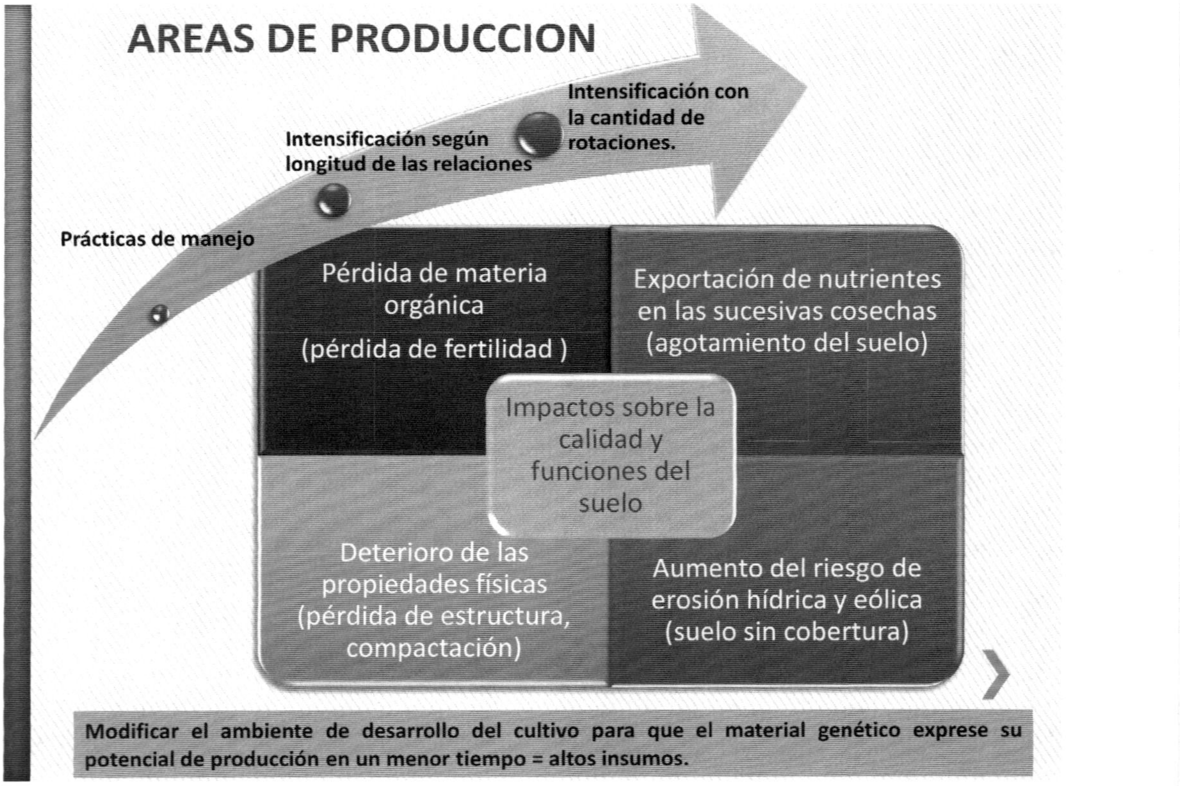
AREAS DE PRODUCCION
Prácticas de manejo
Intensificación según longitud de las relaciones
Intensificación con la cantidad de rotaciones.
Pérdida de materia orgánica (pérdida de fertilidad)
Exportación de nutrientes en las sucesivas cosechas (agotamiento del suelo)
Impactos sobre la calidad y funciones del suelo
Deterioro de las propiedades físicas (pérdida de estructura, compactación)
Aumento del riesgo de erosión hídrica y eólica (suelo sin cobertura)
Modificar el ambiente de desarrollo del cultivo para que el material genético exprese su potencial de producción en un menor tiempo = altos insumos.

IMPACTOS SOBRE RECURSOS HÍDRICOS.

Escasez de datos sobre flujos de agua y consumo en forestaciones para las distintas regiones

Aspecto relevante en regiones con balance hídrico cercano al neutro, o bien en cuencas que abastecen de agua a ciudades o pobladores rurales.

El impacto de forestaciones dependerá del tamaño de la cuenca hidrológica: > probabilidad en cuencas pequeñas-

El impacto dependerá de la posición en el relieve: > en zonas de acumulación y drenaje que en áreas de recarga.

En especies de rápido crecimiento, manejadas a alta densidad, el consumo de agua a nivel de microcuenca puede resultar crítico.

Planificar la proporción de la cuenca pasible de ser destinada a este tipo de plantaciones.

IMPACTOS SOBRE LA BIODIVERSIDAD

Sistema intensivo y simplificado

controlan *per se* la diversidad biológica interna

Cambio Climático

fragmentación y degradación de los hábitats

La magnitud es mayor en la medida que se incremente la superficie plantada y no se planifique la localización de los rodales en el paisaje.

Rusch, Fracassi.

Ambientes Prioritarios de conservación

Especies prioritarias (en peligro)

Áreas de valor para la biodiversidad en Mesopotamia (Zurita y col 2015, preliminar).

Áreas de Valor de la region Chaqueña, Proyungas,2014.

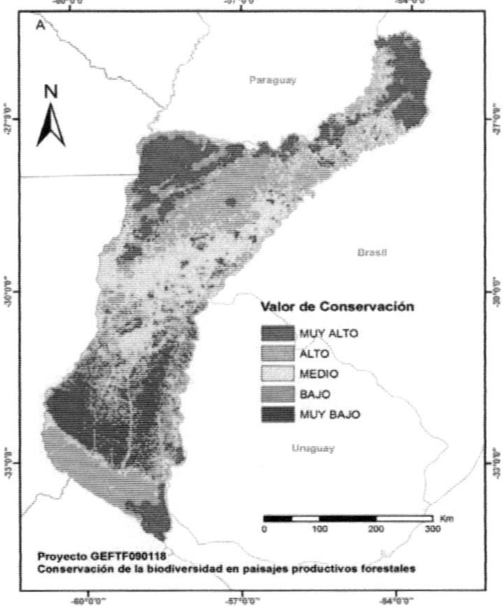

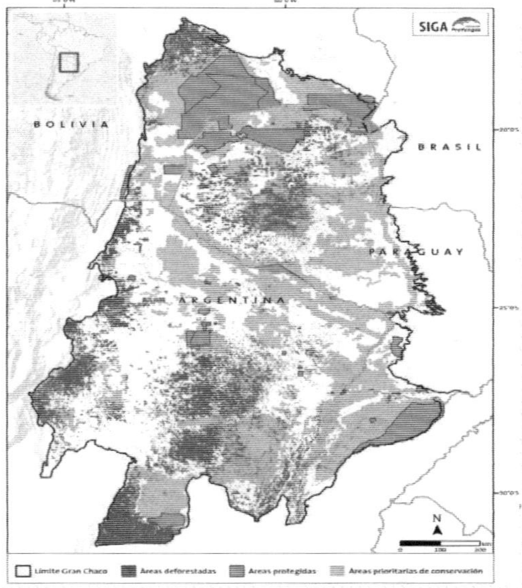

CONSIDERACIONES FINALES

- 6 regiones con potencial de producción de biomasa para energía: Mesopotamia, Pampeana, Central, Noroeste, Delta, Patagonia Andina.
- Los eucaliptos son propuestos como elegibles en 5 de las 6 regiones, clones de *Salicaceas* en 3, y especies de *Pinus* en 2 de ellas.
- Ventaja del empleo de materiales de *Eucalyptus*, *Salix* y *Populus* (manejar los rebrotes, > crecimiento, disminuyendo de esta forma los costos necesarios para contar con otro ciclo productivo).

CONSIDERACIONES FINALES

- Tecnología insuficiente y baja a nula disponibilidad de información (materiales genéticos apropiados, arquitectura de las plantaciones, maquinaria, transporte)
- No se dispone a nivel nacional información sobre producción de biomasa en esquemas de muy altas densidades.
- Son recientes las líneas experimentales en desarrollo particularmente en Delta, Corrientes y NOA.
- Las información suministrada corresponde a especies que no son necesariamente eficientes desde el punto de vista energético.

CONSIDERACIONES FINALES

» Es necesario complementar con balance energético, la evaluación de las huellas de carbono y de agua, y el resultado económico incorporando los costos ambientales "ocultos".

» Es necesario evaluar la liberación de CO_2 durante los procesos del cultivo forestal, del transporte de la madera y de su transformación industrial, ya que pueden implicar altos requerimientos de combustibles fósiles y de sus derivados.

LA ESCALA DEL PROYECTO…

………UN FACTOR DETEMINANATE EN LA MAGNITUD DE LOS IMPACTOS

- Panificar e implementar con una actitud precautoria respecto del cuidado del ambiente.
  - Plantaciones planificadas para responder a demandas de energía pequeñas y mediana industrias locales.
- Los rodales a nivel paisaje se deben implementar con un patrón de distribución disperso y tamaño reducido.
  - Aplicación de buenas prácticas de manejo

SOBRE LA BIODIVERSIDAD

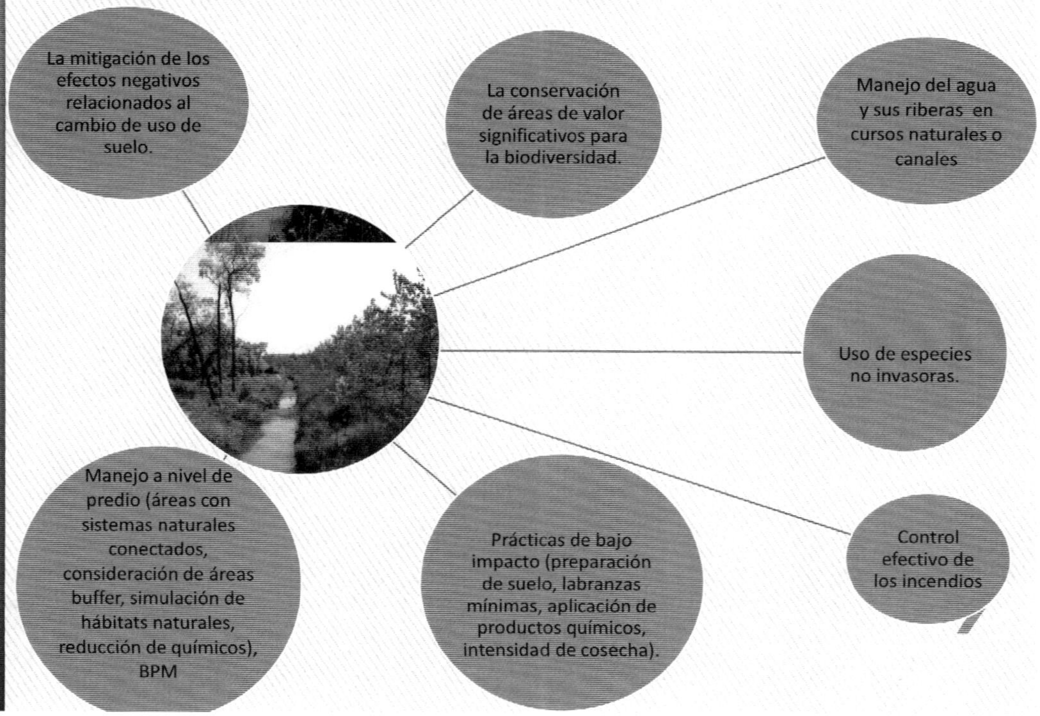

Problemas y riesgos para la sostenibilidad ambiental y productiva

1-Sanitarios.

2- Riesgo hídrico por consumo excesivo. El consumo de agua a nivel de cuenca aumenta cuando las forestaciones en grandes superficies reemplazan pastizales naturales.

3- Erosión hídrica en sectores en relieves montañosos

4- Riesgo de erosión eólica

5- Mayores niveles de exportación de nutrientes en relación a producción de madera de calidad y triturado.

6- Salinización en determinados ambientes

7- Degradación física. Compactación superficial y subsuperficial por ingreso de máquina.

8- Degradación química y de fertilidad

9- Aumento del riesgo de Incendios.

10-Pérdida de especies endémicas de humedales (conectividad)

12-Indispensable e manejo del agua(ingreso-egreso)

»Gracias

Facultad de Ciencias Forestales, Universidad de Concepción

Universidad de Concepción — Facultad de Ciencias Forestales

FOREST PRODUCTIVITY COOPERATIVE

FONDECYT
Fondo Nacional de Desarrollo Científico y Tecnológico

PROYECTO FONDECYT 1140482
"Sustainability of water and nutrient use in short rotation woody crops for bioenergy"

EVALUATION OF SUSTAINABILITY OF DENDROENERGETIC PLANTATIONS IN THE BIOBÍO REGION, CHILE

Rafael Rubilar*, Eduardo Acuña, Edwin Esquivel, Pablo Mena, Tim Albaugh, Chris Maier

2nd RCN Conference on Pan American Biofuels & Bioenergy Sustainability
Buenos Aires, September 13-16, 2016.

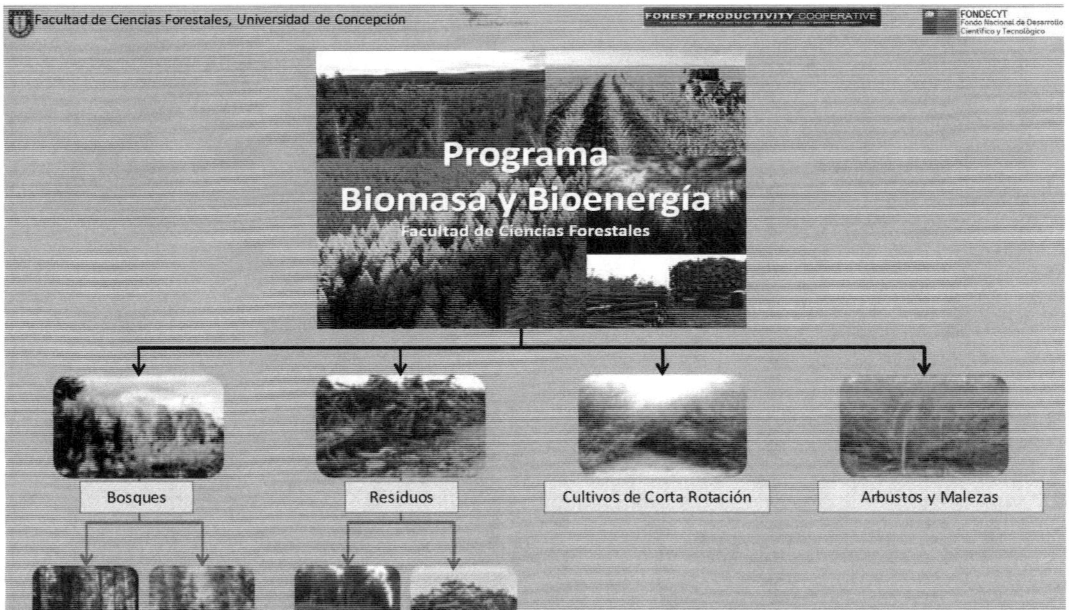

Facultad de Ciencias Forestales, Universidad de Concepción

FOREST PRODUCTIVITY COOPERATIVE

FONDECYT Fondo Nacional de Desarrollo Científico y Tecnológico

Bioenergía Cultivos de Corta Rotación

Escenario

- Agencia Internacional de Energía 25% de la energía primaria de transporte en bioenergía (Smeets, Faaij, & Lewandowski, 2004).
- Chile importa 75% Energía y Biomasa=28% Energía primaria. Creciente demanda de biomasa en Chile (Calor + Pellets).
- Libre de estacionalidad mercado.
- Producción de energía (biomasa/biocombustibles).
- Plan estratégico 2030 Estrategia Dendroenergía (MINAGRI-CONAF, 2015) = 182.000 ha al 2030!

Cultivo

- Altas densidades (> 2500 arboles ha^{-1}).
- Rotaciones cortas (2-4 años).
- Especies de rápido crecimiento.
- Capacidad de rebrote.

https://portfolio.pebblepad.co.uk/cumbria/download.aspx?oid=50262&useroid=0 action=view

Facultad de Ciencias Forestales, Universidad de Concepción

FOREST PRODUCTIVITY COOPERATIVE

FONDECYT Fondo Nacional de Desarrollo Científico y Tecnológico

Conflicto de Uso de la Tierra Cultivos Agrícolas vs Plantaciones Dendroenergéticas

www.carbón-sense.com

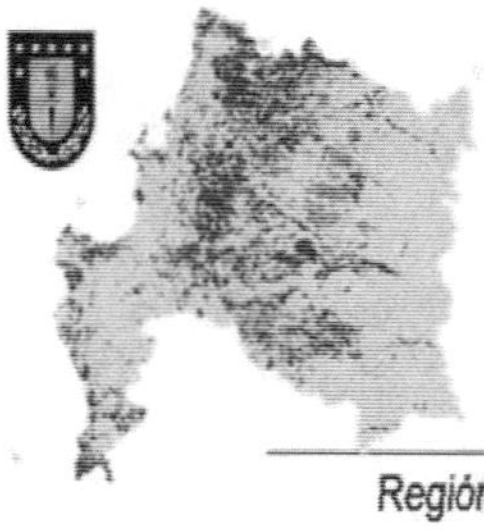

Superficie potencial para plantaciones forestales

Región	Superficie total (ha)[1]	Superficie plantada (ha)[1]	Superficie de uso forestal (ha)[2]	Superficie potencial (ha)[2]
del Libertador B.O.	1.638.700	90.216	593.789	237.516
del Maule	3.029.610	388.123	820.294	328.118
del Bío Bío	3.706.260	791.831	605.253	425.257
de la Araucanía	3.184.230	389.946	711.965	284.786
de los Lagos	6.701.310	208.825	1.783.200	713.262
Total	**18.260.110**	**1.868.941**	**4.514.501**	**1.988.939**

Fuente: [1] INFOR, 2005; [2] Fac. Cs. Forestales, 2005

Ciclo Silvícola de CCR para Biomasa (Bioenergía/Biocombustibles)

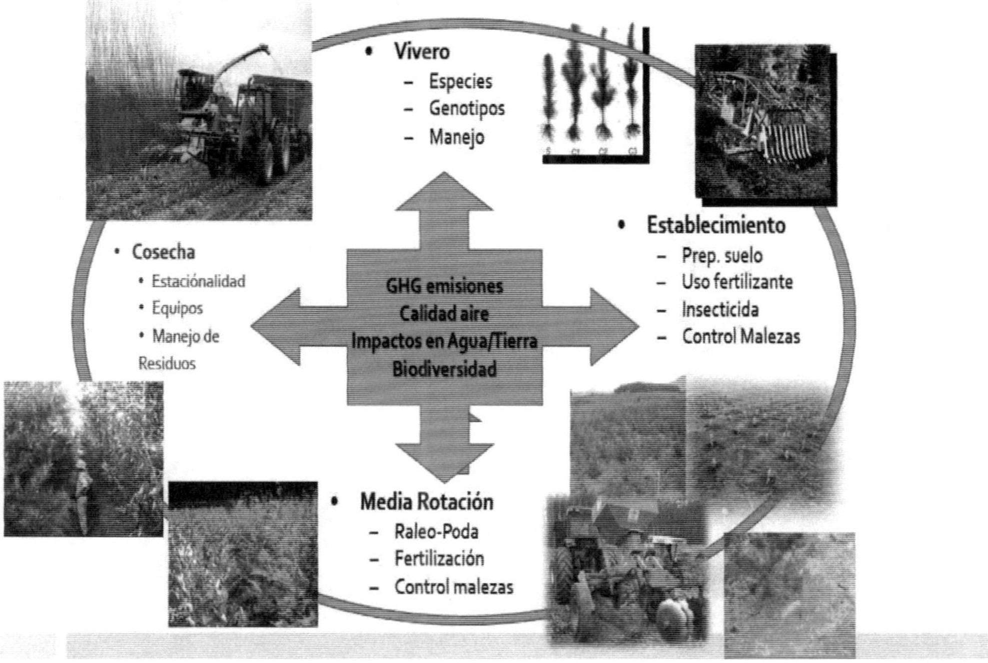

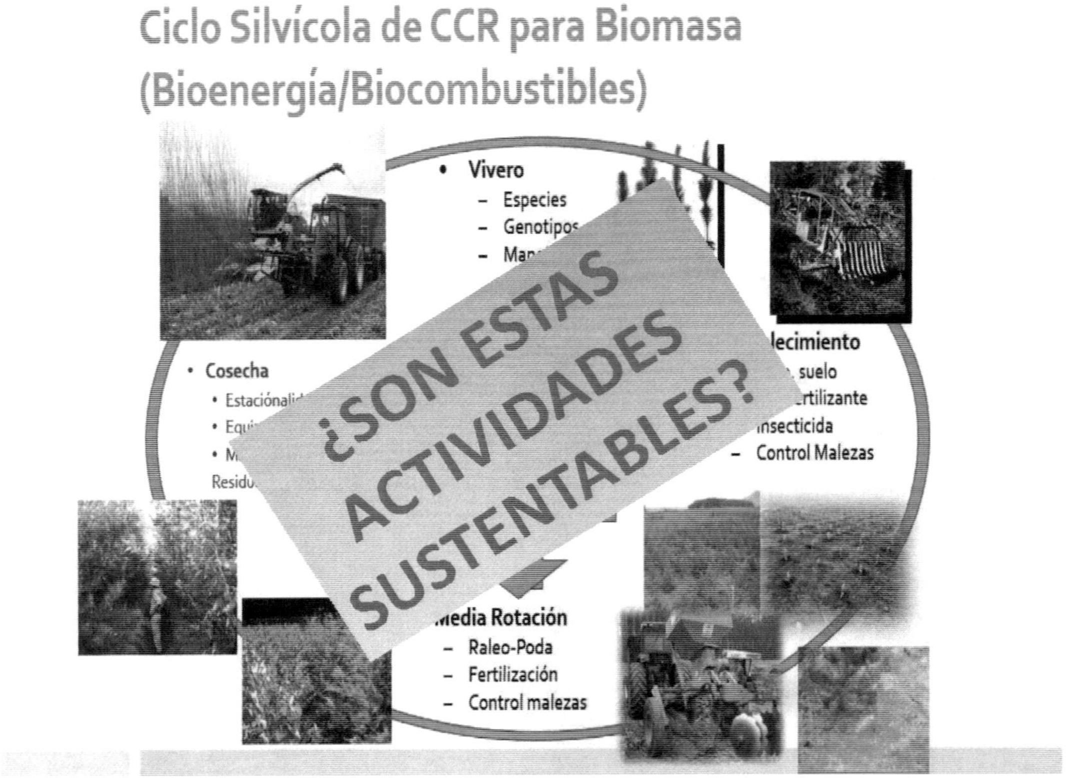
Ciclo Silvícola de CCR para Biomasa
(Bioenergía/Biocombustibles)
• Vivero
– Especies
• Cosecha
lecimiento
suelo
insecticida
– Control Malezas
¿SON ESTAS
ACTIVIDADES
SUSTENTABLES?
– Raleo-Poda
– Fertilización
– Control malezas

Facultad de Ciencias Forestales, Universidad de Concepción
FOREST PRODUCTIVITY COOPERATIVE
FONDECYT
Sitios Potenciales Plantaciones Dendroenergéticas
en Chile de Alta Frágilidad (Hídrica y Nutricional)
Fuente: Fac. Cs. Forestales, 2005
Water Deficit (mm)
-1276 - -1000
-1000 - -800
-800 - -600
-600 - -400
-400 - -200
-200 - 0

Sustentabilidad

La biomasa forestal es una fuente de energía renovable que debe ser manejada sustentablemente.

El mantenimiento del carbono sobre y bajo suelo, y el mantenimiento de la fertilidad del sitio se encuentran entre los principales evaluadores de la sustentabilidad.

El efecto en los recursos hídricos es uno de los principales desafíos en el manejo de cultivos dendroenergéticos.

The Challenge of Lignocellulosic Bioenergy in a Water-Limited World

JOHN S. KING, REINHART CEULEMANS, JANINE M. ALBAUGH, SOPHIE Y. DILLEN, JEAN-CHRISTOPHE DOMEC, REGIS FICHOT, MILAN FISCHER, ZAKIYA LEGGETT, ERIC SUCRE, MIREK TRNKA, AND TERENZIO ZENONE

It is hoped that lignocellulosic sources will provide energy security, offset carbon dioxide enrichment of the atmosphere, and stimulate the development of new economic sectors. However, little is known about the productivity and sustainability of plant cell-wall energy industries. In this study, we used 16 global circulation models to project the global distribution of relative water availability in the coming decades and summarized the available data on the water-use efficiency of tree- and grass-based bioenergy systems. The data on bioenergy water use were extremely limited. Productivity was strongly correlated with water-use efficiency, with C_4 grasses having a distinct advantage in this regard. Our analysis of agroclimatic drivers of bioenergy productivity suggests that relative water availability will be one of the most important climatic changes to consider in the design of bioenergy systems.

(Richardson & Verwijst 2007, Karp y Shield 2008, King et al., 2013)

Facultad de Ciencias Forestales, Universidad de Concepción
FOREST PRODUCTIVITY COOPERATIVE
FONDECYT Fondo Nacional de Desarrollo Científico y Tecnológico

Sustentabilidad Productiva (Nutricional e Hídrica)

- ¿ Cual es el efecto en los pooles de C del suelo?
- ¿Cuál es la magnitud de las remociones de nutrientes asociados a la cosecha de cultivos dendroenergéticos?
- ¿Cuál es el impacto potencial de estas plantaciones dendroenergéticas en la disponibilidad del recurso hídrico?
- ¿Existen diferencias relevantes entre especies, densidades de cultivo y sitios que permitan definir estrategias de sustentabilidad?

Facultad de Ciencias Forestales, Universidad de Concepción

FOREST PRODUCTIVITY COOPERATIVE

FONDECYT Fondo Nacional de Desarrollo Científico y Tecnológico

Estudios Evaluados

- **Sitios** de **Marginales** "Baja Disponibilidad de Recursos" (700 mm, < 1% MO) vs **Fértiles** – "Alta Disponibilidad de Recursos" (>1500 mm, >5% MO).
- **Manejo:** Manejo Extensivo vs. Manejo Intensivo (Fertirrigación).
- **Especies Principales:** Eucalyptus globulus, Eucalyptus camaldulensis, Eucalyptus nitens, Acacia melanoxylon, Acacia dealbata, Populus americana.
- **Densidades** 5000, 10000, 15000 plantas por hectárea.
- **Evaluaciones:** Biomasa producida, Nutrientes Acumulados, Efectos en C del Suelo, Efectos de Disponibilidad de nutrientes, Potencial impacto en recursos hídricos.

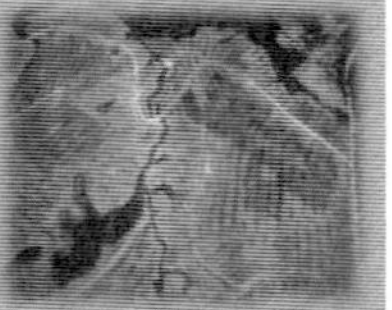

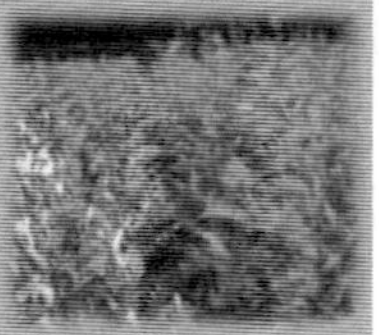

Facultad de Ciencias Forestales, Universidad de Concepción

FOREST PRODUCTIVITY COOPERATIVE

FONDECYT Fondo Nacional de Desarrollo Científico y Tecnológico

Sitios de Estudio Marginales (Manejo Extensivo)

Xeralfs, 800 mm Pp

Psaments, 1100 mm Pp

Facultad de Ciencias Forestales, Universidad de Concepción
FOREST PRODUCTIVITY COOPERATIVE
FONDECYT
Sitios de Estudio Alta Fertilidad & Fertirrigación
3-4 años desarrollo
Paleudand, 1500 mm
Psamments, 1200 mm + 600 mm + NPKmicro

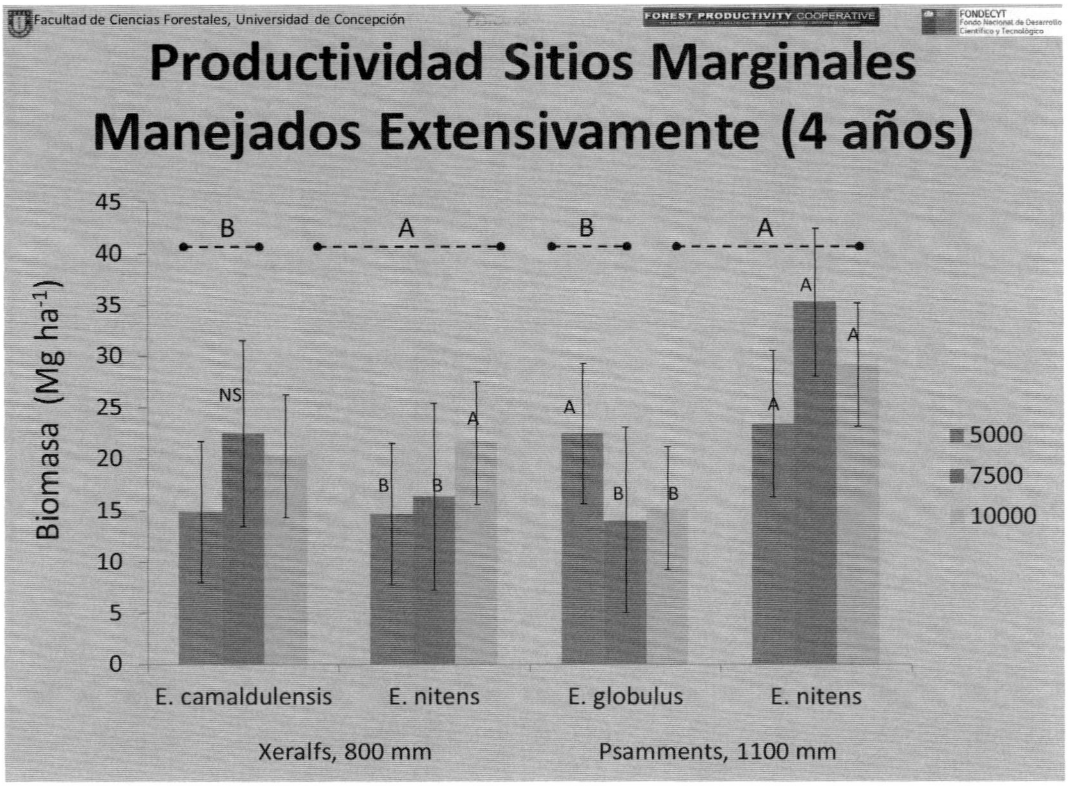
Facultad de Ciencias Forestales, Universidad de Concepción
FOREST PRODUCTIVITY COOPERATIVE
FONDECYT
Productividad Sitios Marginales
Manejados Extensivamente (4 años)
Biomasa (Mg ha-1)
45
40
35
30
25
20
15
10
5
0
B
A
B
A
NS
A
B
B
A
B
B
A
A
A
5000
7500
10000
E. camaldulensis
E. nitens
E. globulus
E. nitens
Xeralfs, 800 mm
Psamments, 1100 mm

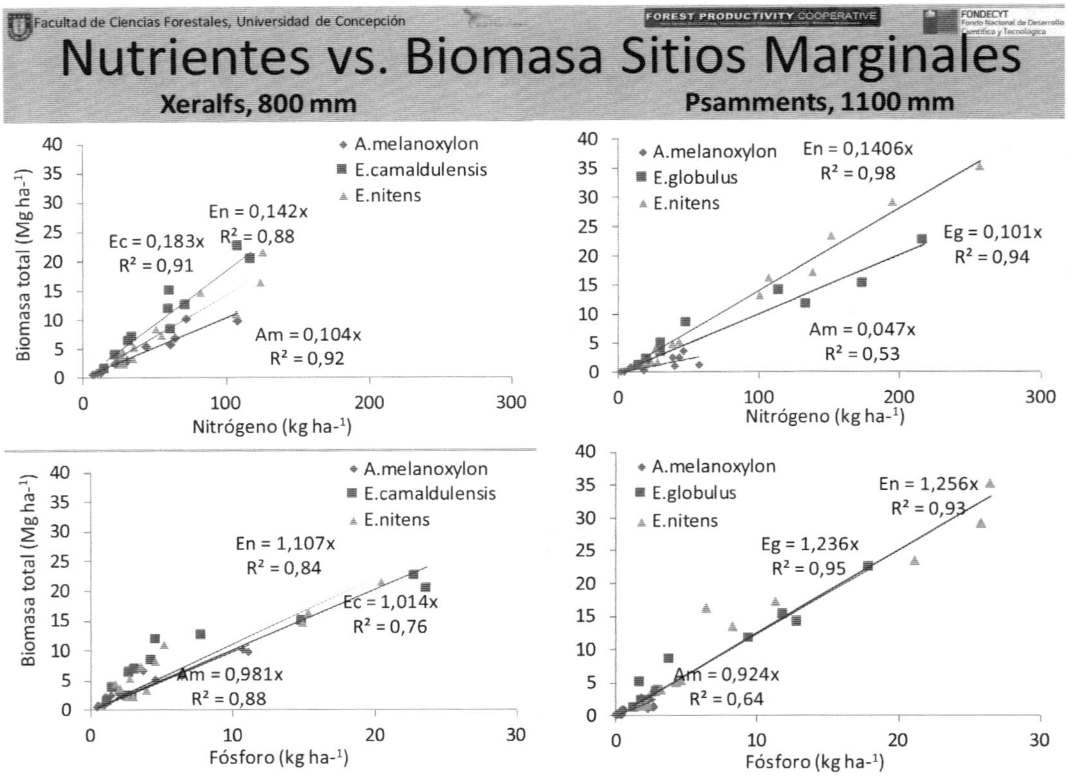

Facultad de Ciencias Forestales, Universidad de Concepción
FOREST PRODUCTIVITY COOPERATIVE
FONDECYT
Nutrientes vs. Biomasa Sitios Marginales
Xeralfs, 800 mm
Psamments, 1100 mm
A.melanoxylon
E.camaldulensis
E.nitens
En = 0,142x
R² = 0,88
Ec = 0,183x
R² = 0,91
Am = 0,104x
R² = 0,92
Biomasa total (Mg ha-1)
Nitrógeno (kg ha-1)
A.melanoxylon
E.globulus
E.nitens
En = 0,1406x
R² = 0,98
Eg = 0,101x
R² = 0,94
Am = 0,047x
R² = 0,53
Nitrógeno (kg ha-1)
A.melanoxylon
E.camaldulensis
E.nitens
En = 1,107x
R² = 0,84
Ec = 1,014x
R² = 0,76
Am = 0,981x
R² = 0,88
Biomasa total (Mg ha-1)
Fósforo (kg ha-1)
A.melanoxylon
E.globulus
E.nitens
En = 1,256x
R² = 0,93
Eg = 1,236x
R² = 0,95
Am = 0,924x
R² = 0,64
Fósforo (kg ha-1)

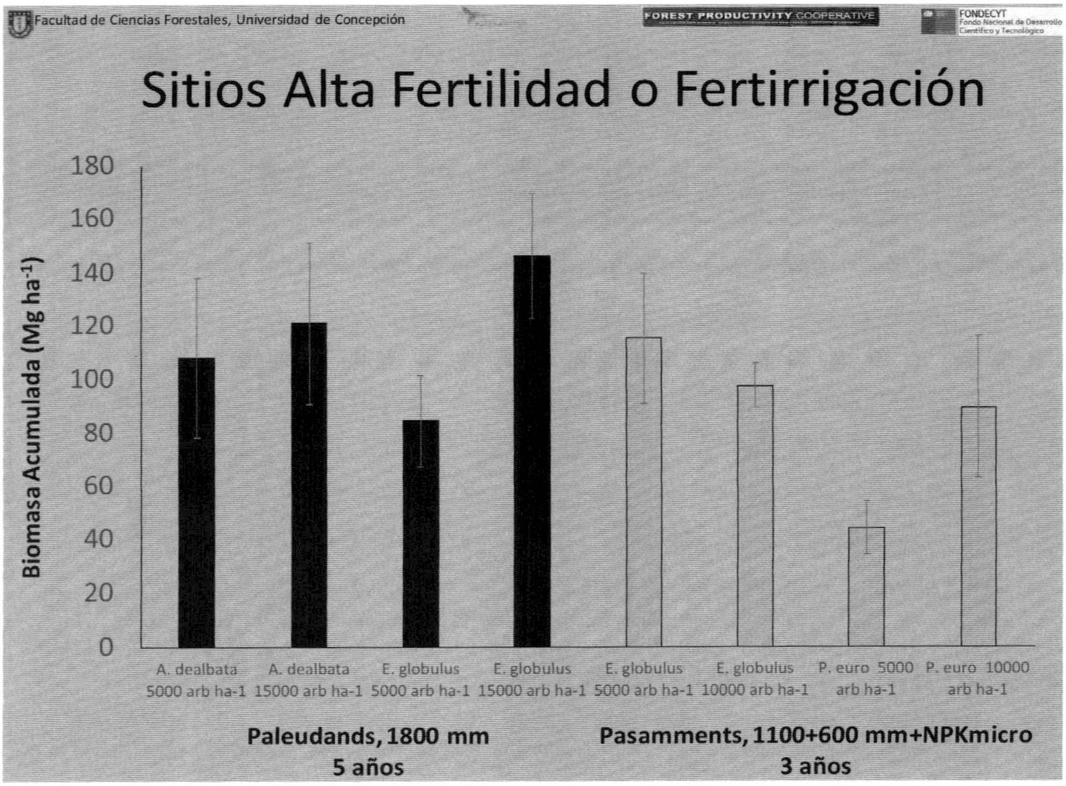

Facultad de Ciencias Forestales, Universidad de Concepción
FOREST PRODUCTIVITY COOPERATIVE
FONDECYT
Sitios Alta Fertilidad o Fertirrigación
Biomasa Acumulada (Mg ha-1)
A. dealbata 5000 arb ha-1
A. dealbata 15000 arb ha-1
E. globulus 5000 arb ha-1
E. globulus 15000 arb ha-1
E. globulus 5000 arb ha-1
E. globulus 10000 arb ha-1
P. euro 5000 arb ha-1
P. euro 10000 arb ha-1
Paleudands, 1800 mm
5 años
Pasamments, 1100+600 mm+NPKmicro
3 años

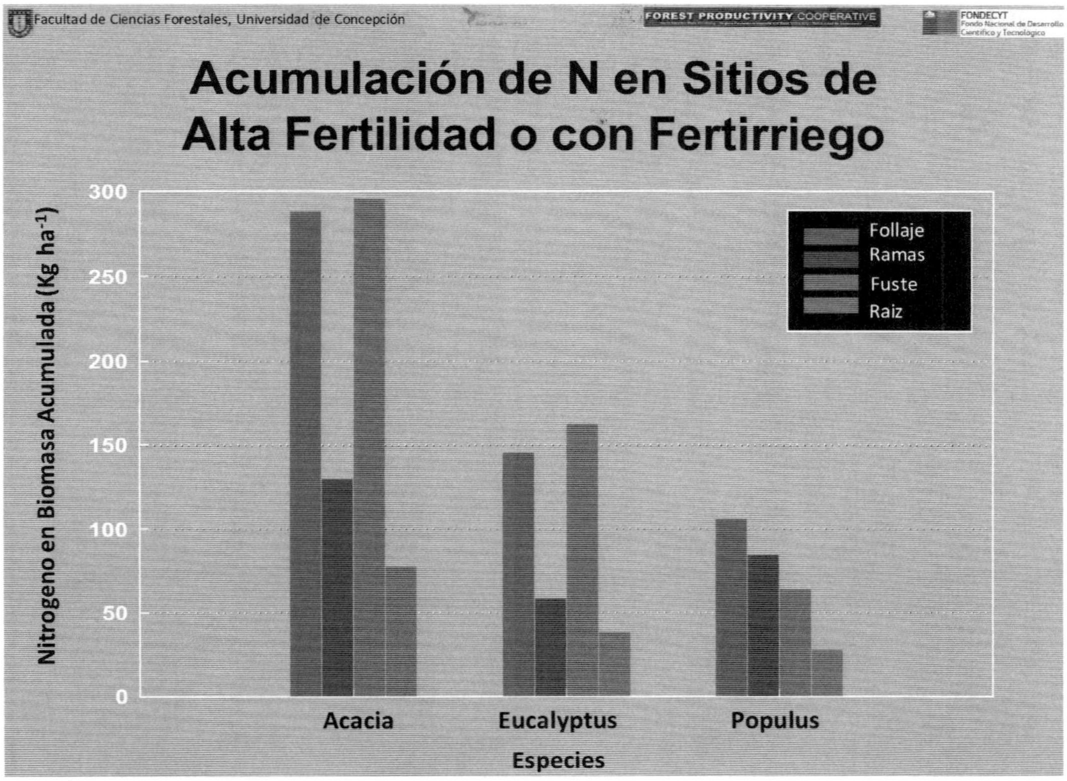
Facultad de Ciencias Forestales, Universidad de Concepción
FOREST PRODUCTIVITY COOPERATIVE
FONDECYT
Acumulación de N en Sitios de Alta Fertilidad o con Fertirriego
Nitrogeno en Biomasa Acumulada (Kg ha-1)
300
250
200
150
100
50
0
Follaje
Ramas
Fuste
Raiz
Acacia
Eucalyptus
Populus
Especies

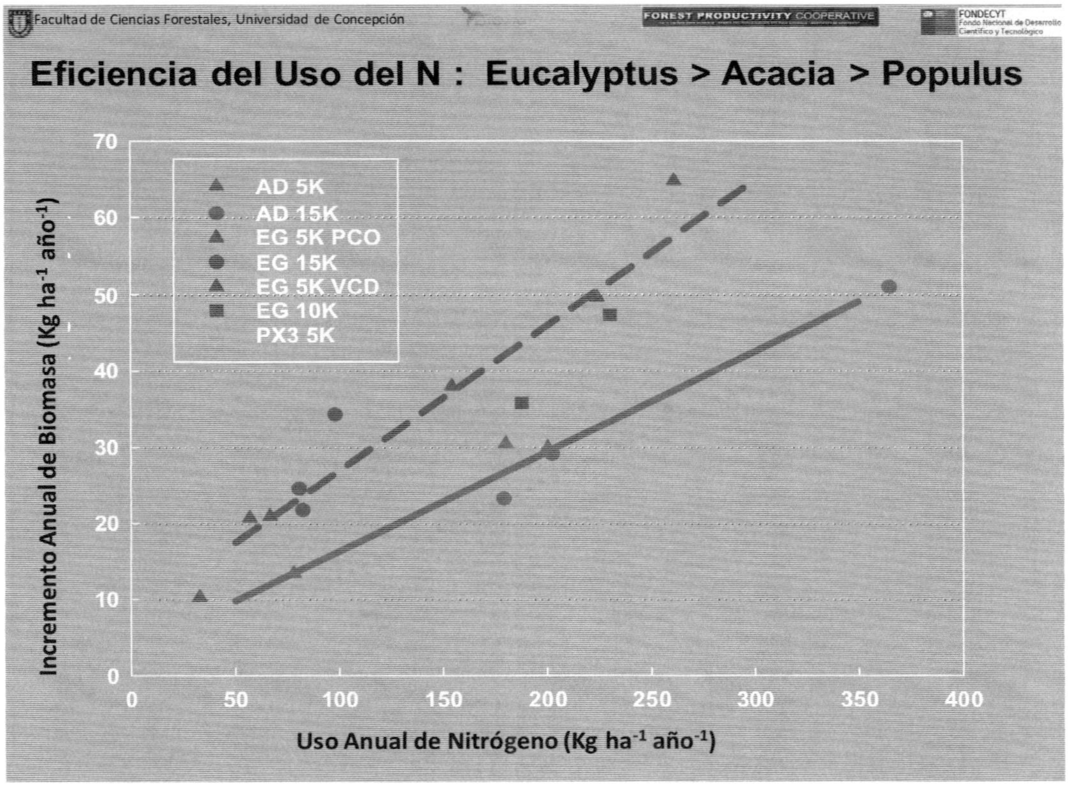
Facultad de Ciencias Forestales, Universidad de Concepción
FOREST PRODUCTIVITY COOPERATIVE
FONDECYT
Eficiencia del Uso del N : Eucalyptus > Acacia > Populus
Incremento Anual de Biomasa (Kg ha-1 año-1)
70
60
50
40
30
20
10
0
AD 5K
AD 15K
EG 5K PCO
EG 15K
EG 5K VCD
EG 10K
PX3 5K
0
50
100
150
200
250
300
350
400
Uso Anual de Nitrógeno (Kg ha-1 año-1)

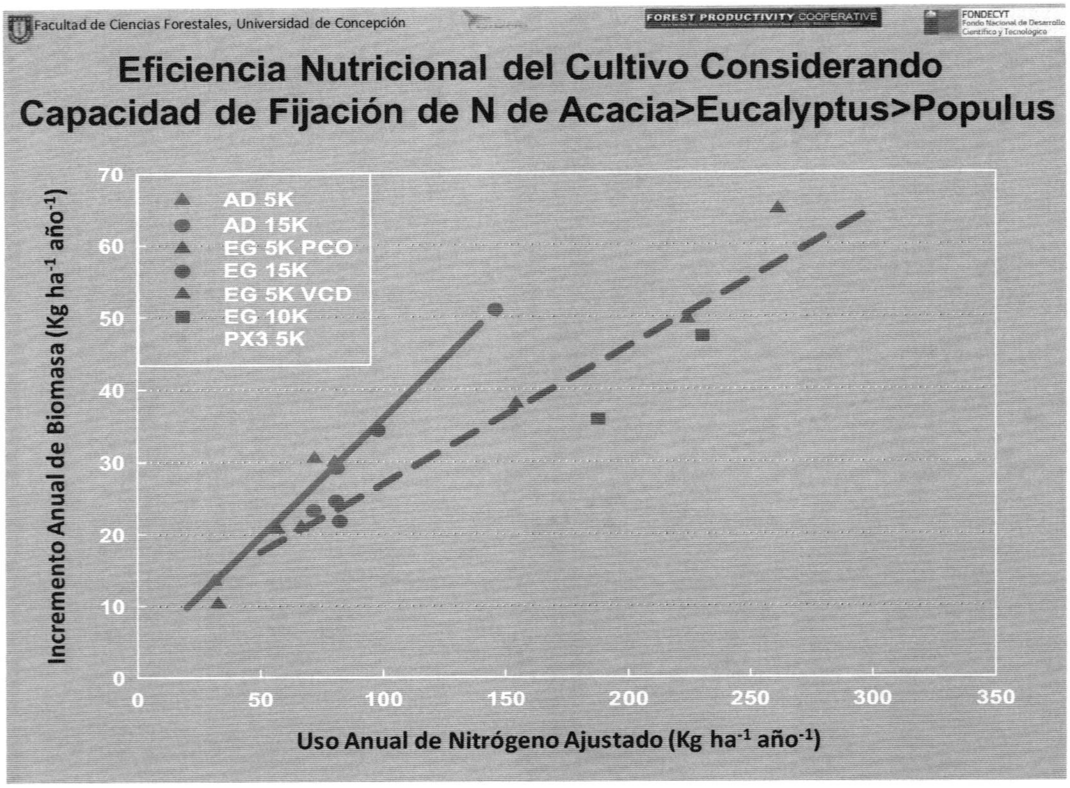

Facultad de Ciencias Forestales, Universidad de Concepción
FOREST PRODUCTIVITY COOPERATIVE
FONDECYT
Eficiencia Nutricional del Cultivo Considerando Capacidad de Fijación de N de Acacia>Eucalyptus>Populus
AD 5K
AD 15K
EG 5K PCO
EG 15K
EG 5K VCD
EG 10K
PX3 5K
Incremento Anual de Biomasa (Kg ha-1 año-1)
Uso Anual de Nitrógeno Ajustado (Kg ha-1 año-1)
0
10
20
30
40
50
60
70
0
50
100
150
200
250
300
350

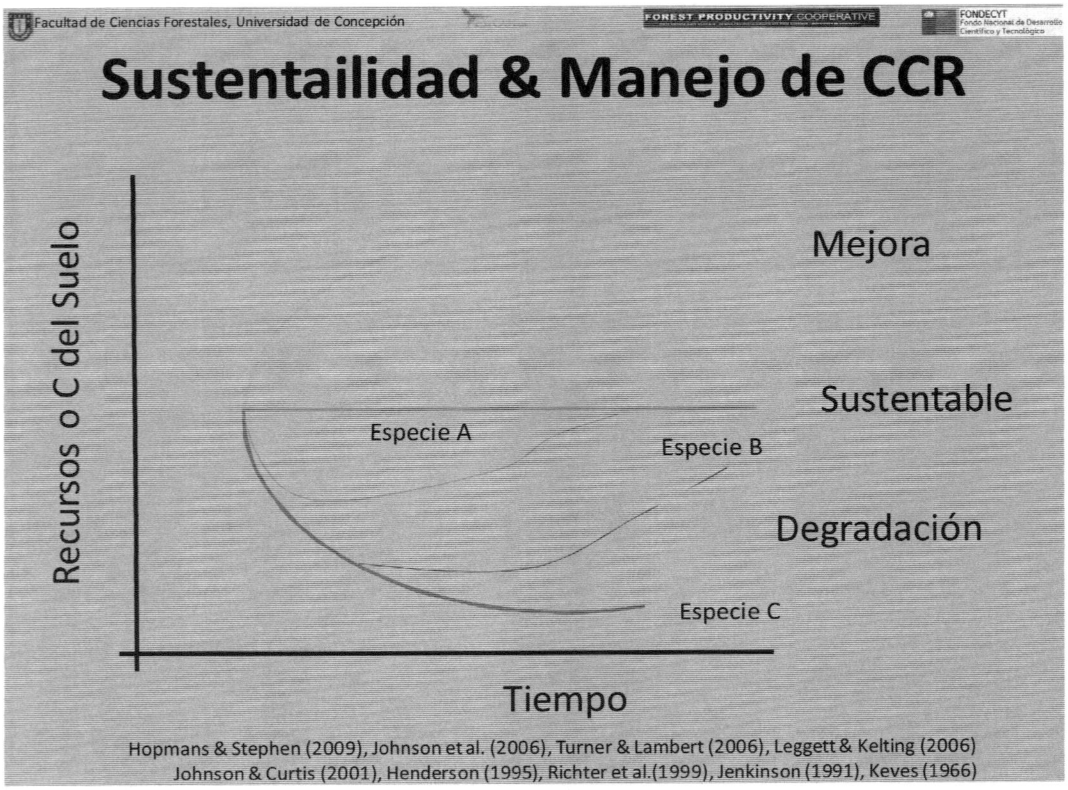

Facultad de Ciencias Forestales, Universidad de Concepción
FOREST PRODUCTIVITY COOPERATIVE
FONDECYT
Sustentailidad & Manejo de CCR
Recursos o C del Suelo
Mejora
Sustentable
Especie A
Especie B
Degradación
Especie C
Tiempo
Hopmans & Stephen (2009), Johnson et al. (2006), Turner & Lambert (2006), Leggett & Kelting (2006)
Johnson & Curtis (2001), Henderson (1995), Richter et al.(1999), Jenkinson (1991), Keves (1966)

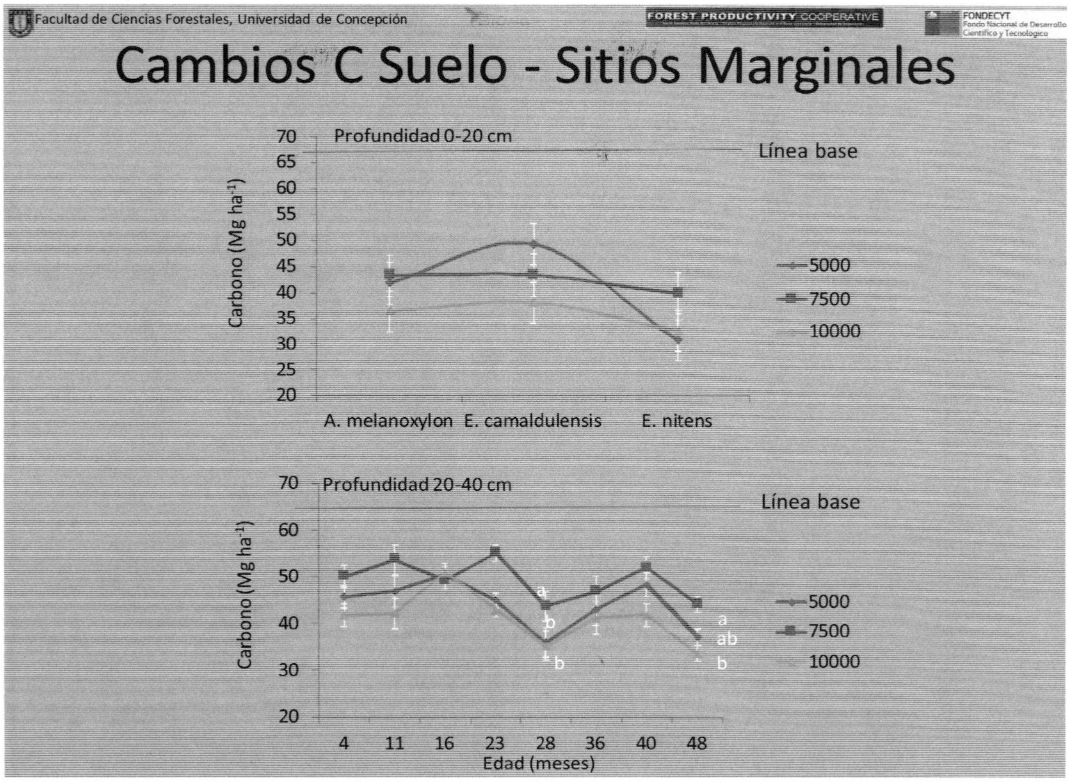
Facultad de Ciencias Forestales, Universidad de Concepción
FOREST PRODUCTIVITY COOPERATIVE
FONDECYT
Fondo Nacional de Desarrollo Científico y Tecnológico
Cambios C Suelo - Sitios Marginales
Profundidad 0-20 cm
Línea base
Carbono (Mg ha-1)
5000
7500
10000
A. melanoxylon
E. camaldulensis
E. nitens
Profundidad 20-40 cm
Línea base
Carbono (Mg ha-1)
a
b
b
a
ab
b
Edad (meses)

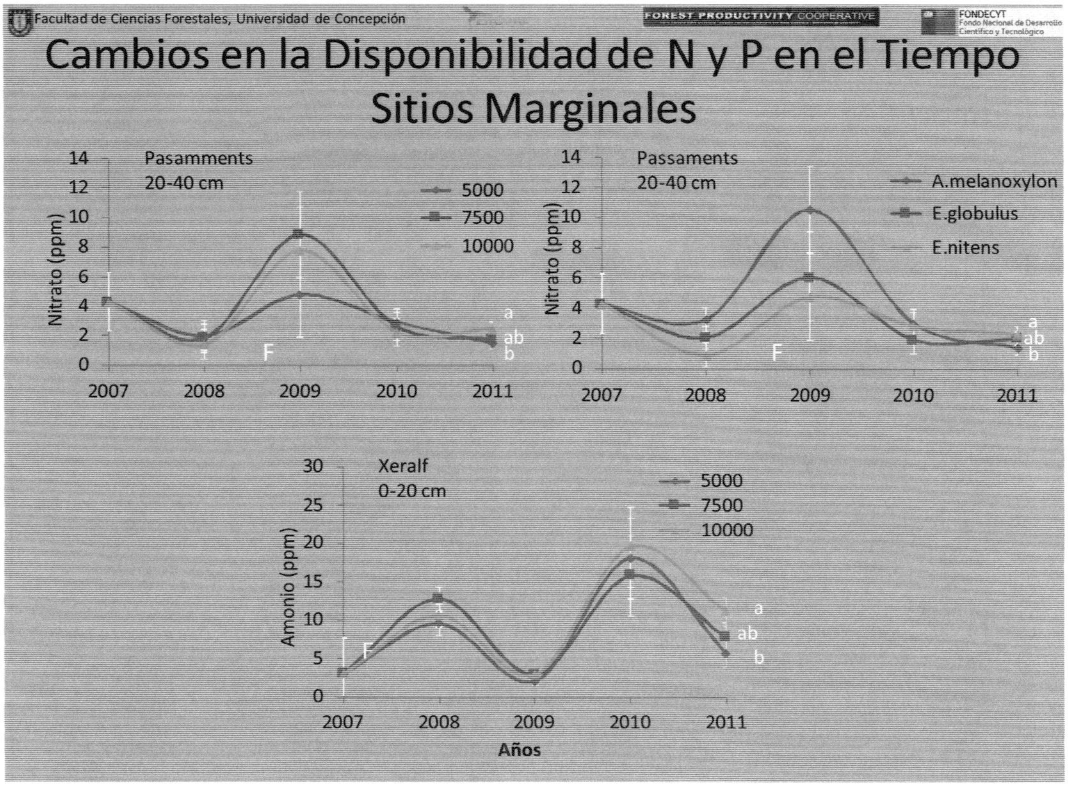
Facultad de Ciencias Forestales, Universidad de Concepción
FOREST PRODUCTIVITY COOPERATIVE
FONDECYT
Fondo Nacional de Desarrollo Científico y Tecnológico
Cambios en la Disponibilidad de N y P en el Tiempo
Sitios Marginales
Pasamments
20-40 cm
5000
7500
10000
Nitrato (ppm)
F
a
ab
b
Passaments
20-40 cm
A.melanoxylon
E.globulus
E.nitens
Nitrato (ppm)
F
a
ab
b
Xeralf
0-20 cm
5000
7500
10000
Amonio (ppm)
F
a
ab
b
Años

Facultad de Ciencias Forestales, Universidad de Concepción | FOREST PRODUCTIVITY COOPERATIVE | FONDECYT Fondo Nacional de Desarrollo Científico y Tecnológico

Sitios de Alta Fertilidad Uso de Agua por Especie y Tratamiento

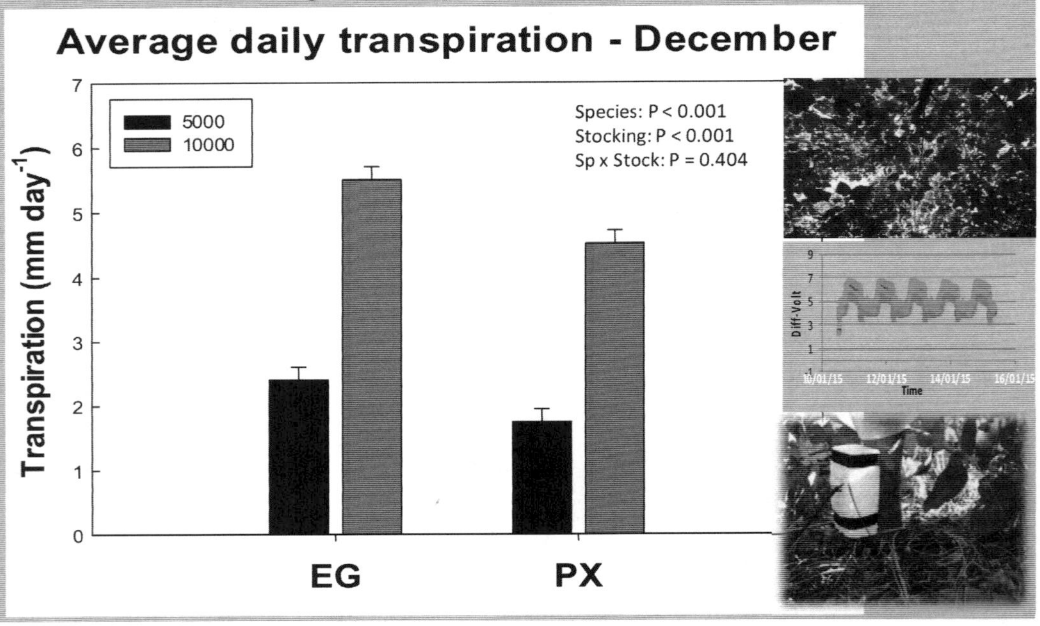

Facultad de Ciencias Forestales, Universidad de Concepción | FOREST PRODUCTIVITY COOPERATIVE | FONDECYT Fondo Nacional de Desarrollo Científico y Tecnológico

Plantaciones Dendroenergéticas Claves Sustentabilidad

- Productividad de PD dependiente del sitio y manejo nutricional para producción de biomasa.
- PD presentan altas tasas de uso nutricional por unidad de biomasa producida (100 kg N/ha/año por 1 ton MS).
- Especies con mayor EN y fijadoras pueden ser mas sustentables pero efectos de acidificación deben ser considerados.
- Cambios en disponibilidad de nutrientes sugieren agotamiento de suelos en rotaciones sucesivas.

Facultad de Ciencias Forestales, Universidad de Concepción FOREST PRODUCTIVITY COOPERATIVE FONDECYT Fondo Nacional de Desarrollo Científico y Tecnológico

Plantaciones Dendroenergéticas Claves Sustentabilidad

- C del suelo no retorna a los niveles de línea base en suelos graníticos y arenales, posible efecto positivo con manejo de rebrotes de largo plazo.
- Uso de Agua de Cultivos de Corta Rotación Altamente Dependiente de Densidad de Cultivo y Efectos en Sustentabilidad Hídrica.
- Se requiere de una reingeniería de Silvicultura Específica al Sitio con Monitoreo de Indicadores Ambientales.

Facultad de Ciencias Forestales, Universidad de Concepción FOREST PRODUCTIVITY COOPERATIVE FONDECYT Fondo Nacional de Desarrollo Científico y Tecnológico

Agradecimientos

FONDECYT
Fondo Nacional de Desarrollo Científico y Tecnológico

PROYECTO FONDECYT 1140482
"Sustainability of water and nutrient use in short rotation woody crops for bioenergy"

Fondo Regional INNOVA Bio Bio 06PC-S133
Energía Verde S.A.
CONAF
OEA
CONSORCIO BIOENERCEL

CORFO
ARAUCO
cmpc
MASISA
Universidad de Concepción
Pontificia Universidad Católica de Valparaíso
Fundación Chile
CONICYT Comisión Nacional de Investigación Científica y Tecnológica
FOREST PRODUCTIVITY COOPERATIVE
BioEnercel S.A.
Universidad de Concepción Facultad de Ciencias Forestales

EXPORTACIÓN DE RESIDUOS DE COSECHA FORESTAL CON FINES ENERGÉTICOS. PÉRDIDA DE NUTRIENTES Y COSTO DE REPOSICIÓN.

Roberto Fernández; Alejandra Von Wallis; Norberto Pahr; Alejandro Friedl; Rodolfo Martiarena; Ana Lupi.

EEA Montecarlo (INTA) - Facultad de Ciencias Forestales (UNaM)

Buenos Aires, 13 – 16 de setiembre de 2016

Trabajo parcialmente financiando por el Proyecto Manejo Sustentable de Recursos Naturales BIRF 7520 AR. Componente 2 Plantaciones Forestales Sustentables.

Unión para el Cambio Rural -UCAR-.

El problema:

En Argentina, la modalidad tradicional de cosecha de Pinus spp, es la extracción exclusiva del fuste. Algunas empresas extraen también residuos de la tala rasa destinados a bioenergía, objetivo para el cual resulta práctico la extracción del árbol entero.

El impacto ambiental y la sostenibilidad son temas centrales en el actual debate internacional sobre bioenergía.

Por ello, es oportuno conocer las implicancias de dicha modalidad de cosecha en la pérdida de nutrientes del sitio.

El problema:

La intensificación en el uso de la tierra puede activar procesos de degradación. Cabe preguntarse entonces hasta qué punto el ingreso económico por venta de bioenergía y/o el volumen de combustible fósil evitado compensan los costos ambientales "ocultos", y en qué plazos.

CICLO DE NUTRIENTES EN SISTEMAS FORESTALES
Entradas
Aportes atmosféricos
Fijación biológica
Meteorización
Reposición de nutrientes
Bosque
almacenaje
Biomasa
Sotobosque
deposición
absorción
Mantillo
Suelo
Suelo
transformaciones
Mineralización y humificación
Procesos de Intercambio. Fijación.
Salidas
Lixiviación
Volatilización
Voladura de cenizas
Cosecha

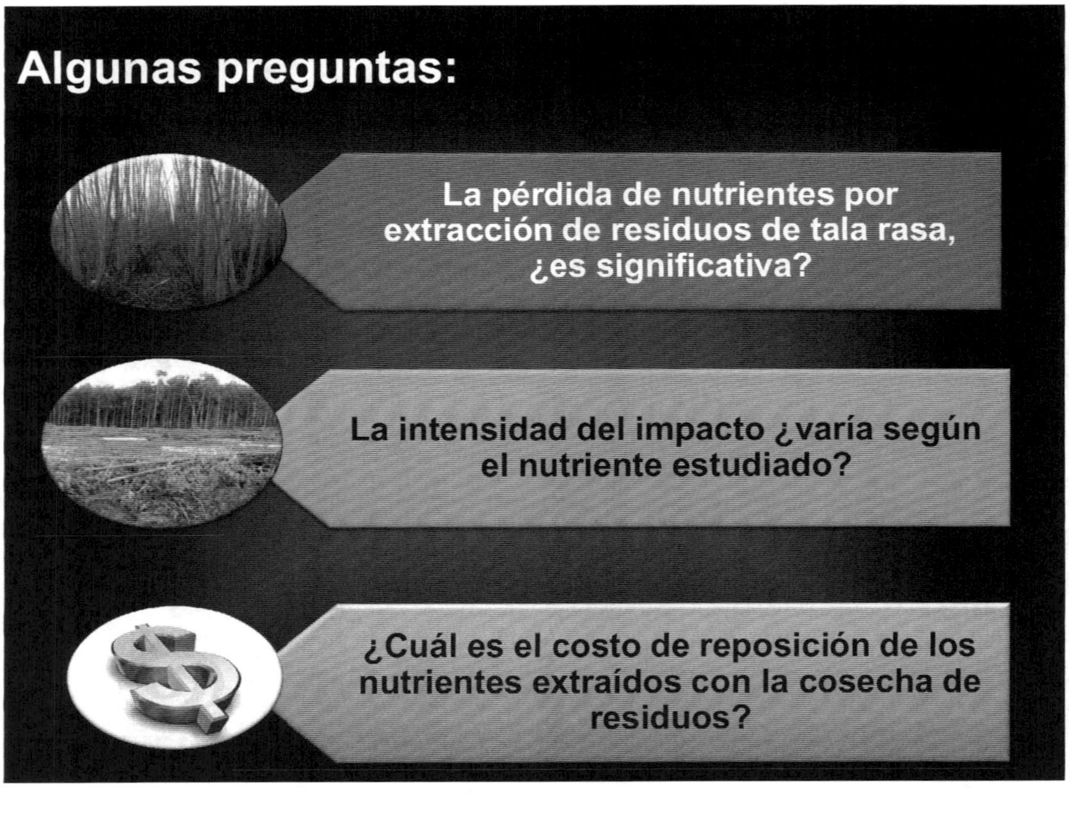
Algunas preguntas:
La pérdida de nutrientes por extracción de residuos de tala rasa, ¿es significativa?
La intensidad del impacto ¿varía según el nutriente estudiado?
¿Cuál es el costo de reposición de los nutrientes extraídos con la cosecha de residuos?

OBJETIVOS:

Cuantificar la biomasa de residuos de cosecha de *Pinus taeda* y la cantidad de N, P y K exportadas como consecuencia de su extracción para bioenergía.

Estimar el costo de reposición de los nutrientes contenidos en los residuos cosechados.

MATERIALES Y METODOS

Localización de la experiencia: NO de la Provincia de Misiones.

Se trabajó en dos rodales de *P. taeda*, de 20 y 26 años, representativos en cuanto a su conducción manejo y producción.

En cada sitio se seleccionó un área homogénea. En ellas se establecieron dos áreas de dos hectáreas cada una.

Sitio 1: 26° 00' 08'' S - 54° 38' 54" O.
Sitio 2: 25° 58´ 52" S - 54° 22´ 65"O.

Empresas PINDO SA y LIPSIA SA.

MATERIALES Y METODOS

Se procedió al inventario forestal y muestreos previos a la cosecha (sotobosque, piso forestal y suelo).

En cada sitio se seleccionó un área homogénea. En ellas se establecieron dos áreas de dos hectáreas cada una, en las cuales se aplicaron los tratamientos.

Tratamientos

- Solo Fuste (SF): consistió en a) Apeo; b) Desrame y despunte del fuste en el sitio del apeo; c) Extracción del fuste comercial hasta cancha de acopio; d) Trozado del fuste d) Carga y transporte.

- Árbol Entero (AE) : consistió en a) Apeo; b) Extracción del árbol completo hasta cancha de acopio; c) Desrame y trozado del fuste; d), Carga y transporte

MATERIALES Y METODOS

Muestreo de residuos de cosecha.

Sistemático, 15 parcelas de 100 m^2, en cada tratamiento.

En cada parcela se instalaron 10 sub-parcelas de 0,5m^2.

Cinco categorías de residuos: 1. Ramas>5; 2. Ramas 1-5; 3. Ramas <1; 4. Acículas y 5. Conos.

MATERIALES Y METODOS

Muestreo de residuos de cosecha.

Cinco categorías

1. Ramas > 5 cm de diámetro
2. Ramas 1-5 cm de diámetro
3. Ramas < 1 cm de diámetro
4. Acículas-hojas
5. Conos

MATERIALES Y METODOS

Análisis estadístico.

Se utilizó como técnica la comparación de las dos poblaciones a través de las medias muestrales.

RESULTADOS.

Materia seca de residuos remanente ($Mg.ha^{-1}$) según modalidades de cosecha. *P. taeda* de 20 años (SITIO 1).

		Diám. de ramas (cm)			Hojas	Conos	Total
		>5	1-5	<1			
SF	$Mg.ha^{-1}$	11,1	26,6	12,7	20,1	3,6	74,0
AE		2,5	16,1	10,6	16,6	3	48,8
Diferencia		**8,6**	**10,5**	**2,1**	**3,5**	**0,6**	**25,2**
FE-AE	%Extraído	77	39	17	17	17	**34**

SF: sólo fuste. AE: árbol entero

La cosecha del árbol entero significó la extracción de 25,2 $Mg.ha^{-1}$ de necromasa.

RESULTADOS.

Materia seca de residuos remanente ($Mg.ha^{-1}$) según modalidades de cosecha. *P. taeda* de 26 años (Sitio 2)

		Diám. de ramas (cm)					
		>5	**1-5**	**<1**	**Hojas**	**Conos**	**Total**
SF		13,9	28,8	11,3	22,4	7,8	84,3
AE	$Mg.ha^{-1}$	7,4	18,4	7,6	16,3	5,3	55,0
Diferencia		6,5	10,4	3,7	6,1	2,5	**29,3**
SF-AE	%Extraído	47	36	33	27	33	**35**

SF: sólo fuste. AE: árbol entero

La cosecha del árbol entero significó la extracción de 29,3 $Mg.ha^{-1}$ de necromasa.

RESULTADOS.

Las mayores diferencias se observaron en "Ramas > 5" y "Ramas 1-5", de las cuales se extrajo con destino a bioenergía el 77 y el 39 % del total de cada estrato, y el 47 y el 36 %, en los Sitios 1 y 2, respectivamente.

RESULTADOS.

Contenido de nitrógeno, en Kg.ha^{-1}, según categoría de residuo y modalidad de cosecha. *P. taeda* de 20 años (Sitio 1)

	Ramas > 5 cm	Ramas 1-5 cm	Ramas <1 cm	Acículas	Conos	Total
SF	30,1	102,4	65,1	210,8	14,9	423,3
AE	6,4	62,5	49,2	175,8	11,7	305,6

SF: sólo fuste. AE: árbol entero

La cosecha del árbol entero significó la pérdida de 117,7 Kg.ha^{-1} de nitrógeno.

RESULTADOS

Contenido de fósforo, en Kg.ha^{-1}, según categoría de residuo y modalidad de cosecha. Tala rasa de *P. taeda* de 20 años (Sitio 1)

	Ramas > 5 cm	**Ramas 1-5 cm**	**Ramas <1 cm**	**Acículas**	**Conos**	**Total**
SF	**1,1**	**4,0**	**5,4**	**16,7**	**0,6**	**27,8**
AE	**0,2**	**2,2**	**4,0**	**13,4**	**0,4**	**20,2**

SF: sólo fuste. AE: árbol entero

La cosecha del árbol entero significó la pérdida de 7,6 Kg.ha^{-1} de fósforo.

RESULTADOS

Contenido de potasio, en Kg.ha^{-1}, según categoría de residuo y modalidad de cosecha. *P. taeda* de 20 años (Sitio 1)

	Ramas > 5 cm	Ramas 1-5 cm	Ramas <1 cm	Acículas	Conos	Total
SF	5,8	17,3	26,7	63,1	5,1	117,9
AE	1,3	10,0	19,7	50,4	3,3	84,7

SF: sólo fuste. AE: árbol entero

La extracción del árbol entero significó la pérdida de 33,3 Kg.ha^{-1} de potasio.

RESULTADOS.

Contenido de N, P y K, en Kg.ha^{-1}, según categoría de residuo y modalidad de cosecha. Tala rasa de *P. taeda* de 26 años (Sitio 2)

		N		P		K	
		SF	AE	SF	AE	SF	AE
Ramas	> 5 cm	28,4	15,2	1,5	0,8	7,3	3,9
	1-5 cm	80,6	51,3	2,9	1,9	30,4	19,3
	<1 cm	63,3	42,7	3,5	2,4	31,0	20,9
Acículas		289	210	12,8	9,3	97,2	70,6
Conos		24,3	16,4	0,7	0,5	11,2	7,5
Total		486	336	21,5	14,8	177,1	122,4
Diferencia SF - AE		150,0		6,6		54,7	

RESULTADOS.

En ambos sitios, la extracción de fósforo debido a la cosecha de residuos resultó equivalente al contenido de fósforo disponible en los suelo.

Al considerar la oferta de los tres nutrientes en el largo plazo es el fósforo el que presenta mayor nivel de criticidad.

RESULTADOS.

El análisis estadístico permitió detectar diferencias significativas entre las modalidades de cosecha,
para los tres nutrientes, en los cinco compartimentos estudiados,
y para ambos sitios.

RESULTADOS.

La compensación de las mineralomasas de N, P y K extraídas con los residuos cosechados requeriría adquirir urea, superfosfato triple y cloruro de potasio por un valor de 177 y 207 U$$.ha^{-1}, para los Sitios 1 y 2, respectivamente (1U$$ = 14,9$)

La estimación del costo de reposición debe incluir también el costo de aplicación y el costo de adquisición de las cantidades de nutrientes necesarias para compensar pérdidas por lixiviación, volatilización, fijación, escorrentía,etc.

CONCLUSIONES.

- La cosecha de residuos con fines energéticos representó una extracción de 25,2 y 29,3 Mg.ha^{-1}, para los Sitios 1 y 2, respectivamente.
- La cosecha de residuos representó una pérdida adicional de 117,7 a 150 kg.ha^{-1} de N, 7,6 a 6,6 kg.ha^{-1} de P, y 33,3 a 54,7 kg.ha^{-1} de K, para los Sitios 1 y 2, respectivamente
- La extracción adicional de fósforo debido al tratamiento árbol entero resultó equivalente al contenido de fósforo disponible en el suelo en los dos sitios.
- La extracción adicional de N, P y K representó un costo oculto por hectárea cosechada, de 177 y 207 U$$.ha^{-1}, para los Sitios 1 y 2, respectivamente (1U$$ = 14,9$)

CONSIDERACIONES FINALES.

La implementación de proyectos bio-energéticos en base a residuos de cosecha requiere de mayor conocimiento, en varios aspectos, entre ellos:

- Información sobre la huellla de carbono y los costos - energéticos y económicos - del conjunto de las operaciones necesarias para la utilización en procesos de transformación en otras formas de energía.
- Determinar la calidad nutritiva de los residuos del proceso de transformación energética –cenizas- y evaluar técnicas para su aplicación a campo, como mecanismo de compensación de reposición de nutrientes.

Gracias por su atención

STEVENS INSTITUTE of TECHNOLOGY THE INNOVATION UNIVERSITY

CES

The cycle of nutrients at industrial plants: valorization of energetic-laden wastewater streams using microalgae *Scenedesmus obliquus* to produce bioenergy

Juliana Abraham[1], Abhishek Roychowdhury[1], Christos Christodoulatos[1], Yanxia Lin[1], Per Arienti[2], Benjamin Smolinski[2], and Washington Braida[1]

[1]Stevens Institute of Technology, [2]RDCOM-ARDEC, Picatinny Arsenal

2nd RCN Conference on Pan American Biofuels & Bioenergy Sustainability

Buenos Aires, Argentina
September 13-16, 2016

Distribution A: Approved for public release; distribution is unlimited

STEVENS INSTITUTE of TECHNOLOGY THE INNOVATION UNIVERSITY

CES

Presentation outline

- Net Zero Waste, Water and Energy at industries
- Why unicellular algae?
- Biomass conversion
- Characterization of industrial water streams
- Toxicity assessment to microalgae
- Nutrient uptake by microalgae
- Reuse of waste streams
- Scale-up (100 L raceway reactors)

Distribution A: Approved for public release; distribution is unlimited

Net Zero challenge for sustainable industries

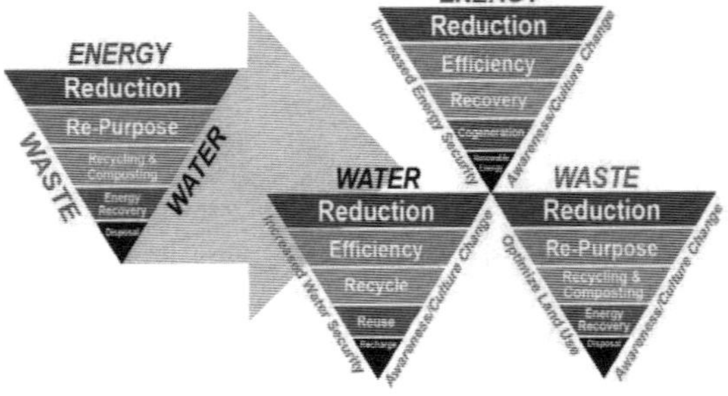

- Net Zero Energy → generate energy on site as much as it is consumed.
- Net Zero Water → limiting the consumption of freshwater resources & recycling.
- Net Zero Waste → reduce, reuse, and recover waste streams, converting them to valuable resource, eliminate landfilling.

Why unicellular algae?

Algae can be found in several environments such as fresh or marine water as well as in the surface of moist soils and rocks.

Fastest growing photo-synthesizers. Macronutrients required: C, N and P

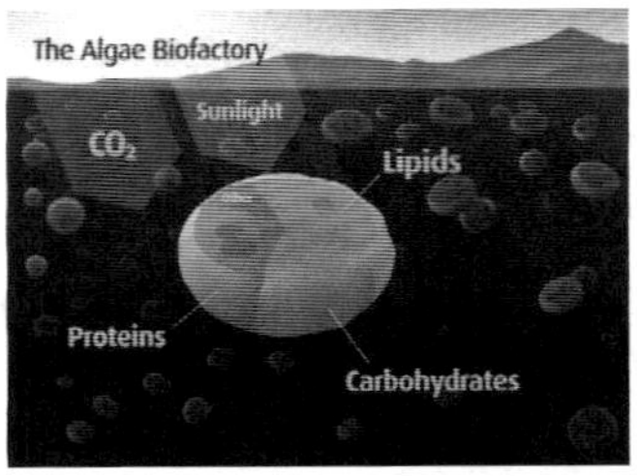

$$6CO_2 + 6H_2O + \text{Solar energy} \rightarrow C_6H_{12}O_6 + 6O_2$$

Carbon dioxide Water Glucose Oxygen

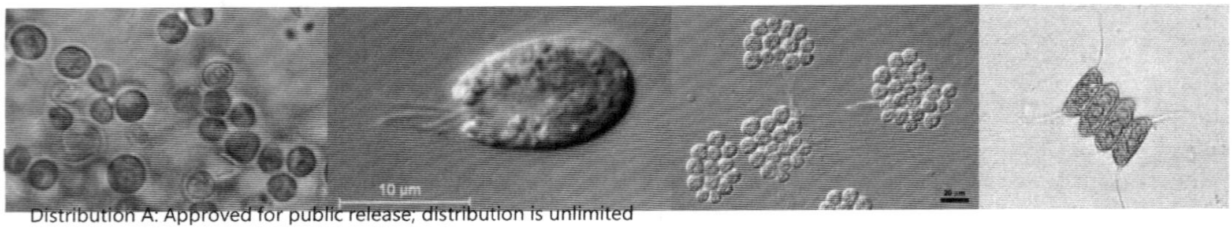

Biomass Conversion

Microalgae grow faster than plants and occupy less space.

Several types of bioenergy can be produced from algae biomass (biofuel from lipids, H_2, and biogas by anaerobic digestion.)

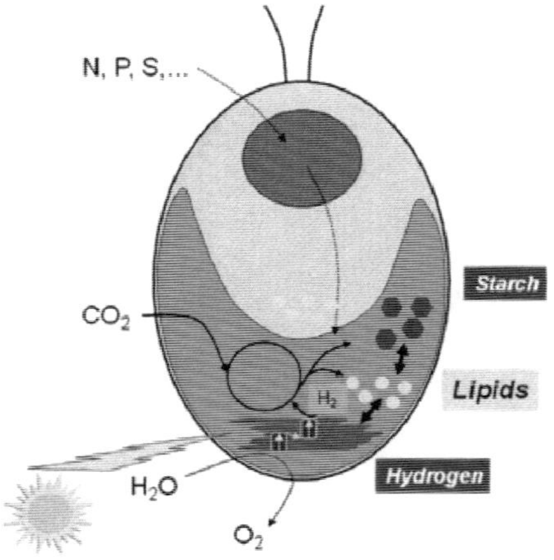

Algal oil yield is higher than the obtained from other plant crops.

Crop	Oil yield (L ha^{-1})
Corn	172
Soybean	446
Sunflower	952
Coconut	2,689
Microalgae (30% oil)	58,700
Microalgae (70% oil)	136,900

(Source: Trivedi et al. 2015)

Utilization of Waste Streams for Algae Growth

Characterization of waste streams (WS):
- Nutrients' content (Hach® tests)
- EC content (HPLC/ HEPI MS)
- Toxicity tests (OECD)

↓

Use of WS for algae growth (laboratory and pilot scales)

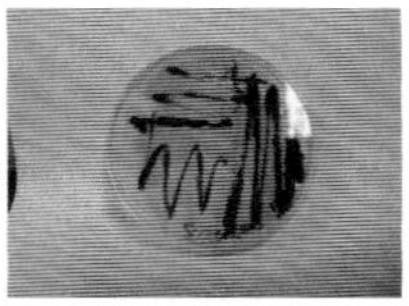

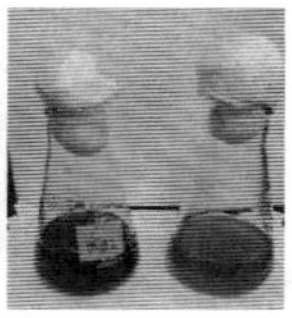

Algae Growth and Scale Up for Fuel Production

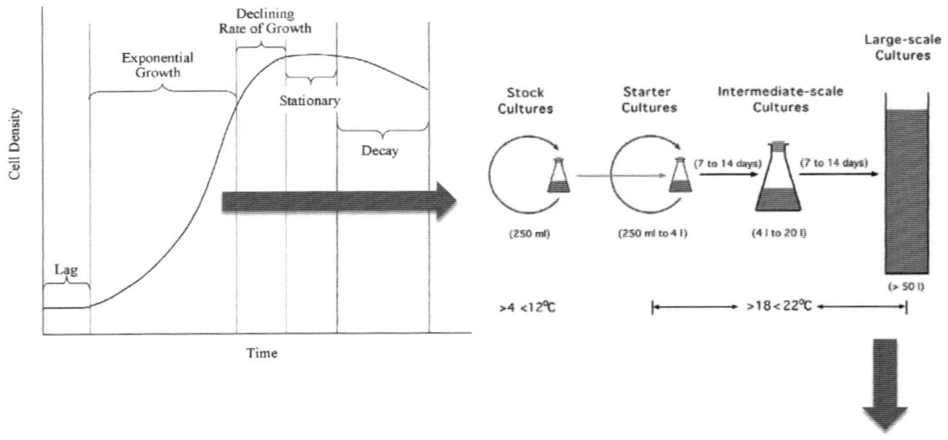

Biofuel & Biogas

Toxicity Assessment of Waste Streams

- Strain used: *Scenedesmus obliquus ATTC®* 11477
- Protocol adapted to microplates, OECD (2011)
- Adjust pH of samples= 7.00, if needed
- Incubation at 25°C, 3700 lux and 14:10 h light: dark cycle for 6 days
- 5 % CO_2 during lighting period
- Measure algae growth (Fluorescence) and color changes

Protocol

Percentage of waste stream tested	60%	40%	20%	10%	0% (C+)	Blank
Volume of wastewater (μL)	1800	1200	600	300	0	0
Volume of Medium 1X (μL)	600	1200	1800	2100	2400	2700
Volume of Medium 10X (μL)	300	300	300	300	300	300
Algal cell Suspension (μL)	300	300	300	300	300	0

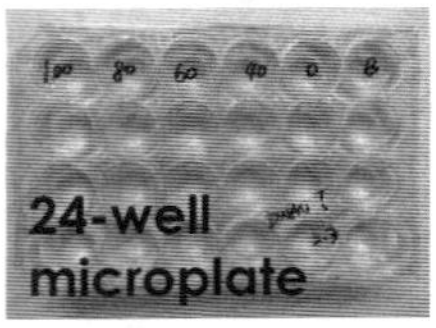

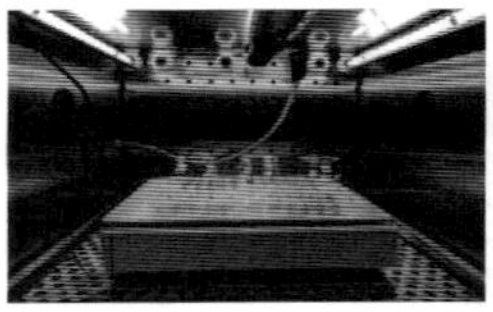

Characterization of Waste streams from Industrial Plant

Waste Streams	pH	TN (ppm)	$N\text{-}NH_3$ (ppm)	$N\text{-}NO_3^-$ (ppm)	$N\text{-}NO_2^-$ (ppm)	TP (ppm)	TOC (ppm)	Toxicity due to
1	0.66	10200	B.D.L	8874	B.D.L.	0.01	40.01	NQ
2	0.67	6910	B.D.L.	6481	0.06	B.D.L.	6.01	NO_3^-
3	5.12	287000	82400	124645	B.D.L.	B.D.L.	192	NH_3, NO_3^-
4	3.95	295	14.20	42.00	B.D.L.	0.04	30216	-
5	6.75	33.70	0.53	1.21	1.41	0.04	57.1	-
6	12.15	4870	19.80	1680	1.07	B.D.L.	569	NTO
7	7.08	47.00	B.D.L.	0.70	B.D.L.	B.D.L.	194	DNAN
8	4.61	16.30	28.30	B.D.L.	B.D.L.	0.04	175	-
9	6.59	221	12.70	2.84	70.86	0.12	2132	RDX, NTO, NQ
10	7.67	4.23	B.D.L.	B.D.L.	B.D.L.	0.04	4.89	-

B.D.L: below detection limit, TN: total nitrogen, TP: total phosphorous, TOC: total organic carbon.

Good source of nutrients for microalgae growth

Waste Stream #5 – Toxicity Assessment

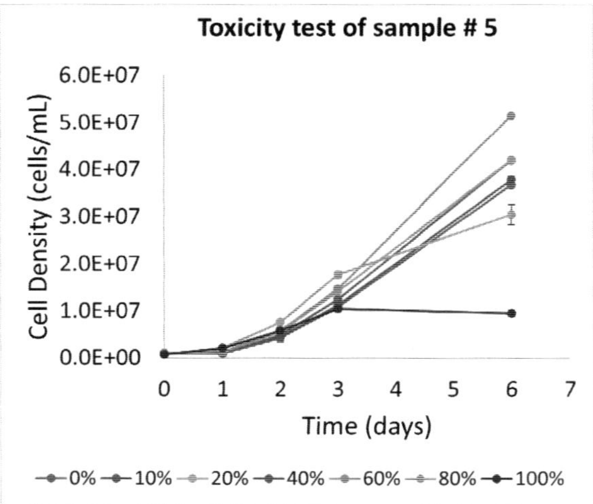

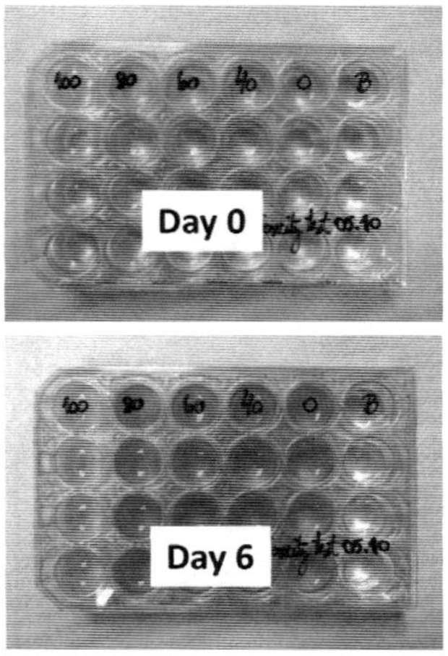

No toxicity evidenced, good development of color

Growth decreased due to lack of nutrients in 80 and 100% samples

Waste Stream #2 – Toxicity Assessment

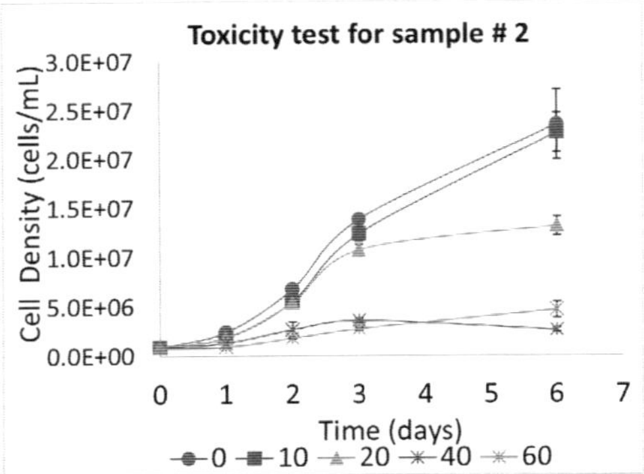

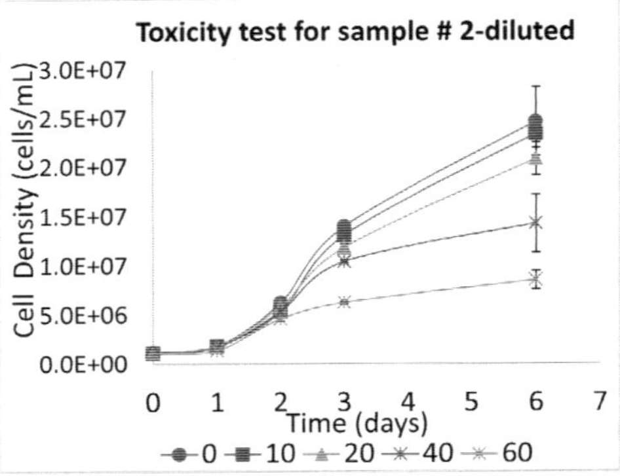

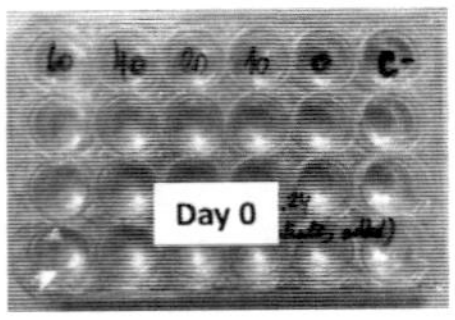

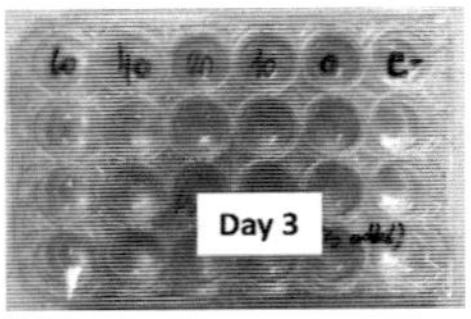

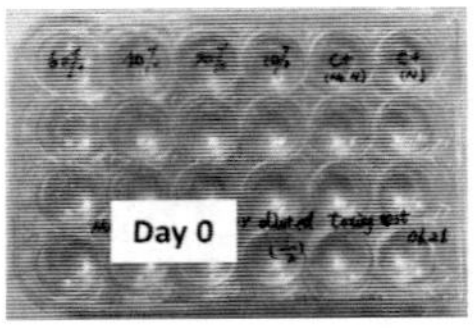

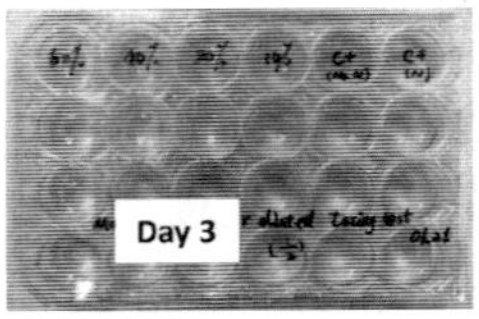

Toxicity decreased when diluted the NO_3^- concentration that was inhibiting growth

Waste Stream #7 – Toxicity Assessment

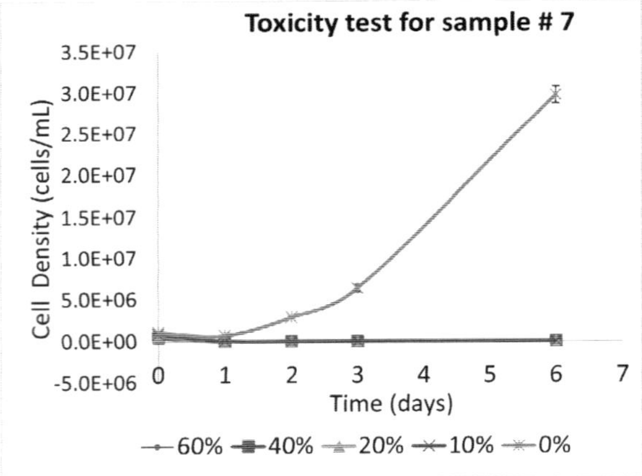

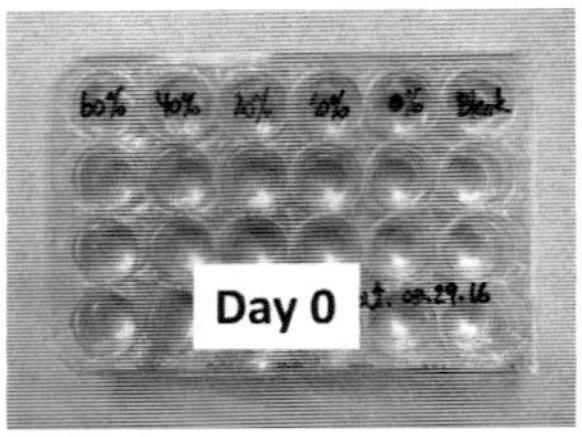

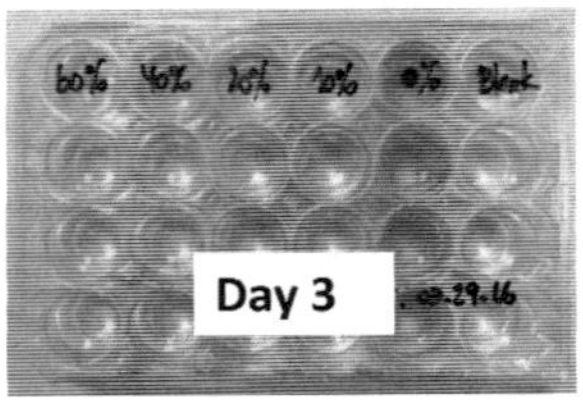

High Toxicity due to EC, WS needs to be treated before recycling
The majority of the wastewaters are not highly toxic for *S. obliquus* growth and thus could be used as nutrient media without pretreatment.

Nutrient uptake by microalgae

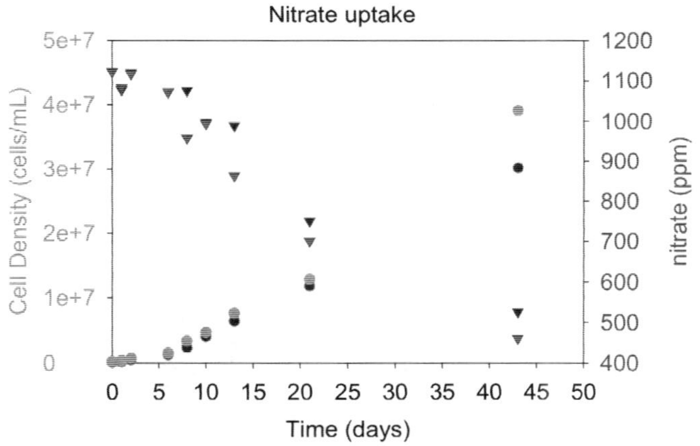

At laboratory scale, microalgae showed an uptake of 35% of initial nitrate concentration over time.

The WS can be used as nutrient media for microalgae growth

Algae Growth In Mixtures of Waste Streams

Mixture of WS performed: WS # 3+ WS# 9+ micronutrient traces
Experiments performed in triplicates + 2 control (mixture alone)

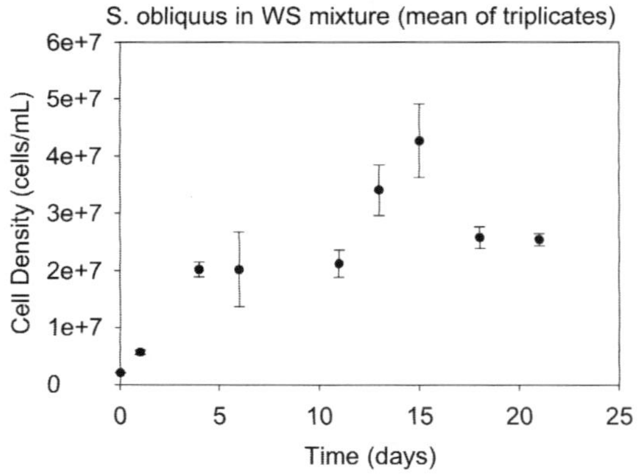

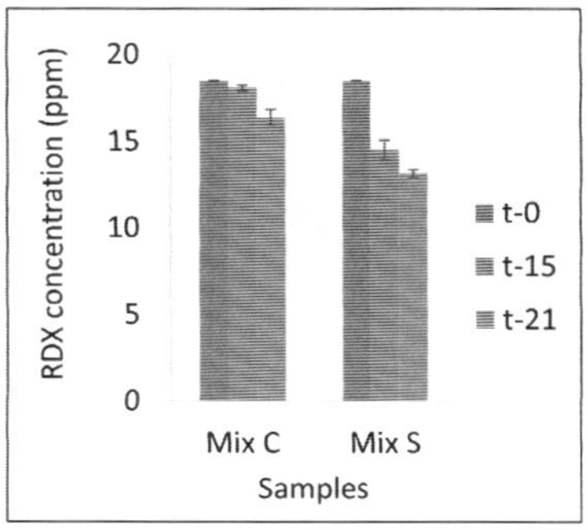

The dry weight of algal culture was increased from 0.17 g/L to 2.33 g/L from day 0 to day 21

An overall 28% decrease of RDX concentration was observed over time.

Scale-up to 100 L raceway reactors

On-line Control of parameters

Scale-up to 100 L raceway reactors

Scale-up to 100 L raceway reactors

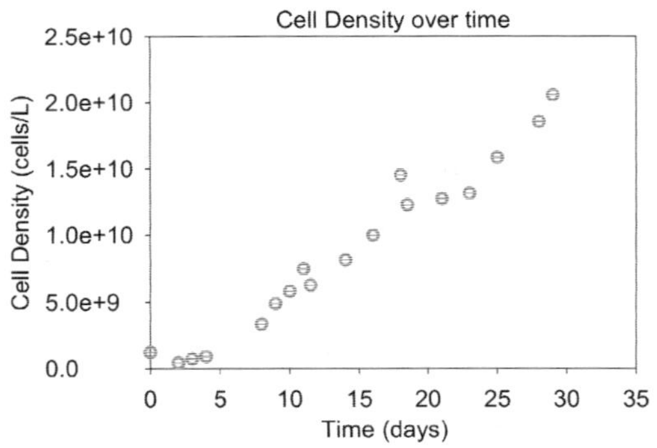

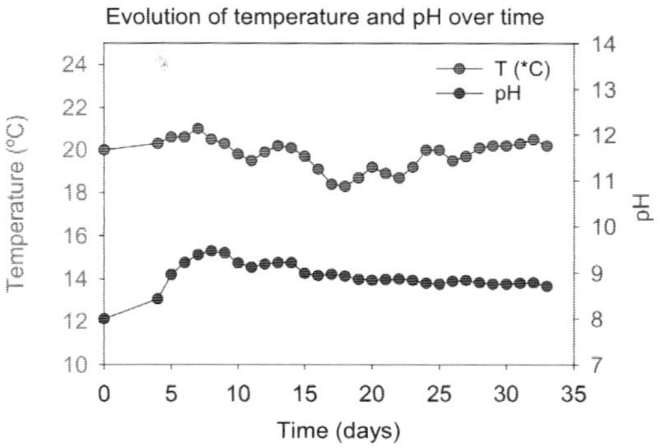

0.3 g/L dry weight was obtained in the reactor after 25 days

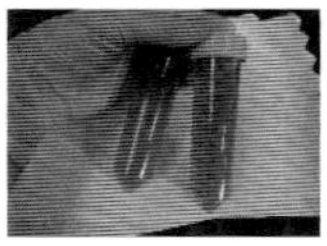

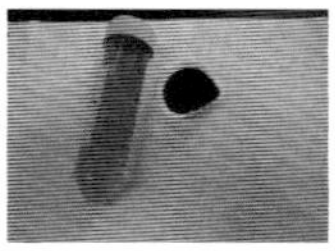

Acknowledgements

This project is financially supported by the Consortium for Energy, Environment and Demilitarization (CEED) contract number SINIT-15-0013

The research is performed at Center for Environmental Systems (CES)- Stevens Institute of Technology

STEVENS INSTITUTE of TECHNOLOGY
THE INNOVATION UNIVERSITY

Thank you for your attention

jabraham@stevens.edu

Supplementary information

NTO DNAN

RDX NQ

The correct names are 1) 3-nitro-1*H*-1,2,4-triazol-5(4*H*)-one (NTO); 2) 1-methoxy-2,4-dinitrobenzene (2,4-dinitroanisole, DNAN); 3) 1,3,5-trinitro-1,3,5-triazinane (RDX); 4) nitroguanidine

Natural Resources Canada / Ressources naturelles Canada

Renewable natural gas (RNG) from forestry and agricultural biomass in Canada

B.D. Titus[1], F. Preto[2], B. Joss[3], D. Paré[1], D. Lee[4], Brianna Kelly[4], D. Auger[1], M. Bock[1]

1. Natural Resources Canada, Canadian Forest Service
2. Natural Resources Canada, CanmetENERGY
3. Natural Resources Canada, Canadian Forest Service, Canadian Wood Fibre Centre
4. Agriculture and Agri-Food Canada

RNG from anaerobic digestion and direct thermochemical conversion (gasification) of biomass can be used as a drop-in fuel to displace natural gas. The Canadian Gas Association (CGA) has set a distribution target of 5% RNG-blended natural gas by 2025 rising to 10% by 2030, resulting in a reduction of 14 Mt of GHGs (= 3 million cars).

85% of RNG in Canada could come from gasification because of feedstock availability and process economics. Forestry residues are priority feedstock because of higher energy density, more stable and consistent sustainable supply, and fewer technical challenges; technology for agriculture residues will follow.

Two essential underpinnings of an RNG sector are (i) **inventories** of potential sustainable feedstock availability that will determine the size of the sector, and (ii) appropriate **conversion technologies** for the feedstocks available. Other issues to be addressed include economic and supply-chain analyses, and development of appropriate policies, regulations and environmental standards to ensure social license.

Annual estimated availability of biomass was estimated in circles from 25- to 150-km radius in 25-km increments around 288 major compressor stations along transmission pipelines for forest harvesting and mill residue, agricultural residue, and forest salvage after fire. 50-75 km is considered an economic haul distance.

G4 Insights is planning two RNG plant sizes:

Large: 10,000 GJ/day requiring ~275,000 odt biomass/year

Small: 400 GJ/day requiring ~11,000 odt biomass/year

Next steps: Complete biomass analyses along major transmission and distribution pipelines; scale up conversion plants to operational size; supply chain analyses; enabling policies and regulations.

Schematic of G4 Insights RNG conversion process

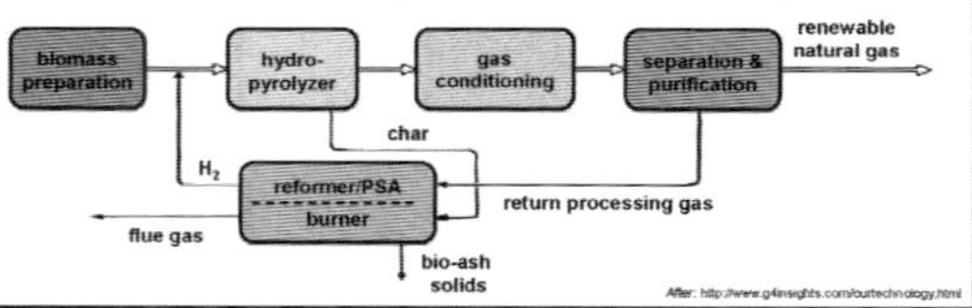

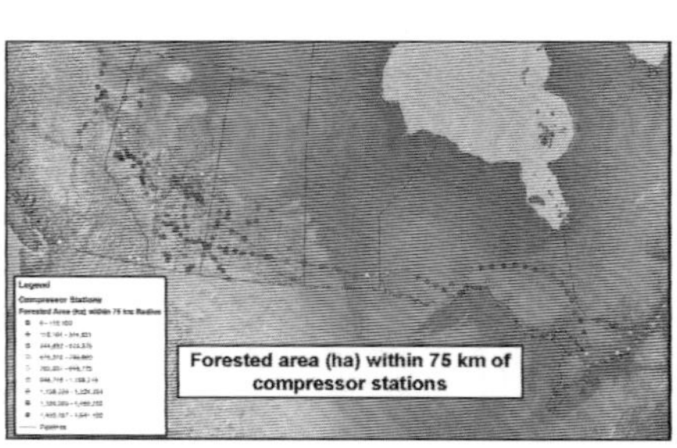

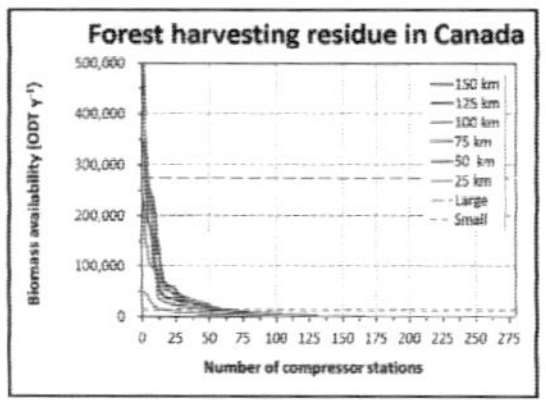

Feedstock source	Plant size	Radius of circular fuelsheds (km) 25	50	75	100	125	150
Forest mill residue	Large	13	33	62	104	133	158
	Small	33	88	131	157	188	210
Forest harvesting residue	Large	0	2	32	58	91	119
	Small	77	125	168	196	210	226
Total forestry biomass	Large	13	46	97	121	166	192
	Small	90	137	179	203	213	229
Agricultural harvesting residue	Large	0	75	117	148	179	202
	Small	141	183	205	222	231	240
Total forestry and agricultural biomass	Large	19	124	198	231	255	267
	Small	219	254	268	274	277	277

Contact

Project co-leads: Brian.Titus@Canada.ca or Mattthew.Bock@Canada.ca
Conversion technology: Fernando.Preto@Canada.ca
BIMAT inventory: Brent.Joss@Canada.ca (forestry) or David.Lee3@Canada.ca (agriculture)
Remote sensing inventory: David.Pare@Canada.ca (forestry)

Acknowledgements: Colleagues in NRCan and the Canadian Gas Association and member companies; research funds from the Forest Innovation Program (FIP) and the Environmental Innovation Program (EIP) of the Government of Canada; travel funds from the U.S. National Science Foundation; and organizers of the 2nd *RCN Conference on Pan American Biofuels & Bioenergy Sustainability*, Buenos Aires, Argentina, September 13-16, 2016.

Canada

2ª Conferencia Panamericana sobre Sustentabilidad en Biocombustibles y Bioenergía Buenos Aires, Argentina - Septiembre 13-16, 2016

An overall assessment of the technologies in use for the production of renewable fuels and energy.

Eduardo A. Lombardo
INCAPE (CONICET, FIQ-UNL)
Santa Fe, ARGENTINA

FIQ Facultad de Ingeniería Química
UNL Universidad Nacional del Litoral

Outline

- **World energy outlook**
- **Biofuels: Commercial scale projects and costs**
- **Renewable chemicals**
- **American continent panorama: Renewable energies and biofuels.**
- **Conclusions**

Consumo mundial de energía primaria

(BP annual review 2016)

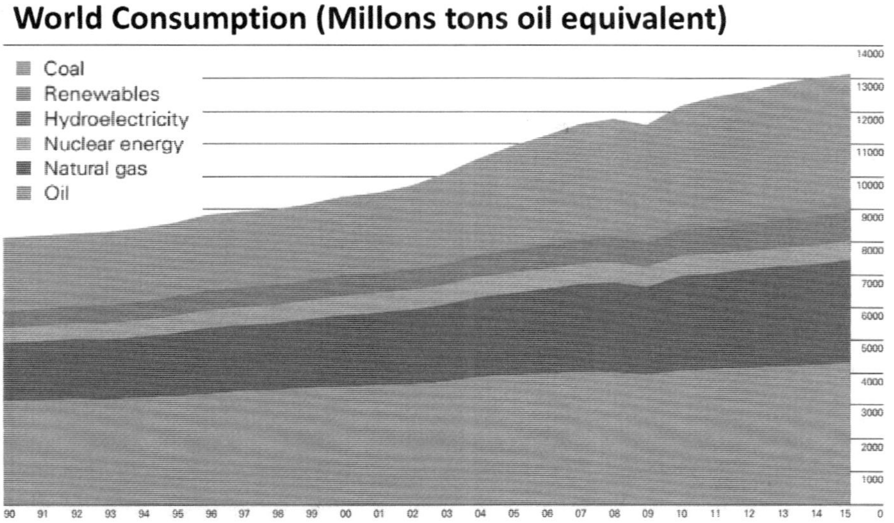

World primary energy consumption grew by a below-average 1.0% in 2015, the slowest rate of growth since 1998 (other than the decline in the aftermath of the financial crisis). Growth was below average in all regions except Europe & Eurasia. All fuels except oil and nuclear power grew at below-average rates. Oil remains the world's dominant fuel and gained global market share for the first time since 1999, while coal's market share fell to the lowest level since 2005. Renewables in power generation accounted for a record 2.8% of global primary energy consumption.

Non-hydro renewables share of power generation

(BP annual review 2016)

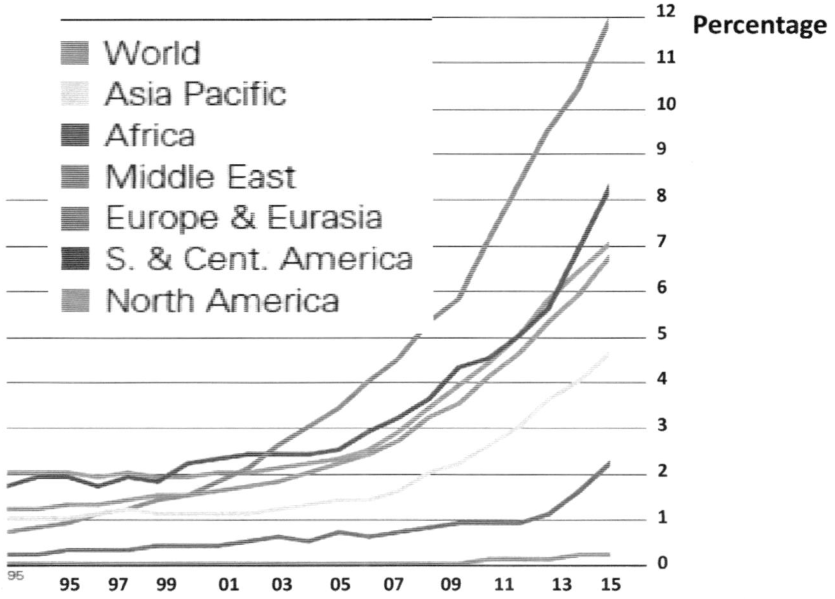

Renewable energy grew by 15% (2015). Wind had the largest growth increment, solar had the highest growth rate (33%). N-h accounted for 6.7% of global power generation up from 2% in 2005.

Solar Electricity

Two routes: a) **Photovoltaic** b) **Thermal**

a) **In the last decade:**

i) Thinner PV silicon wafers (400→150 μm) became available.

ii) The electrical contacts are now printed with conducting ink on the wafers.

These two factors significantly decreased the cost of the PV option.

5 years ago the installed Kw cost was €3500 now €1000.

Solar Electricity

Two routes: a) **Photovoltaic** b) **Thermal**

a) **In the last decade:**

i) Thinner PV silicon wafers (400→150 μm) became available.

ii) The electrical contacts are now printed with conducting ink on the wafers.

These two factors significantly decreased the cost of the PV option.

5 years ago the installed Kw cost was €3500 now €1000.

b) **Thermal main advantage:**

Affordable heat accumulation that allows 24 h operation of the electricity generation unit.

Wind electricity

- In the last decade advances in the turbine design and the use of novel composite materials made eolic power use far cheaper, e.g.blades >100m and composites with higher strength/weight (Enhanced use of carbon fibers).

Wind electricity

- In the last decade advances in the turbine design and the use of novel composite materials made eolic power use far cheaper, e.g.blades >100m and composites with higher strength/weight (Enhanced use of carbon fibers).

In brief, ***for the first time the annual (2015) investment in new eolic and photovoltaic electricity generation has been higher than that reported for thermoelectric units burning coal and natural gas.***

HC Biofuels and 2nd generation ethanol production pathways

Biomass

- Acid hydrolisis → Fermentation → Ester → Ethanol
- Enzimatic hydrolisis → Fermentation → Ethanol
- Consolidated bioprocessing (CBP)** BLUFF → Ethanol
- Catalytic pyrolisis → Hydroprocessing → HC Biofuels
- Gasification*
  - →Patented bacteria (HCN) → Ethanol
  - → FT synthesis → HC Biofuels
  - → MTG → HC Biofuels

*PROBLEMS: Buildup of tars, impurities in the syngas, undesirable H_2/CO ratios, finding the right catalyst.

** Saccarolysis, hydrolisis, fermentation C5/C6 sugars in one reactor!!!

Commercial scale cellulosic biofuel projects (2014-16)

Company	Location	Nameplate capacity (MGY)	(Product)	At present
Mossi Ghisolfi	Crescentino, Italy	20	Ethanol	Operating
Gran Bio	Alagoas, Brasil	20	Ethanol	Operating
Raizen (Shell)	Piracicaba, Brasil	10	Ethanol	Operating
Poet DSM	Emmetsburg, IA, USA	20	Ethanol	Operating
DuPont	Nevada, IA, USA	25	Ethanol	Operating
Abengoa	Hugoton, KS, USA	25	Ethanol	Shut(Ch 11)
Sunliquid	Bavaria, Germany	36	Ethanol	Operating
INEOS	Vero Beach, Fl, USA	8	Ethanol	Poisoned

Commercial scale cellulosic biofuel projects (2014-16)

Company	Location	Nameplate capacity (MGY)	(Product)	At present
Mossi Ghisolfi	Crescentino, Italy	20	Ethanol	Operating
Gran Bio	Alagoas, Brasil	20	Ethanol	Operating
Raizen (Shell)	Piracicaba, Brasil	10	Ethanol	Operating
Poet DSM	Emmetsburg, IA, USA	20	Ethanol	Operating
DuPont	Nevada, IA, USA	25	Ethanol	Operating
Abengoa	Hugoton, KS, USA	25	Ethanol	Shut (Ch 11)
Sunliquid	Bavaria, Germany	36	Ethanol	Operating
INEOS	Vero Beach, Fl, USA	8	Ethanol	Poisoned
Clear Fuels	Collinwood, TN, USA	20	HC biofuel	Pilot plant
Kior	Narchez, MS, USA	40	HC biofuel	Bankrupt
Sundrop Fuels	Alexandria, LA, USA	**50**	HC biofuel	Const (2020)

Cellulosic ethanol production costs (2016)*

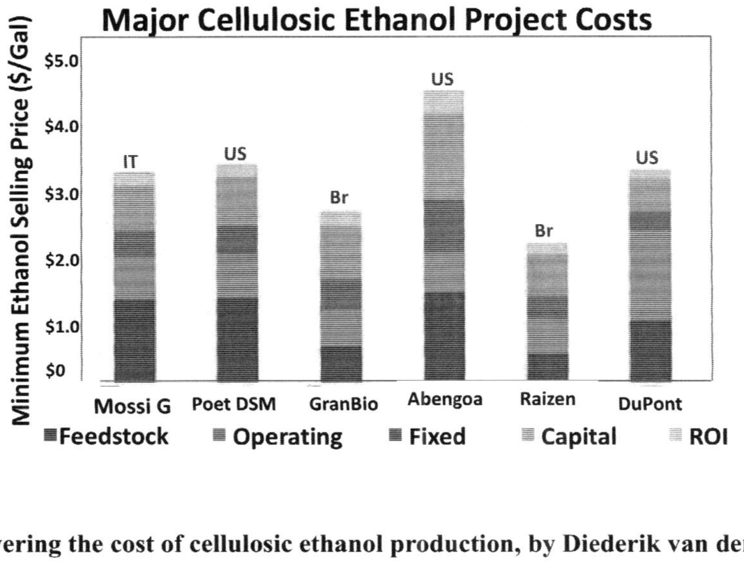

***Uncovering the cost of cellulosic ethanol production, by Diederik van der Hoeven**
www.luxresearch.com (2/28/2016)

Process Sunliquid-Clariant

Commercial plant operating in Bavaria

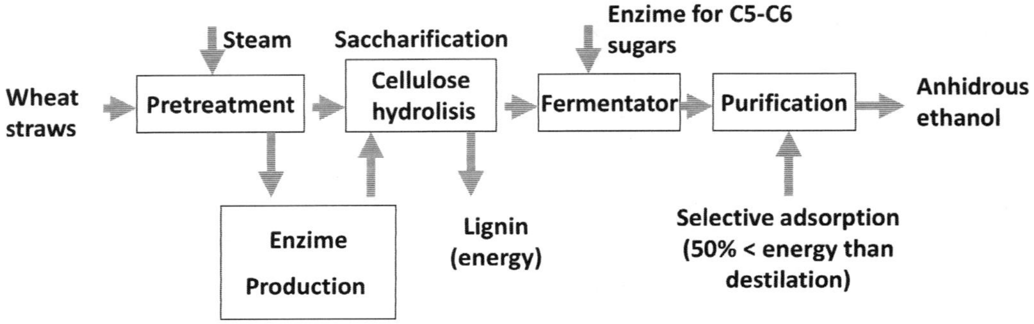

5 tons of wheat straws → 1 ton of anhydrous ethanol

Third generation algae biofuels

1) Algae produce an oil that can easily be refined into diesel or even certain gasoline components.

2) It can be genetically manipulated to produce everything from ethanol and butanol to gasoline, jet and diesel fuels.

3) It has produced up to 9000 gallons/acre (85000 liters/Ha) of fuel.

4) Can use a diverse array of carbon sources (e.g. CO_2)

Third generation algae biofuels

1) Algae produce an oil that can easily be refined into diesel or even certain gasoline components.

2) It can be genetically manipulated to produce everything from ethanol and butanol to gasoline, jet and diesel fuels.

3) It has produced up to 9000 gallons/acre (85000 liters/Ha) of fuel.

4) Can use a diverse array of carbon sources (e.g. CO_2)

Drawback: Algae require very large amounts of nitrogen and phosphorous to grow.

Third generation algae biofuels

1) Algae produce an oil that can easily be refined into diesel or even certain gasoline components.

2) It can be genetically manipulated to produce everything from ethanol and butanol to gasoline, jet and diesel fuels.

3) It has produced up to 9000 gallons/acre (85000 liters/Ha).

4) Can use a diverse array of carbon sources (e.g. CO_2)

Drawback: Algae require very large amounts of nitrogen and phosphorous to grow.

Warning flags

EXXON-Mobil: Abandoned the field (spent 200 millon dollars)!

Algenol (leader in algae biofuel) switched in 2015 to carbon capture and water treatment!

Fourth generation biofuels: Advanced solar biofuels

They maybe produced by*:

i) Designer photosynthetic microorganisms to produce photo-biological solar fuels.

ii) Combining photovoltaics and microbial fuel production (Electrobiofuels).

iii) Synthetic cell factories or organelles specifically tailored for the production of desired high value chemicals and biofuels.

***Eva-Mari Aro, Ambio, Jan 2016, 45(Suppl 1), 24-31**
Royal Swedish Academy of Sciences

Fourth generation biofuels: Advanced solar biofuels

They maybe produced by*:

i) Designer photosynthetic microorganisms to produce photo-biological solar fuels.

ii) Combining photovoltaics and microbial fuel production (Electrobiofuels).

iii) Synthetic cell factories or organelles specifically tailored for the production of desired high value chemicals and biofuels.

Closest approach: Audi-Joule technology

$CO_2 + H_2O$ (CNB)$\rightarrow$ Ethanol or HC fuels

CNB = tailored cianobacterias

***Eva-Mari Aro, Ambio, Jan 2016, 45(Suppl 1), 24-31**
Royal Swedish Academy of Sciences

Renewable Organic Acids (2016)

Company	Product	Potential Markets
Myriant	Succinic, fumaric, acrylic	Resins, plasticisers, polyesters
DSM N.V.	Several	Feed and food products
BASF	Several	Same
Cargill	Citric acid	Preservative, flavoring, detergents pharmaceuticals, stabilizer
Novozimes	New enzimes / lactic acid	Enable for various fermentations
Natureworks/Corbion	Lactic acid	Polylactic acid
Metabolix	Polyhidroxyalcalonates	Fully biodegradable use in plastics
Dow chemicals	Acetic acid / vynil acetate	Plastics
DuPont/ADM Avantium Technologie	2,5-Furandicarboxilic acid	PTF replacement for PET
Rivertop renewables	Glucaric acid	Detergents, corrosion inhibitors

Renewable Organic Acids (2016)

Company	Product	Potential Markets
Myriant	Succinic, fumaric, acrylic	Resins, plasticisers, polyesters
DSM N.V.	Several acids	Feed and food products
BASF	Several acids	Same
Cargill	Citric acid	Preservative, flavoring, detergents pharmaceuticals, stabilizer
Novozimes	New enzimes / lactic acid	Enable for various fermentations
Natureworks/Corbion	Lactic acid	Polylactic acid
Metabolix	Polyhidroxyalcalonates	Fully biodegradable use in plastics
Dow chemicals	Acetic acid / vynil acetate	Plastics
DuPont/ADM Avantium Technologie	2,5-Furandicarboxilic acid	**PTF replacement for PET**
Rivertop renewables	Glucaric acid	Detergents, corrosion inhibitors
BioAmber and Mitsui	**Succinic acid**	**Plastics, resins, poliesters**
Renovia	**Adipic acid**	**Nylon and polyurethanes**

Bioadipic acid (Johnson Mathey-Rennovia)
Huge global demand (U$S 6000 x 10^6)

Two steps

1) Glucose + O_2 → Glucaric acid

(Miniplant in operation since 07-19-2015)

Bioadipic acid (Johnson Mathey-Rennovia)
Huge global demand (U$S 6000 x 10^6)

Two steps

1) Glucose + O_2 → Glucaric acid

(Miniplant in operation since 07-19-2015)

2) Glucaric acid + H_2 → Adipic acid (Under construction)

Both plants located at J-M R&D center, Stockton, England

Renewable energy and fuels in America (I)

- **Solar and eolic energies in Argentina.** There are regions of high insolation (NW) and windy areas (Patagonia), ranked among the best in the world.
- In 2006 the law 26190 was approved and in 2010 the goverment announced once more the development of eolic and photovoltaic parks. **But nothing happened.**
- The present government has given a new push to this option. **On 09/05 the first international call opened with offers of energy capacity 6 times larger than expected (6000 Mw).**

Renewable energy and fuels in America (II)

- In the last 20 years **active programs have been executed in other American countries** (Uruguay, Chile, Perú, Brasil, México, USA, Canada) as shown in the next slide.
- **Second generation biofuels.** USA and Brasil are the leaders. Brasil has been using 1G ethanol as car fuel since the late seventies. Now is by far the main fuel in use (> 80% of vehicles).

Non-hydro renewables share of power generation

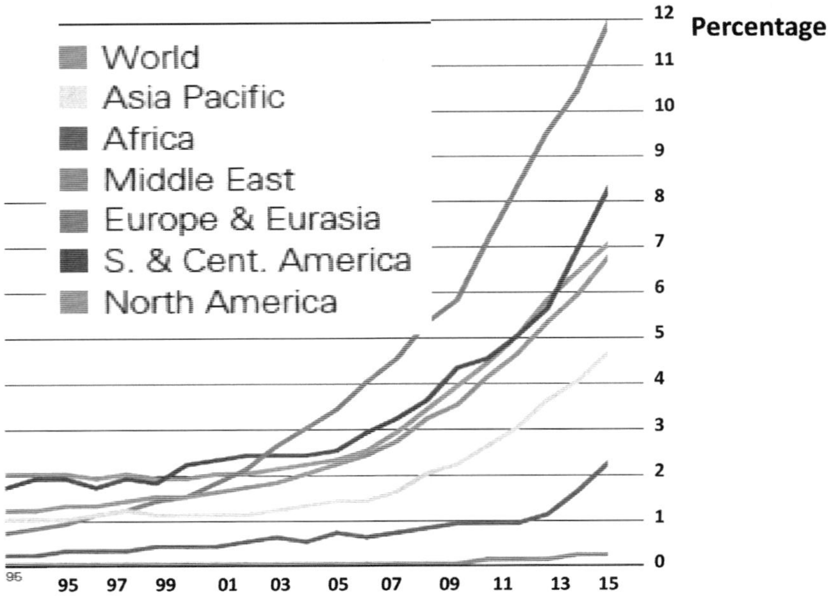

Renewable energy grew by 15%. Wind had the largest growth increment but solar had the highest growth rate (33%). N-h accounted for 6.7% of global power generation up from 2% a decade ago.

Conclusions

- Sharp and continuous increase in the last decade of **wind and solar electricity**. Still represents 7% of total world energy. Technology is still improving. A good investment bet.

Conclusions

- Sharp and continuous increase in the last decade of **wind and solar electricity**. Still represents 7% of total world energy. Technology is still improving. **A good investment bet.**
- ***Biofuels*.** **Lignocellulosic ethanol** most promising. Its production next to 1st generation ethanol plants, best option.
- **Innovations needed** in both HC and ligno ethanol biofuels.

Conclusions

- Sharp and continuous increase in the last decade of **wind and solar electricity**. Still represents 7% of total world energy. Technology is still improving. A good investment bet.
- ***Biofuels*.** **Lignocellulosic ethanol** most promising. Its production next to 1st generation ethanol plants, best option.
- **Innovations needed** in both HC and ligno ethanol biofuels.
- Third (**Algae**) and fourth (**Solar fuels**) generation are still green. **A good promise Audi-Joule approach.**

Conclusions

- Sharp and continuous increase in the last decade of **wind and solar electricity**. Still represents 7% of total world energy. Technology is still improving. A good investment bet.
- ***Biofuels***. **Lignocellulosic ethanol** most promising. Its production next to 1st generation ethanol plants, best option.
- **Innovations needed** in both HC and ligno ethanol biofuels.
- Third (**Algae**) and fourth (**Solar fuels**) generation are still green. **A good promise Audi-Joule approach.**
- **Chemical products from renewable feedstock**: Early stage. A few large demonstration-comercial plants are in operation, e.g. succinic & glucaric acid.

Conclusions

- Sharp and continuous increase in the last decade of **wind and solar electricity**. Still represents 7% of total world energy. Technology is still improving. A good investment bet.
- ***Biofuels***. **Lignocellulosic ethanol** most promising. Its production next to 1st generation ethanol plants, best option.
- **Innovations needed** in both HC and ligno ethanol biofuels.
- Third (**Algae**) and fourth (**Solar fuels**) generation are still green. **A good promise Audi-Joule approach.**
- **Chemical products from renewable feedstock**: Early stage. A few large demonstration-commercial plants are in operation, e.g. succinic & glucaric acid.
- The **simultaneous production of biofuels and chemicals** is a desirable option to increase rentability of novel technologies.

Suspension Bridge, Santa Fe (Argentina)

Thanks for your attention

Backup Slides

Producción de etanol a partir de residuos lignocelulósicos (INEOS at Vero Beach, Fl)

INEOS Group compró desde 1998 plantas de Dow, Rhodia, BASF, Chevron, Philips, Monsanto, Hoesch y BPchemicals (ventas 45 x 10^9 u$s)

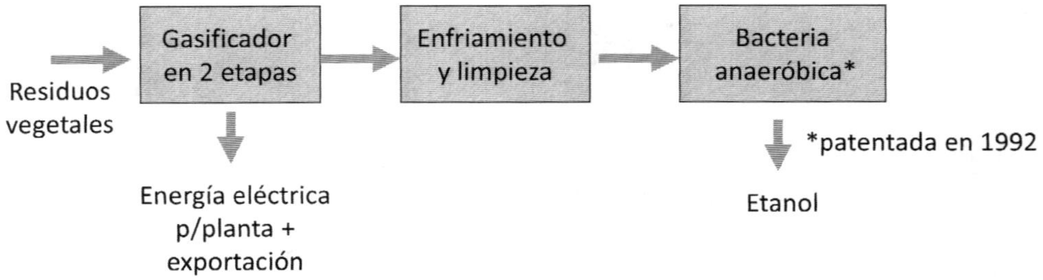

La planta produce 300-330 l de etanol por tonelada seca de residuo.

Las bacterias anaeróbicas convierten 95-98% del residuo vegetal.

Problema: Al ir de escala piloto a pionera se detuvo la producción de alcohol

Causa: Bajísima concentración de HCN que mata las bacterias
Solución posible: Absorber selectivamente el HCN de la corriente de gas

El scrubber fue instalado en 1915 pero la planta sigue sin funcionar

FODA de biorefinerías

Fortaleza: **Gran disponibilidad de materia prima.**

Oportunidad: **Se requiere gran variedad de catalizadores y de procesos.**

Debilidad: Depende de decisiones políticas poco predecibles (Argentina).

Amenaza: Producción de shale gas y oil nacional abundante (Argentina posee la 2ª reserva más grande de shale gas del mundo: 22 x 10^{12} m^3)

Argentina: Total primary energy supply (2000-2012)

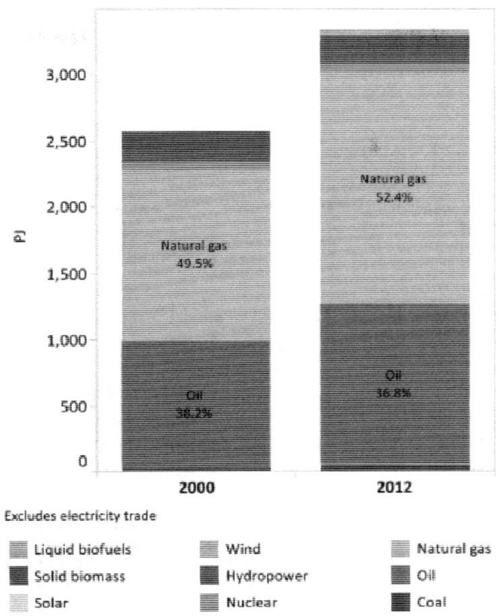

Argentina: Total primary energy supply (2000-2012)

2. Statistics

Total Primary Energy Supply

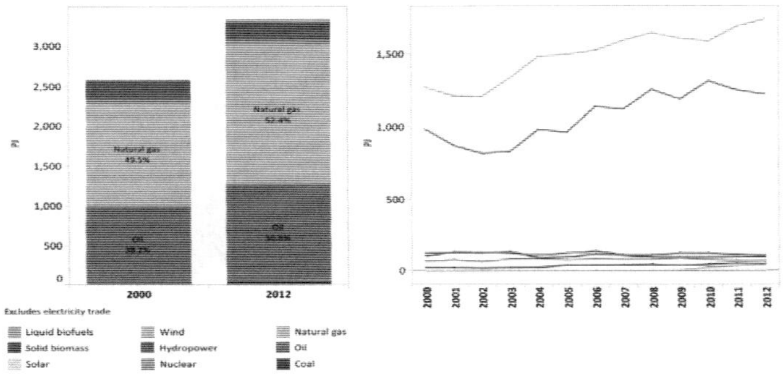

		Total Primary Energy Supply	Share of renewables
2000	Total	2,578.1 PJ	
	Of which renewables	227.5 PJ	8.8%
2012	Total	3,359.3 PJ	
	Of which renewables	245.3 PJ	7.3%

Total includes electricity trade

		Total Primary Energy Supply	Share in total renewables
2012	Wind	1.3 PJ	0.5%
	Solar	0.0 PJ	0.0%
	Liquid biofuels	40.1 PJ	16.4%
	Solid biomass	98.4 PJ	40.1%
	Hydropower	105.4 PJ	43.0%

Mercado mundial de combustibles

Producción mundial de petróleo	**4600 x 10^6 Ton**
Producción mundial de etanol	**75 x 10^6 Ton**
Producción mundial de biodiesel	**47 x 10^6 Ton**
Producción de básicos petroquímicos:	**255 x 10^6 Ton**

La producción de biocombustibles y de petroquímicos a partir de biomasa solo se justifica por mejorar la calidad ambiental

Los gobiernos de muchos países han dictado leyes y creado organismos que catalizan la producción de biocombustibles

- **Brasil durante años ha subsidiado la industria alcoholera**
- **USA lo mismo aplicado al alcohol de maíz y ahora también al alcohol lignocelulósico**
- **Unión Europea también tiene leyes promocionales para desarrollo de fuentes de energía limpias (Residuos agrícolas y forestales, eólica, geotérmica y solar)**

WTW(1) reducción de emisiones GHG(2) para 5 materias primas (relativas a gasolina) calculadas con el modelo GREET(3)

Reducción de emisiones WTW(1)–GHG(2)	Maíz	Caña de azúcar(5)	Marlo maíz	Switchgrass (6)	Miscanthus (7)
Incluyendo LUC(4)	34%	51%	96%	88%	108%
Excluyendo LUC(4)	44%	68%	94%	89%	95%

(1) Well to wheel

(2) GHG: gases de efecto invernadero (ej. CO_2)

(3) GREET: Greenhouse gases, Regulated Emissions and Energy use in transportation

(4) LUC: Land use change

(5) Atención se calculó usando azúcar de Brasil. Si este balance se hiciera en Brasil se acentuaría la diferencia a favor de la caña de azúcar vs. maíz

(6) Switchgrass: pasturas altas silvestres

(7) Otra variedad de pasturas silvestres con mejor rendimiento de etanol

Outline

- Biorefining opportunities
- Example of lignin
- Some reference cases
- Challenges and opportunities

Bioproduct pulp mill

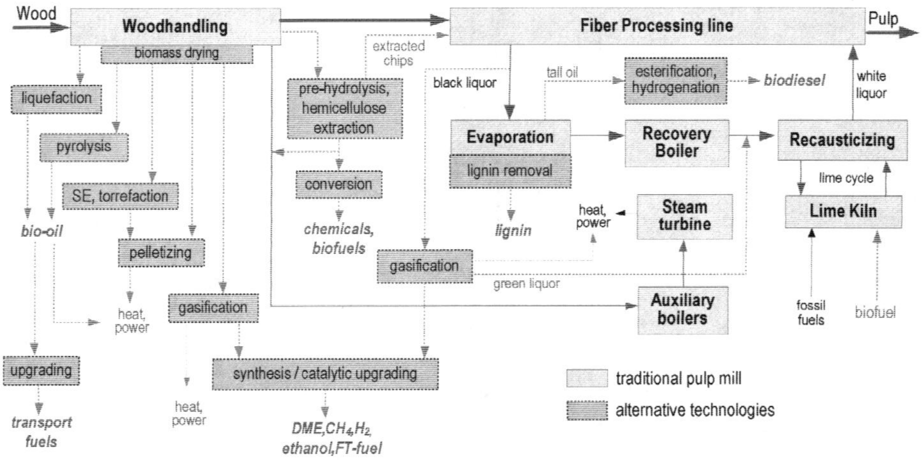

Bioproduct pulp mill

Metsä Fibre Äanekoski, Finland (Start-up Q3 2017)

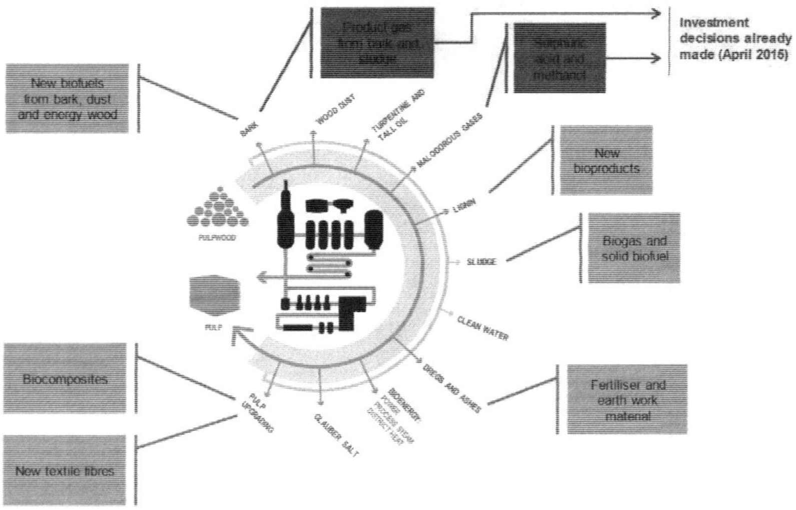

Source: Weymarn 2015, Metsä Fibre

Heat and Power

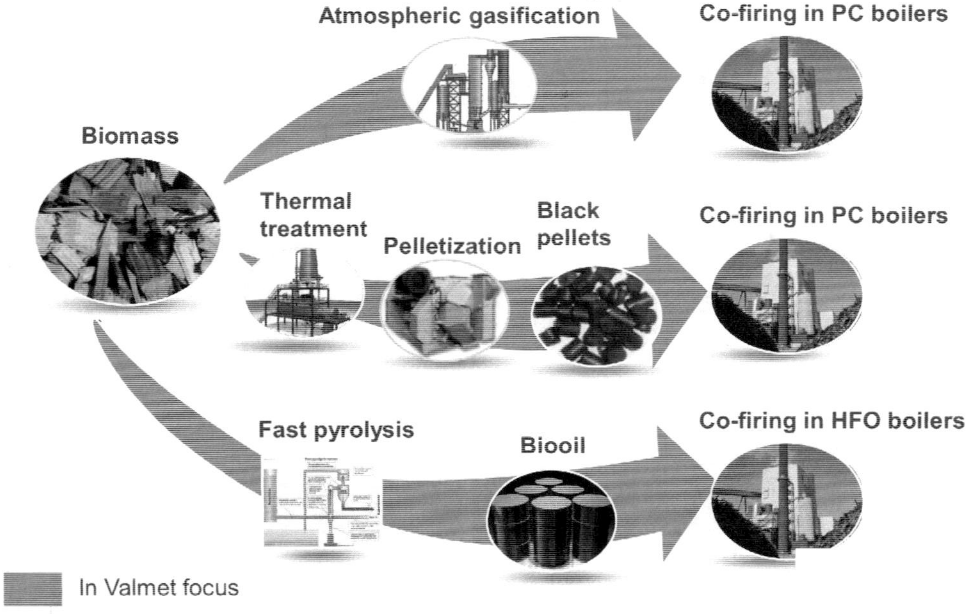

Transportation fuels

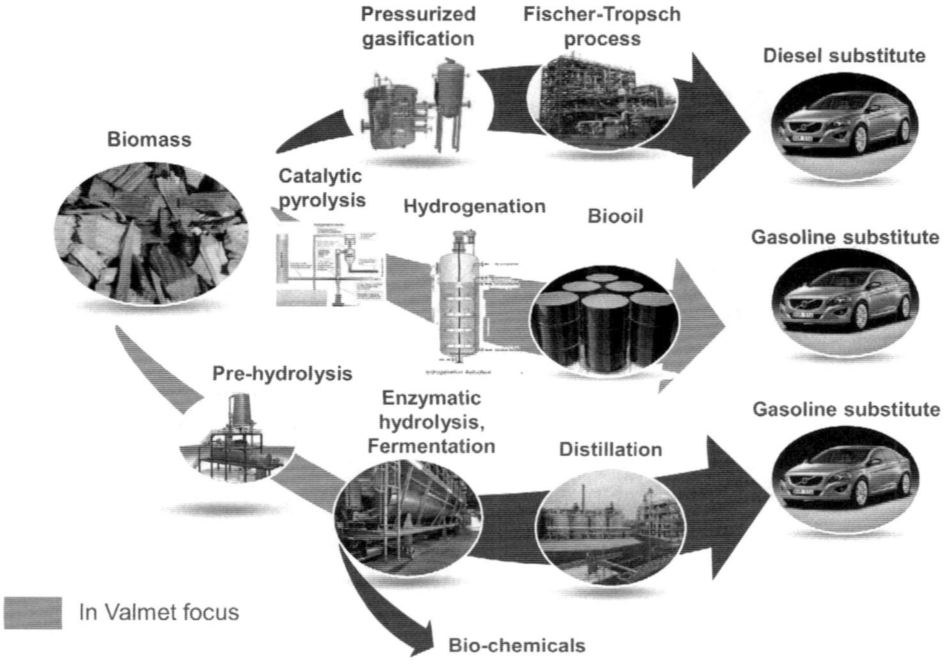

Biomaterials

LignoBoost and focus on lignin

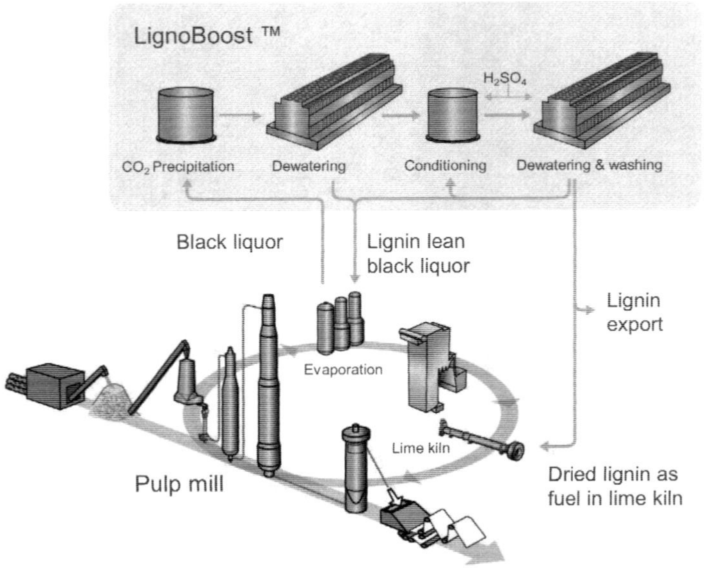

Valmet

Lignin

Nature's own binding agent

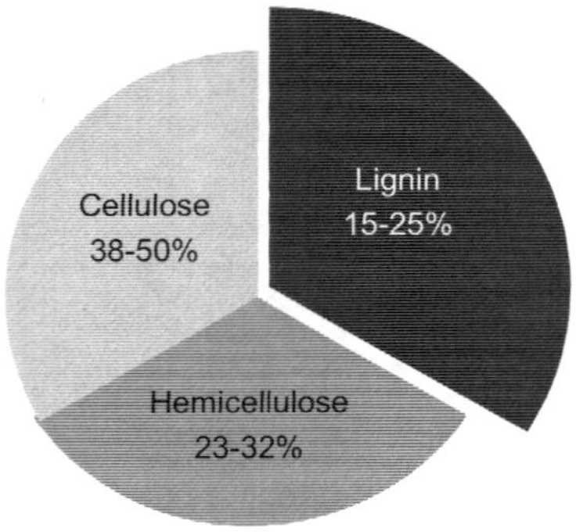

Approximate amount of lignin in black liquor, kg/ton pulp

- Spruce 510
- Birch 340
- Eucalyptus 340

> 50,000,000 t/a of lignin in world wide kraft pulp mills

C	64–66%
O	26–27.5%
H	5.7–5.8%
N	0.1%
Na	0.1–0.3%
S	1–3% (typically 2–2.5%)
HHV	25–26.5 MJ/kg

Lignin content: 92–98%

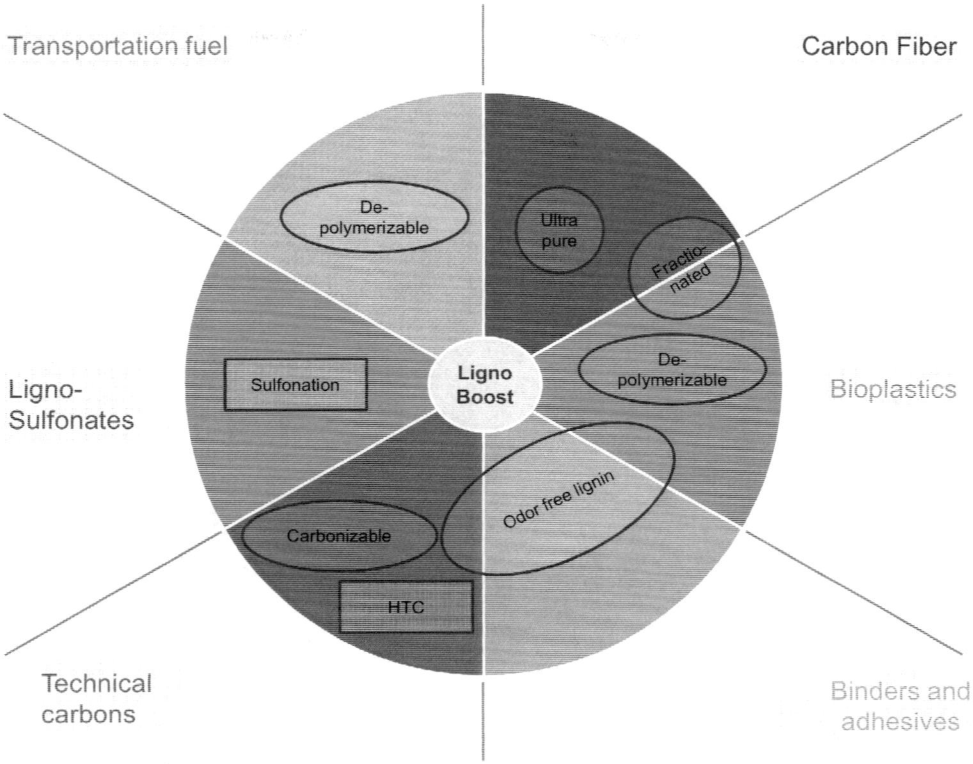

LignoBoost plants in operation

LignoBoost Demo AB, Bäckhammar, Sweden, 2006

- Annual capacity of 10,000 tons as dry lignin (65% DS)
  - Demonstration and research plant

Domtar Plymouth, USA, 2013

- Annual capacity of 25,000 tons as dry lignin (65% DS)
  - Increase of pulp production capacity by 5% and efficiency by off-loading the recovery boiler
  - Domtar sells lignin as BioChoice™ product

Stora Enso Sunila, Finland, 2015

- Annual capacity of 50,000 tons as dry lignin (95% DS)
  - Decreased use of fossil fuel
  - First step towards a Biorefinery
    - Sell lignin to new bio market

Replacing Fossil Fuels with Renewable Energy

Circulating fluidized bed gasifier plant that converts biomass into product gas

Gasifier added to an existing power plant to replace 40% of coal by biomass

Total investment cost ca. 40 M€.

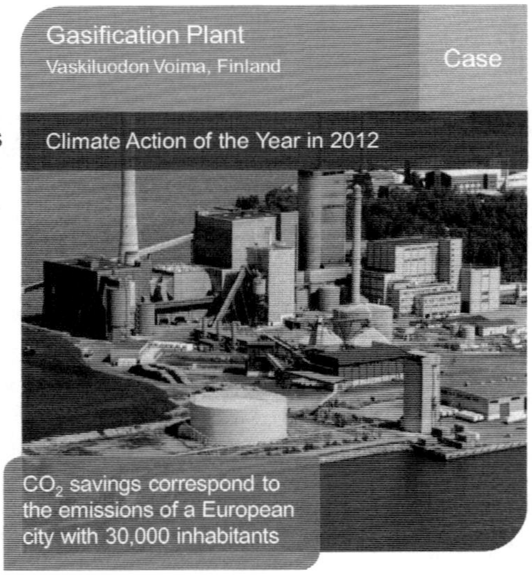

Key results

- Fossil fuels can be partly or completely replaced with renewable energy
- Existing assets utilized and power plant life-time significantly extended

Additional Revenue from Bio-oil Production (1/2)

Fast pyrolysis of biomass integrated to a fluidized bed boiler

Production of bio-oil by pyrolysis from forest residue and other biomass

Renewable bio-oil production of 50,000 t per annum

Investment of ~30M EUR

Key results

- Results in significant reductions in CO_2 emissions
- Bio-oil can replace heavy fuel oil in heat and power generation and might be further refined to transportation fuel

Additional Revenue from Bio-oil Production (2/2)

- Heating value 13-18 MJ/kg (LHV)
- Water content 20-30 wt%
- Viscosity: between light and heavy fuel oils
- High ignition temperature
- Acidic, pH 2-3
- Flash point 40-65 °C
- Density 1.2 kg/l
- Polymerizes slowly
- Immiscible with mineral oils

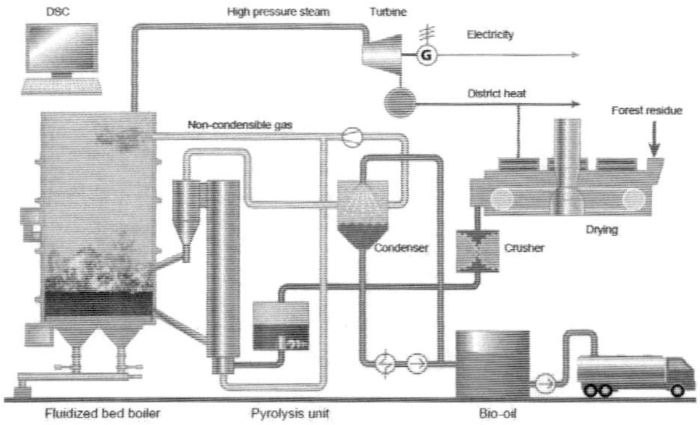

Gasification Plant for Energy Production (1/2)

- Gasifier plant that combusts waste into product gas
  - Reduction of fossil fuels and emissions
  - 170,000 tons of coal replaced by waste-fuelled energy
  - 50 MW of electricity
  - 90 MW of district heat
  - Total investment ~€160M

Key results

- Highest possible electrical efficiency achieved from waste to energy
- Reduction of use of fossil fuels and CO_2 emissions

Gasification Plant for Energy Production (2/2)

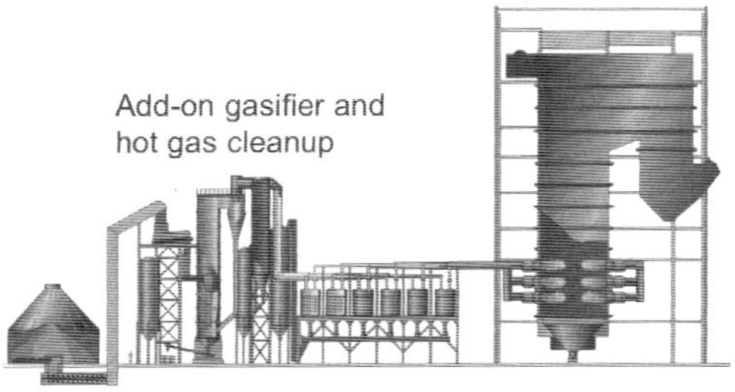

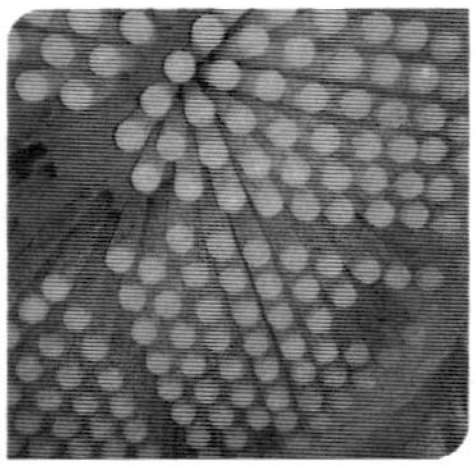

Not allowed:

- Diapers and similar hygiene products
- Packaging that contains food
- Leather, e.g. shoes, bags
- Videotapes, DVDs and CDs, PVC, etc
- Waste that contains aluminum, metals.

Fuel Production Technology for Biogas Vehicles

Indirect gasification of forest residues and wood pellets to produce synthetic natural gas

A 20MW demonstration plant based on indirect gasification

Key results

- Synthetic natural gas is comparable to natural gas and can also be used to replace fossil vehicle fuels

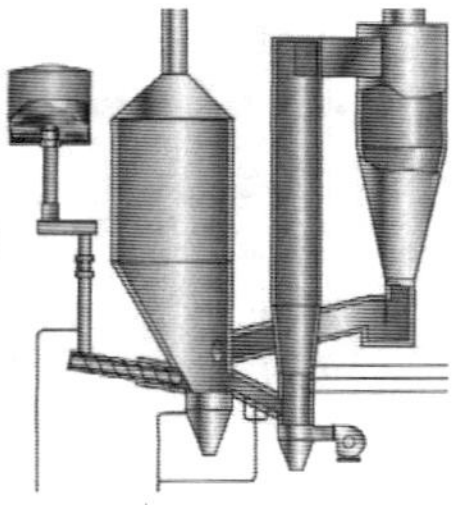

Challenges and Opportunities

- Non conventional technologies have reached good levels of maturity.
- Some resistance to new technologies
  - Need for more consistent government policies, but also people's attitude towards a more sustainable economy
  - Powerful fossil fuel lobby
- High investment cost for novel technologies (risk)
  - Feasibility also influenced by feedstock quality, electricity/fossil sales price, scale and process maturity/efficiency.
- Important to work with partners/end users for developing and expanding the biorefining market
- Availability of biomass: planted forests and agriculture residues

"Everything that's made with fossil material today can be made from a tree tomorrow"

Karl-Henrik Sundström, CEO, Stora Enso

THANK YOU!
Valmet
FORWARD
MEMBER OF
Dow Jones
Sustainability Indices
In Collaboration with RobecoSAM

Second-Generation Ethanol: The Need is Becoming a Reality

Conference: RCN Conference on Pan American Biofuel and Bioenergy Sustainability

Year: 2016

Proceeding: 2nd RCN Conference on Pan American Biofuels & Bioenergy Sustainability

Group: Abstracts

Session: Session 7: The Future of the Bioenergy Industry/El Futuro De La Industria De La Bioenergía

Time: Friday, September 16, 2016 - 9:20am-9:45am

Authors:

Pereira, G., GranBio

Bioethanol is currently one of the most important sources of renewable fuel and has the potential to partially replace petroleum-based liquid fuels. The United States is the largest ethanol producer in the world, with an average production of 54 billion liters per year. Most of its production derives from cornstarch. The Carbon Intensity Index (CI) for corn starch derived ethanolâ75.97g of carbon dioxide equivalent (CO2e) emitted per megajoule (MJ) of energy producedâis close to the gasoline index (95.86g of CO2e/MJ), and the use of corn starch derived ethanol has little impact on the emission of greenhouse gases. Compared to ethanol from cornstarch, production using sugarcane has a lower CI (56.66g of CO2e/MJ), and greater emission reductions can be achieved by replacing gasoline or other fuels with ethanol obtained from sugarcane. Brazil, the second largest world producer of ethanol, produces about 30 billion L/y of ethanol and has the largest flex-fuel vehicle fleet in the world. Nevertheless, the Brazilian ethanol sector is in a state of economic stress caused mainly by the high operational costs and the fact that sugarcane productivity seems to have reached a plateau. In this scenario, conversion of the sugars present in lignocellulosic biomass into ethanol emerges as a promising technology with the potential to revolutionize the renewable fuels sector. In this talk, we will present the advances and challenges in producing and fermenting lignocellulosic sugars, demonstrating the huge potential that this technology represents for the renewable fuels sector.

Quantifying the Challenges of Adopting Sustainability Criteria for Biofuel Production

Written by Damiana Serafini
IISD-International Institute for Sustainable Development

March 13th, 2016
This paper was commissioned by IATA

Publication link: http://www.iisd.org/library/quantifying-challenges-adopting-sustainability-criteria-biofuel-production

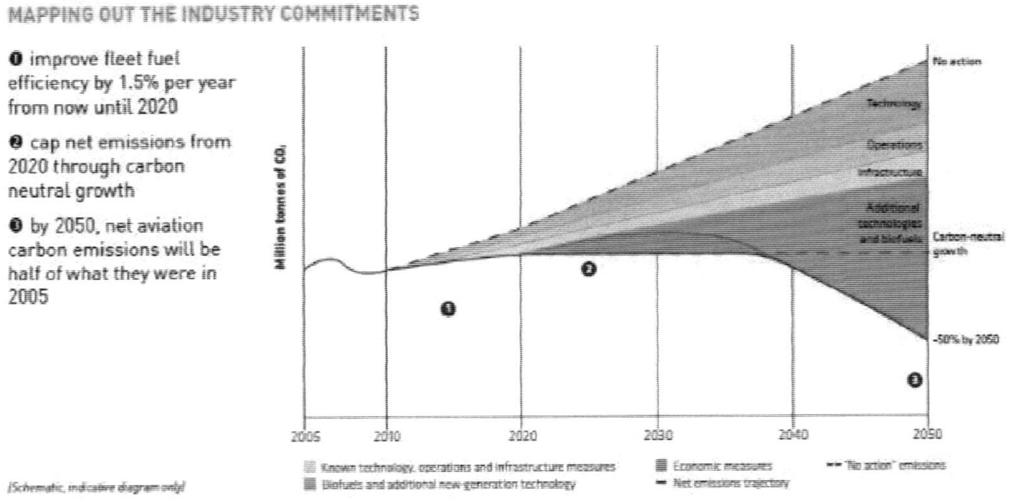

Operators Certified by RSB				
Group	Entity	Region	Product	Feedstock
1	GreenWood Tree Farm Fund, LP	USA	Biomass Producer	Wood Chips-Hybrid Poplar
	Biomass Supplies Pvt Ltd	Asia-Sri Lanka	Biomass Producer	Wood chips-Gliricidia trees
	Sunchem Biofuel Development South Africa	South Africa	Feedstock Producer	Solaris Seed Tobacco
2	Camelina Company Spain	EU-Spain	Biodiesel and aviation kerosene producer	Camelina
	Shoalhaven Starches Pty Ltd (Manildra Group of Companies)	Oceania-Australia	Ethanol Producer	Waste starch from wheat processing
	Beijing Shougang LanzaTech New Technology Co., Ltd.	Asia-China	Ethanol Producer	CO feed gases via microbial fermentation
3	SkyNRG	EU-Netherlands	Supply Chain & Logistics for Biokerosene/Jetfuel	Used cooking oil and other feedstocks
4	Dynamic Fuels LLC	North America-USA	Renewable Diesel / Biojet Mix	Wastes, animal by-products, greases and vegetable oils

Round Table On Sustainable Materials Principles (www.rsb.org)

Principle 1: Legality

Principle 2: Planning, Monitoring and Continuous Improvement

Principle 3: Greenhouse Gas Emissions

Principle 4: Human and Labour Rights

Principle 5: Rural and Social Development

Principle 6: Local Food Security

Principle 7: Conservation

Principle 8: Soil

Principle 9: Water

Principle 10: Air

Principle **11:** Use of Technology, Inputs and Management of Waste

Principle 12: Land Rights

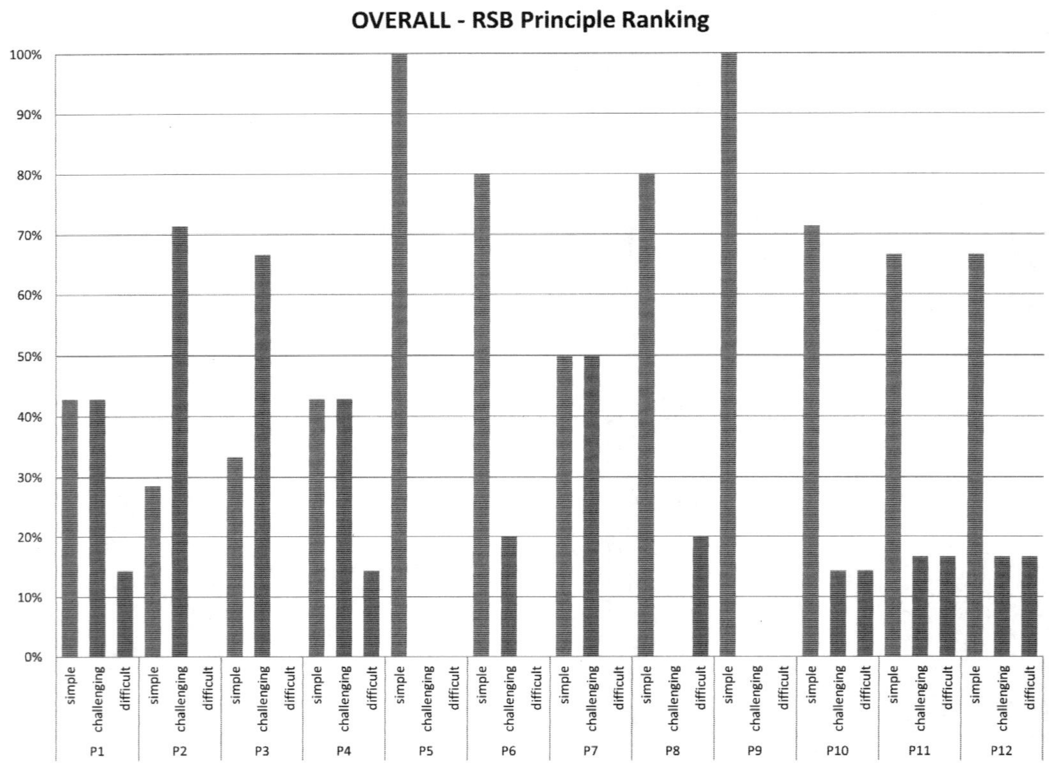

Thank you for your attention!

Muchas gracias por su atencion!

Damiana Serafini
Expert on Renewable Energies and Sustainable Development
damiana.serafini@gmail.com

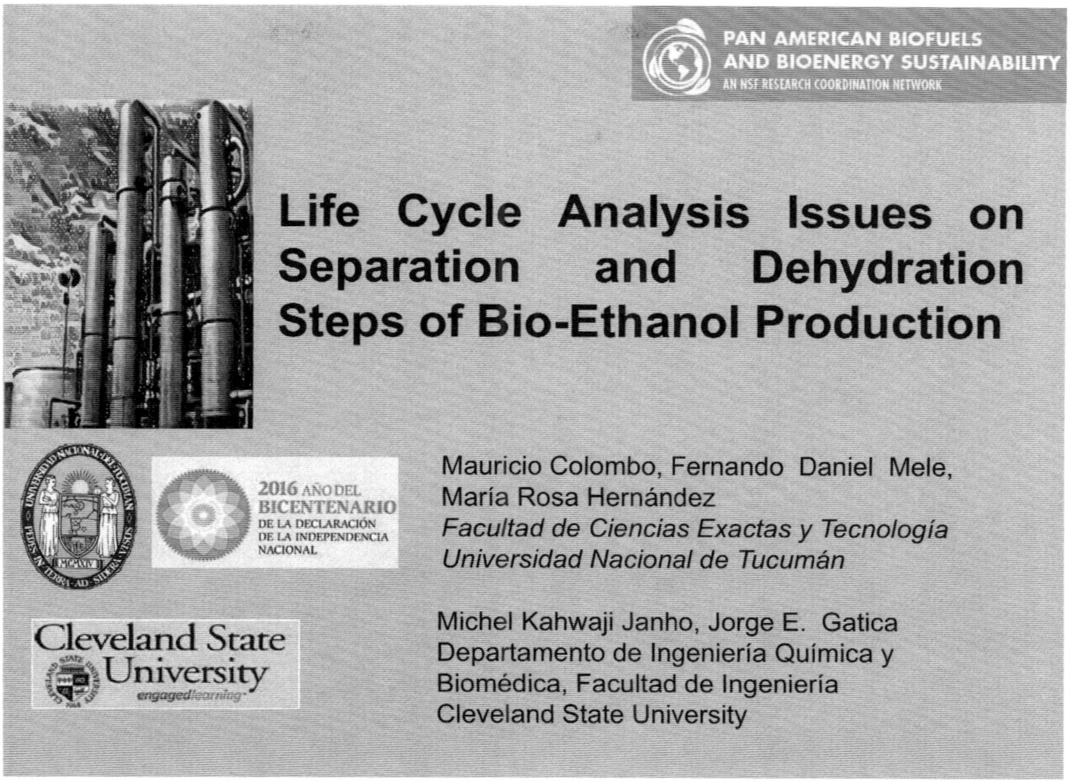
PAN AMERICAN BIOFUELS
AND BIOENERGY SUSTAINABILITY
AN NSF RESEARCH COORDINATION NETWORK
Life Cycle Analysis Issues on Separation and Dehydration Steps of Bio-Ethanol Production
2016 AÑO DEL BICENTENARIO DE LA DECLARACIÓN DE LA INDEPENDENCIA NACIONAL
Mauricio Colombo, Fernando Daniel Mele, María Rosa Hernández
Facultad de Ciencias Exactas y Tecnología
Universidad Nacional de Tucumán
Cleveland State University
engagedlearning
Michel Kahwaji Janho, Jorge E. Gatica
Departamento de Ingeniería Química y Biomédica, Facultad de Ingeniería
Cleveland State University

Cadena de suministros de la caña de azúcar en la Argentina
300,000 hectáreas
Cosecha, 21 millones ton
23 ingenios y 19 destilerías
ETANOL
220 millones litros
AZÚCAR
2,3 millones ton
BAGAZO
ELECTRICIDAD
combustible
alimentos
medicamentos
alco-química
alim.
sucro-química
papel

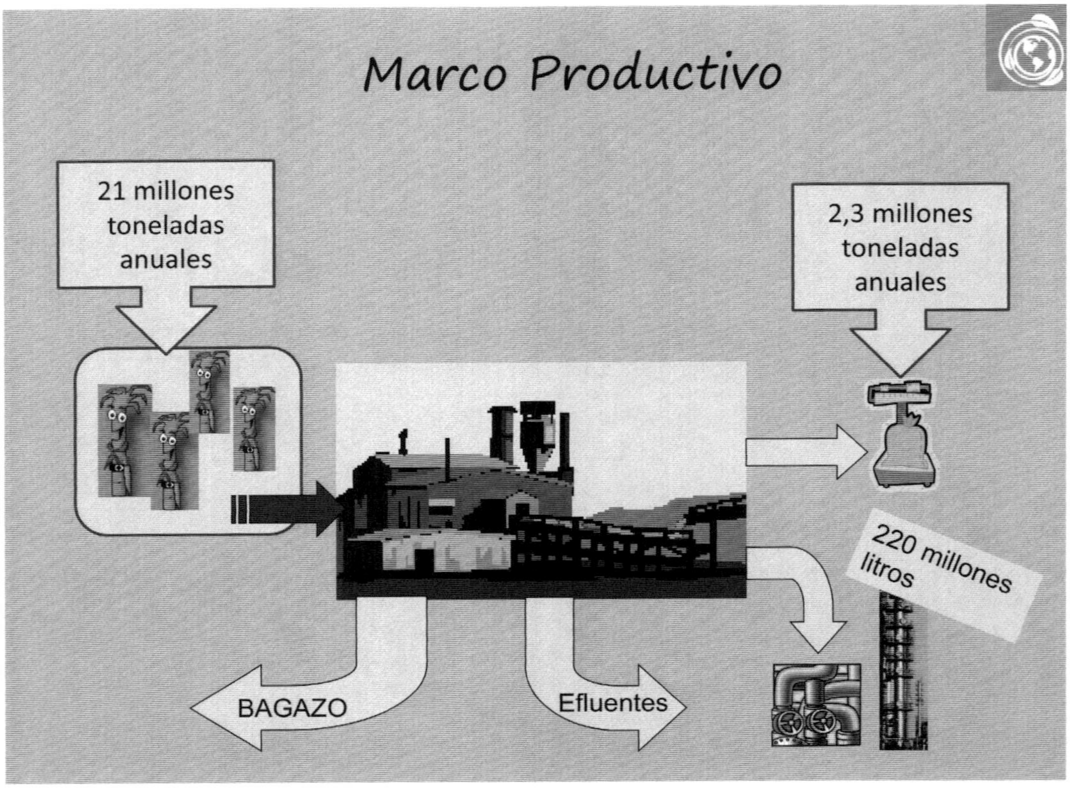
Marco Productivo
21 millones toneladas anuales
2,3 millones toneladas anuales
220 millones litros
BAGAZO
Efluentes

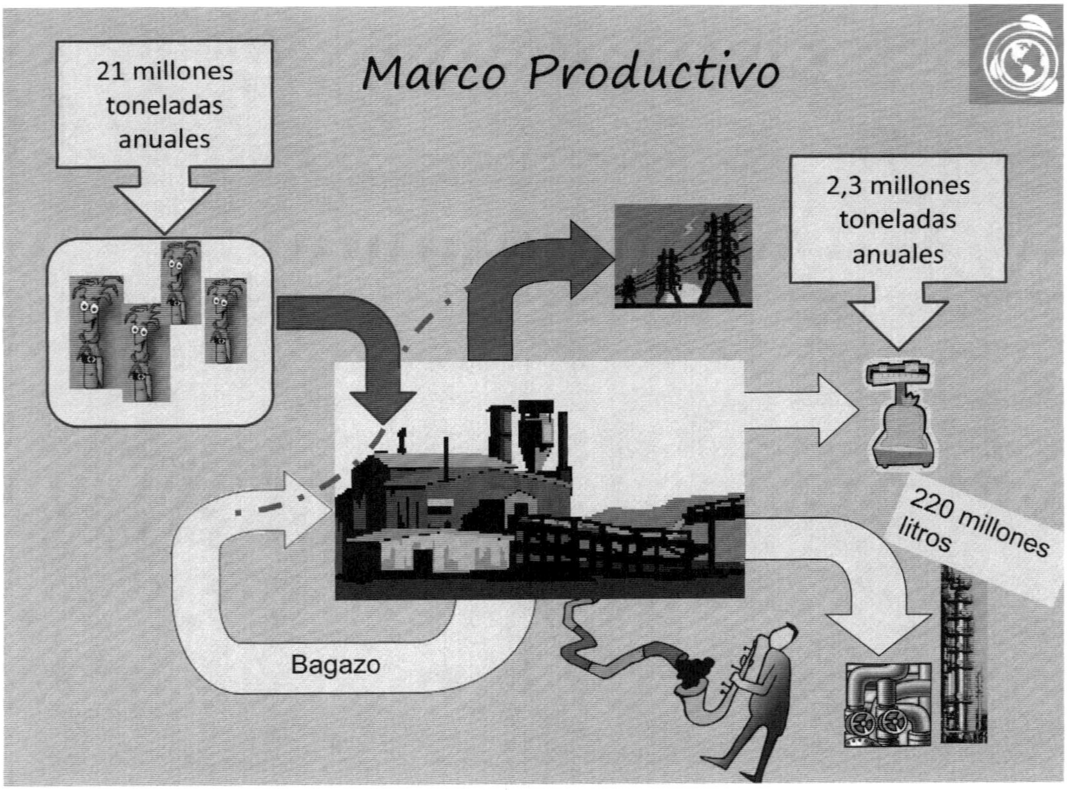
Marco Productivo
21 millones toneladas anuales
2,3 millones toneladas anuales
220 millones litros
Bagazo

Diagrama de flujo de un complejo sucro-alcoholero

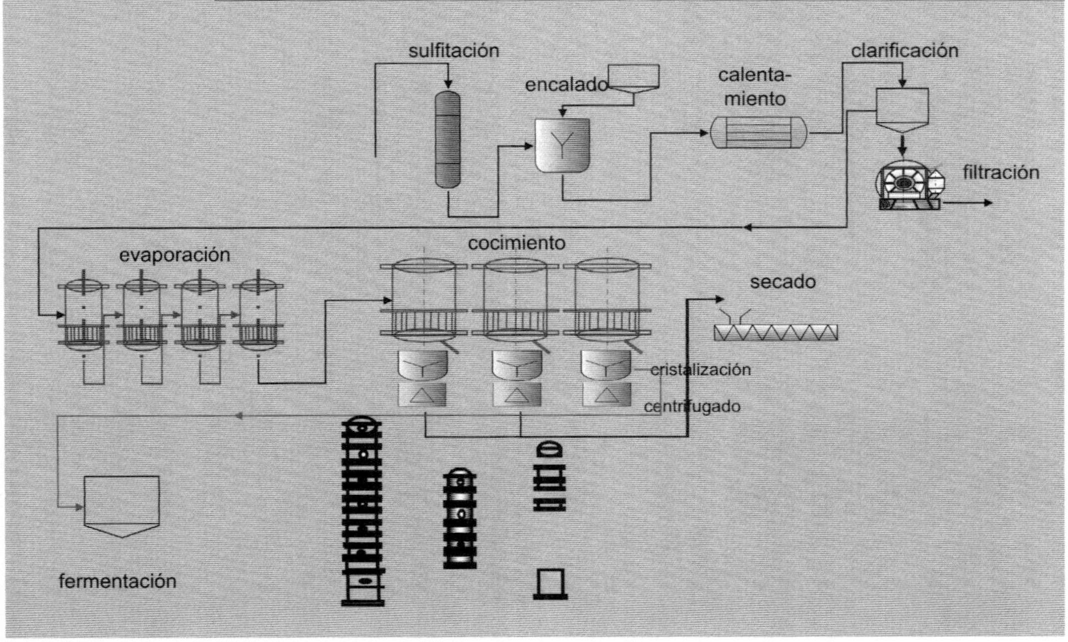

Formas posibles de eliminación de vinaza

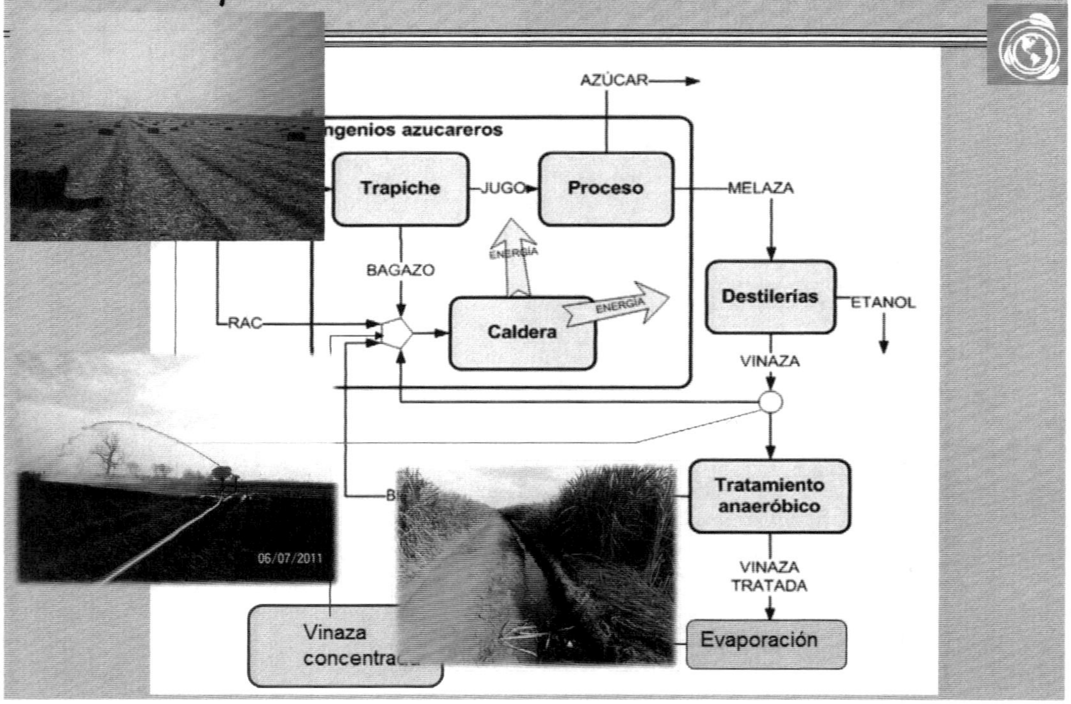

Componente	kmol/h
Acetato de etilo	0.00953
Etanol	40.20
2- Propanol	0.0928
1- Propanol	0.0928
Agua	1215.8
1- Pentanol	0.023

Simulación caso base

Paquete termodinámico : UNIQUAC, UNIFAC-HOC

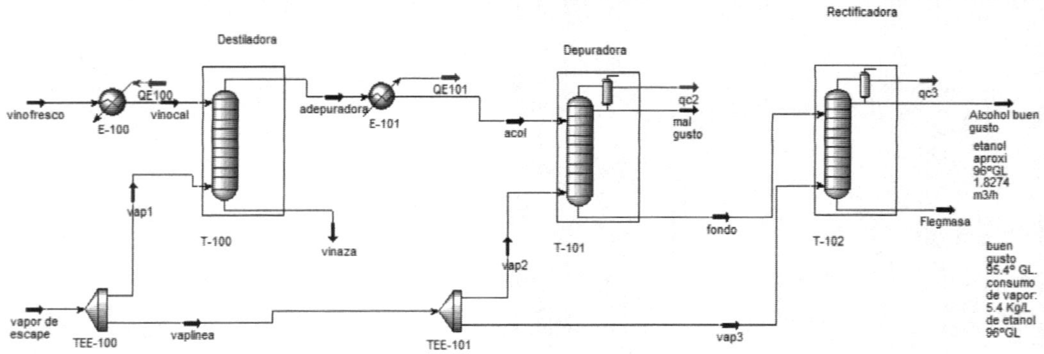

OBJETIVO : Disminuir el consumo de vapor y la generación de vinaza

$$\varphi = \frac{\sum_i V_i[kmol/h] + Vinaza[kmol/h]}{ABG[kmol/h]}$$

Cómo lo conseguimos?

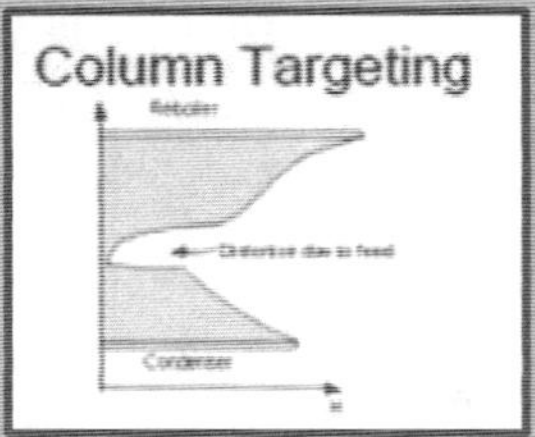

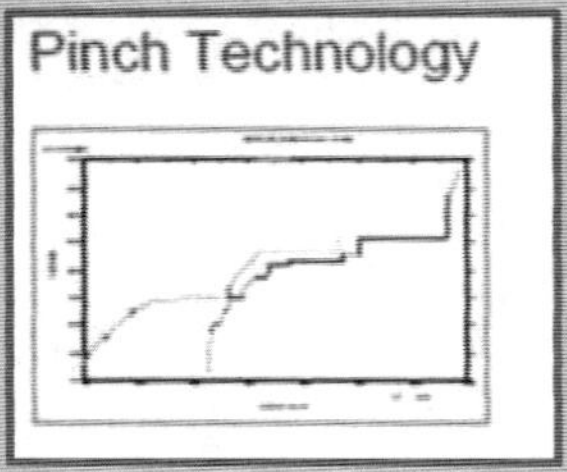

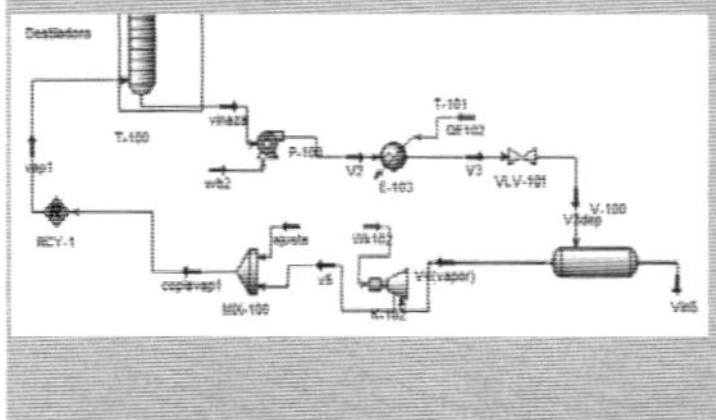

1. Minimizando el consumo de energía de cada columna en particular.
2. Integrando energéticamente el tren de destilación para disminuir el consumo de vapor.
3. Evaporando parcialmente vinaza persiguiendo a la vez la disminución del caudal de vinaza y la disminución de consumo de vapor a través de recirculación.

Recirculación de vinaza

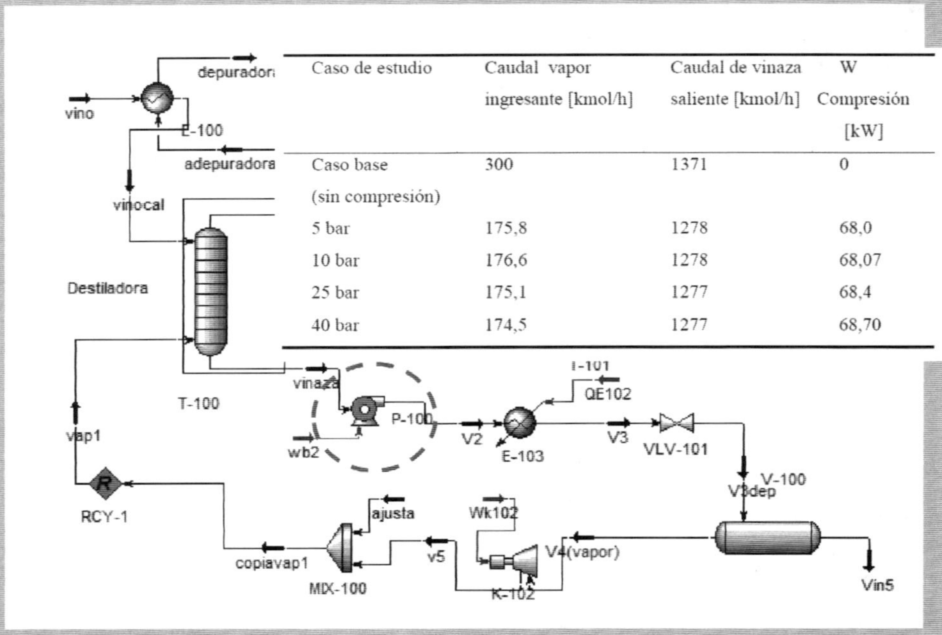

Caso de estudio	Caudal vapor ingresante [kmol/h]	Caudal de vinaza saliente [kmol/h]	W Compresión [kW]
Caso base (sin compresión)	300	1371	0
5 bar	175,8	1278	68,0
10 bar	176,6	1278	68,07
25 bar	175,1	1277	68,4
40 bar	174,5	1277	68,70

Recirculación de vinaza

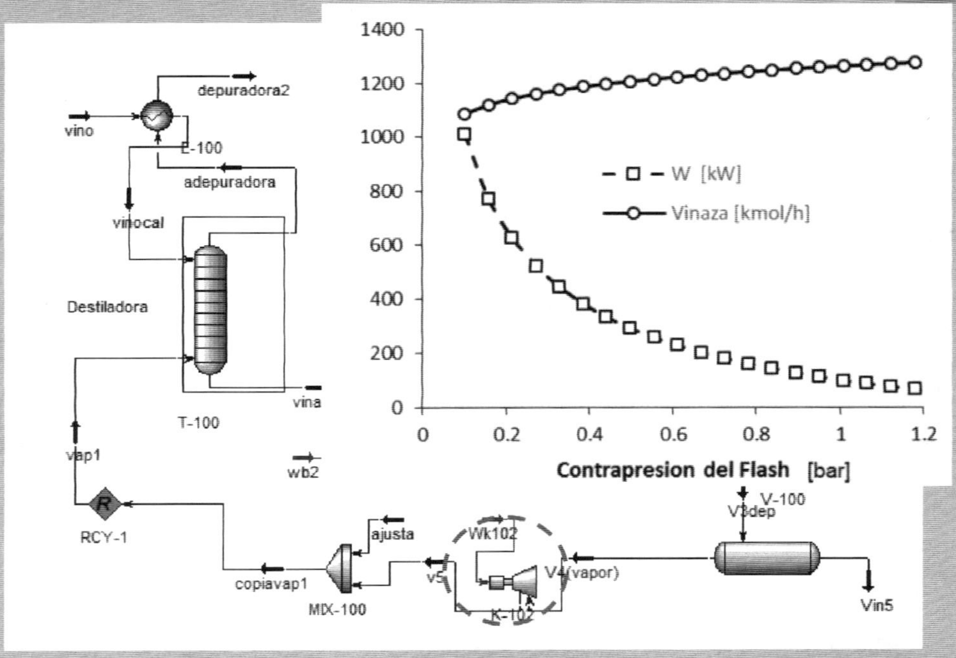

Recirculación de vinaza

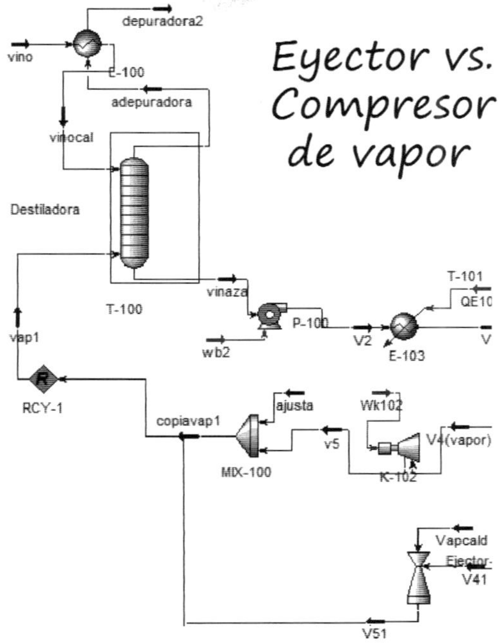

Eyector vs. Compresor de vapor

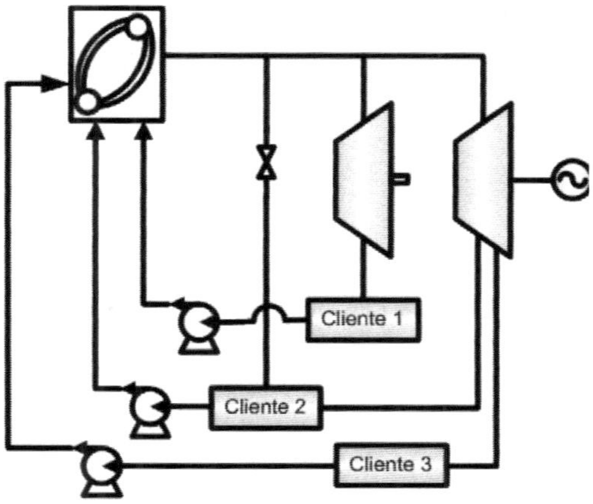

Recirculación de vinaza

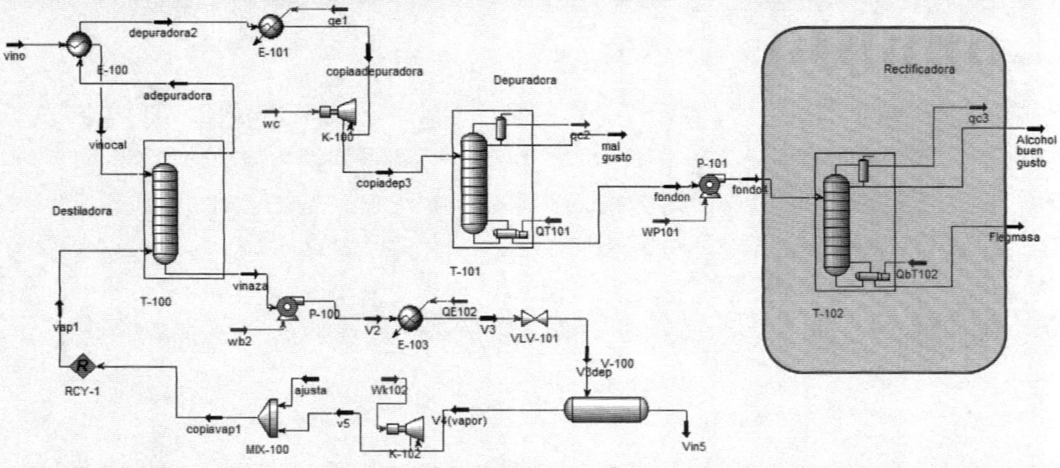

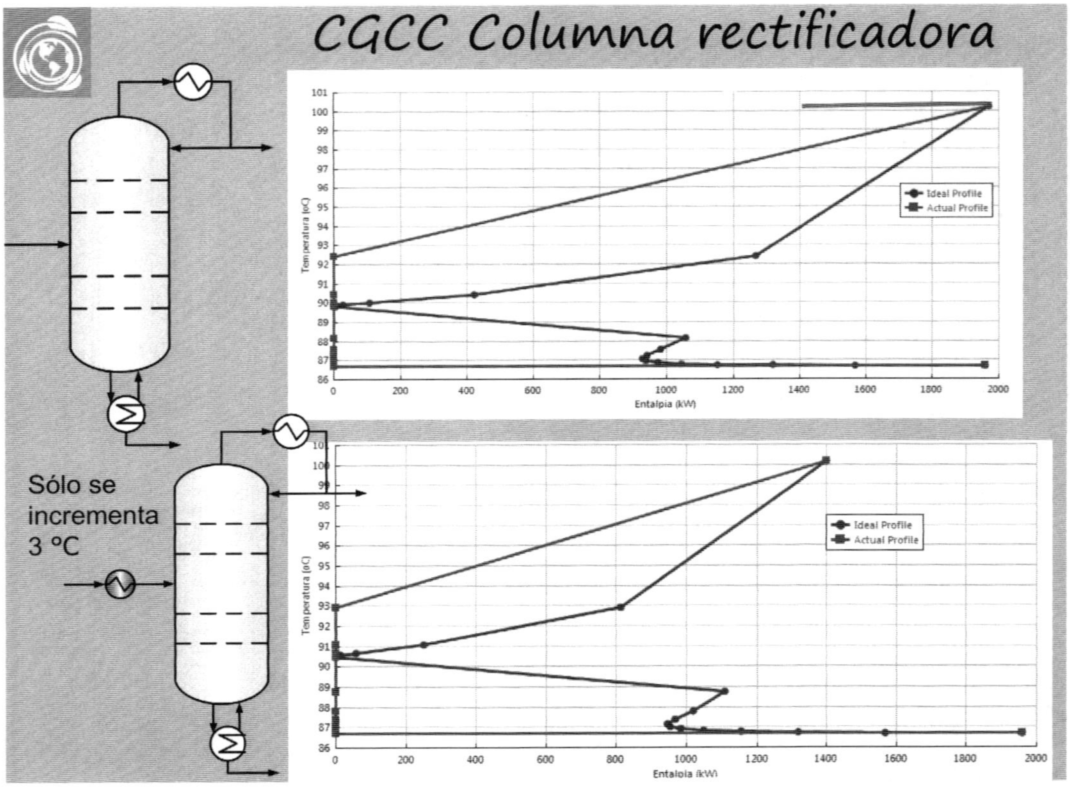
CGCC Columna rectificadora
Ideal Profile
Actual Profile
Temperatura (oC)
Entalpia (kW)
Sólo se
incrementa
3 °C
Ideal Profile
Actual Profile
Temperatura (oC)
Entalpia (kW)

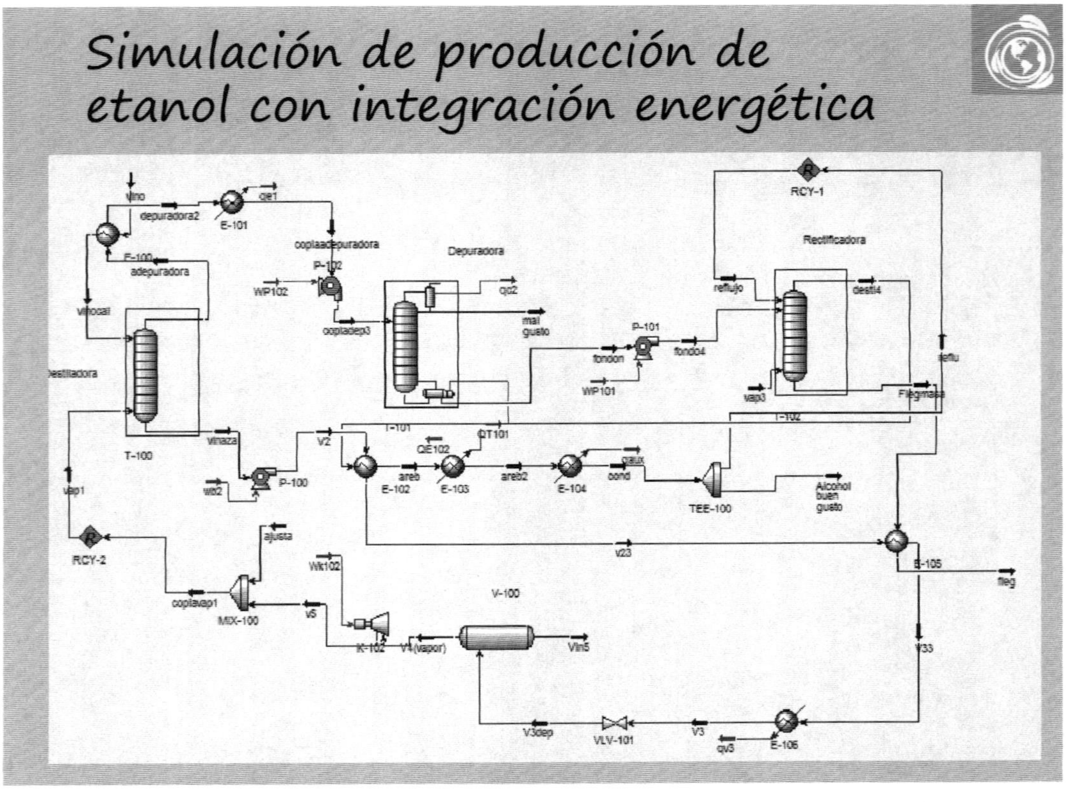
Simulación de producción de
etanol con integración energética
RCY-1
Rectificadora
Depuradora
E-101
P-102
P-101
T-100
P-100
E-102
E-103
E-104
TEE-100
Alcohol
buen
gusto
E-105
RCY-2
MIX-100
V-100
K-102
VLV-101
E-106

Resultados de la simulación

Caso	$\sum V_i$ [kmol/h]	Vinaza [kmol/h]	Etanol [kmol/h]	Φ
Base	549	1371	33,27	57,7
Evaporación parcial vinaza	455,46	1095	32,76	47,32
Integración energética	387,3	1095	32,71	45,3

Se reduce la generación de vinaza en un 25 % aproximadamente

Se reduce el consumo de vapor en un 30 % aproximadamente

LCA de los 3 casos estudiados

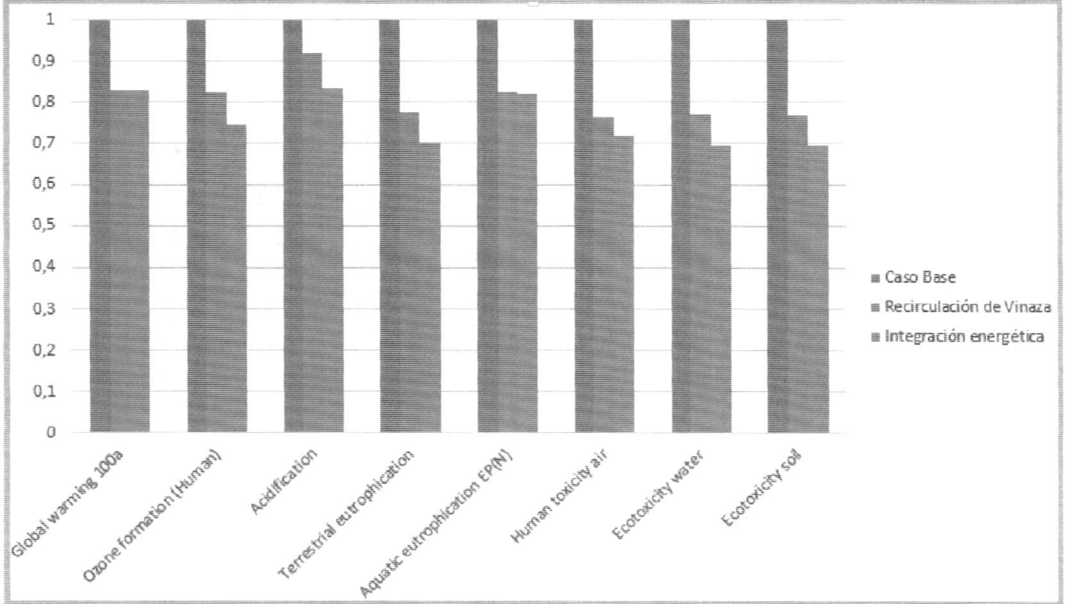

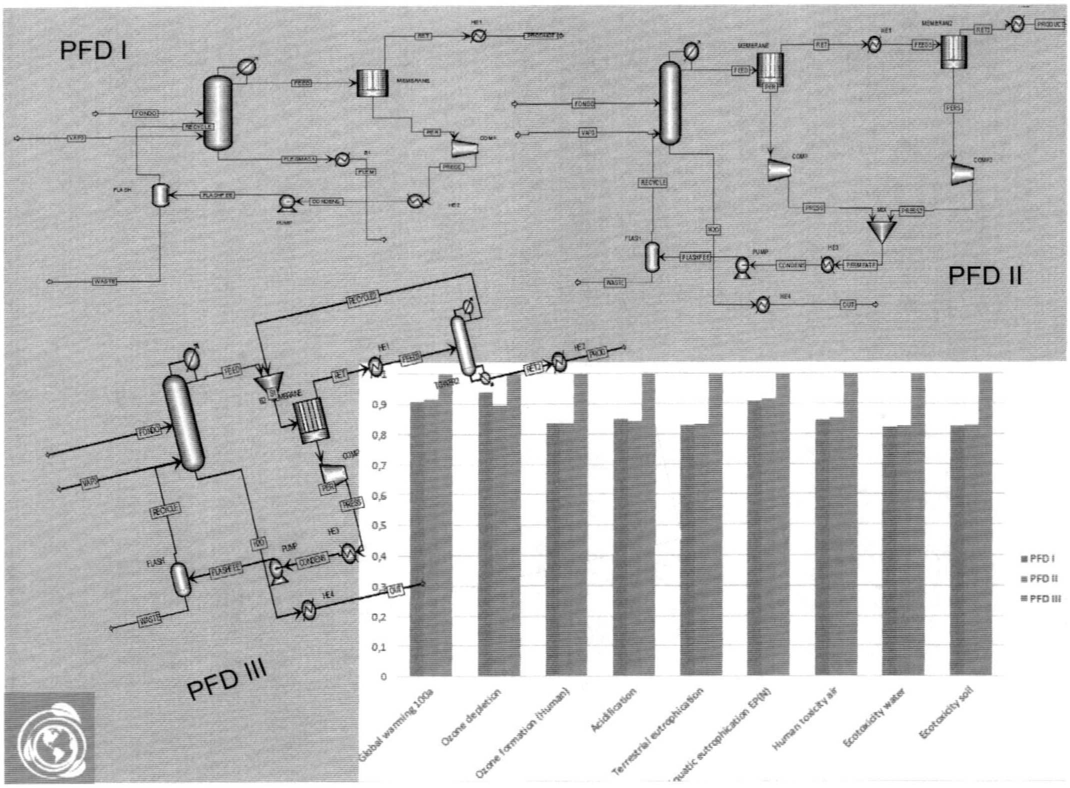

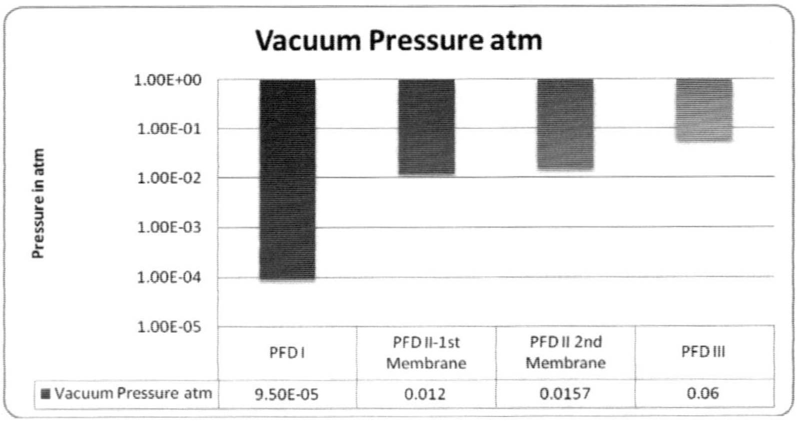

	PFD I	PFD II-1st Membrane	PFD II 2nd Membrane	PFD III
Vacuum Pressure atm	9.50E-05	0.012	0.0157	0.06

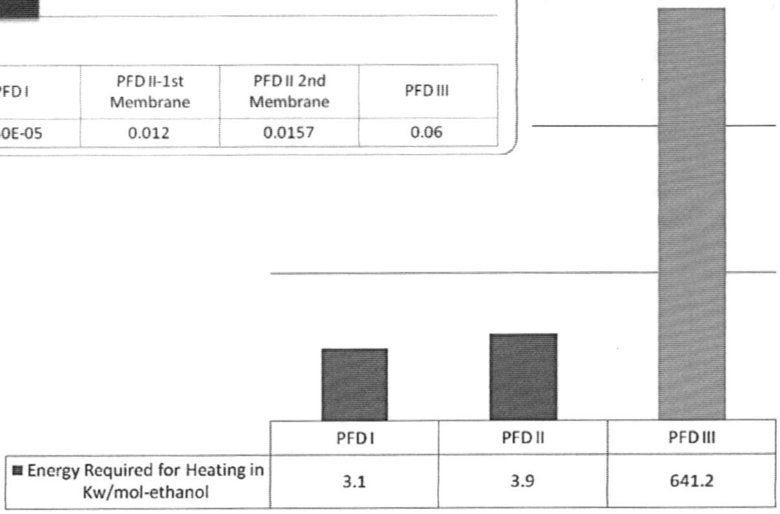

	PFD I	PFD II	PFD III
Energy Required for Heating in Kw/mol-ethanol	3.1	3.9	641.2

LG
LA GACETA

La Gaceta Tucumán

Sábado 10 de Septiembre 2016 · Desde 1912

La vinaza, un riesgo latente y preocupante

Las mejores perspectivas que se presentan en la matriz sucroalcoholera de Tucumán, promisorias y optimistas, enfrenta sin embargo una sombra que los factores productivos y la administración provincial deben resolver más temprano que tarde: qué hacer con la vinaza. Este es un subproducto de la fabricación de alcohol a partir de la caña de azúcar, que en su composición química registra alta salinidad y materia orgánica, y sus volúmenes de producción le confieren un elevado poder contaminante de diques, ríos y arroyos, pudiendo afectar la flora y la fauna que en estos cursos de agua se desarrolla.

El debate sobre el destino de este subproducto se reactualizó en los últimos días, luego de que un alto funcionario del Ministerio de Ambiente de la Nación advirtiera que "Tucumán no ha alcanzado un buen sistema de gestión" para revertir la posibilidad daño ambiental y ratificó lo que cañeros, industriales, expertos, funcionarios y gran parte del mundo vinculado a la actividad ya sabían: el aumento dispuesto por el Gobierno Nacional de 10% a 12% del corte de bioetanol de caña de azúcar en las naftas generará más vinaza en los ingenios. .

Encuentro de la Mesa Nacional para la Gestión Sustentable de la Vinaza Sucro-alcoholera

Panel 2: Tecnologías

Coordinadora: Dra. Ing. Mi Ra KIM

Secretaría de Planeamiento y Políticas

Ministerio de Ciencia, Tecnología e Innovación Productiva
Presidencia de la Nación

7 y 8 de Septiembre de 2016

Secretaría de Planeamiento y Políticas

Ministerio de Ciencia, Tecnología e Innovación Productiva
Presidencia de la Nación

Temas discutidos en el Panel de Tecnologías:

- Optimización del proceso de gestión de la vinaza
- Fortalecimiento en el monitoreo y análisis de la vinaza
- Aprovechamiento y/o remediación de la vinaza:

  - Tratamientos con aprovechamiento agrícola
  - Tratamientos Biológicos
  - Tratamientos por concentración y aprovechamiento para otros usos

CONCLUSIONES

- ☑ Se definió un índice de performance que combina aspectos económicos con aspectos ambientales que permite fácilmente realizar el análisis y valoración de distintos escenarios.
- ☑ Se desarrolló un modelo ad-hoc para conseguir simular el proceso de pervaporación en Aspen Plus.
- ☑ Las herramientas de ingeniería de procesos (en este caso simulación e integración energética) se convierten en un asistente eficaz para la evaluación de distintas alternativas evitando cálculos engorrosos, que a veces obscurecen la solución del problema propuesto, permitiendo el análisis de numerosas alternativas con relativa facilidad de cómputo.

AGRADECIMIENTOS

Las actividades desarrolladas se enmarcan dentro del Convenio Internacional de Colaboración Académica y Científica al que suscriben la Universidad Nacional de Tucumán (UNT, Argentina) y la Universidad Estatal de Cleveland (CSU, EEUU).

Los autores del presente trabajo agradecen el financiamiento otorgado por la Secretaría de Ciencia, Arte e Innovación Tecnológica de la UNT.

Gracias por su atención !!

UTN FACULTAD REGIONAL MENDOZA
UNIVERSIDAD TECNOLÓGICA NACIONAL

CONICET

2nd RCN Conference on Pan American Biofuels & Bioenergy Sustainability
September 13-16, 2016. Buenos Aires, Argentina

Are CAM crops a sustainable bioenergy feedstock? The case of the prickly pear cactus for ethanol production in semi-arid regions of Argentina

Roxana Piastrellini, Alejandro Pablo Arena, Bárbara Civit

Grupo CLIOPE

www.cliope.org

Grupo CLIOPE

CLIOPE - Lines of work

- Life Cycle Assessment
- Sustainability indicators
- Renewable energy devices
- Thermoeconomic analysis
- Sustainable building

UTN FACULTAD REGIONAL MENDOZA
UNIVERSIDAD TECNOLÓGICA NACIONAL

Red Argentina de Ciclo de Vida

RAHH
RED ARGENTINA DE HUELLA HÍDRICA

RICV Red Iberoamericana de Ciclo de Vida

PRé

Are CAM crops a sustainable bioenergy feedstock?

CAM – Crassulacean Acid Metabolism

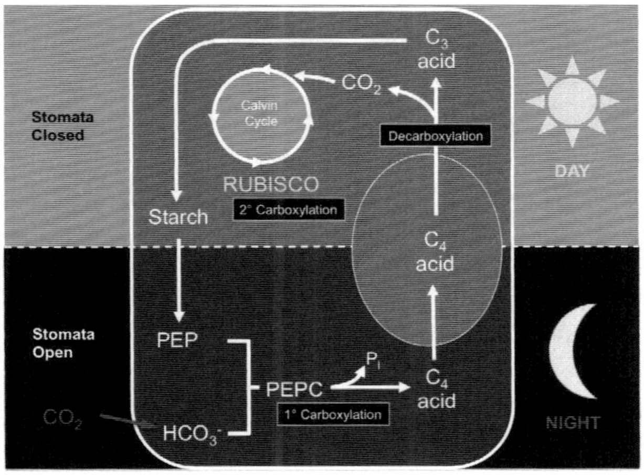

www.cliope.org

Crops for bioenergy

	C3 crops	C4 crops	CAM crops
WUE (g CO_2/kg H_2O)	1-3	2-5	1-8
Examples	✓ wheat, ✓ rape, ✓ soybean	✓ corn, ✓ sugarcane	✓ cacti, ✓ agave

This greater efficiency in water use, represent a lower environmental impact?

www.cliope.org

Case study:

Grupo CLIOPE

Prickly pear cactus for ethanol production in the Argentine Northwest region

WF of 1 MJ of energy

www.cliope.org

Water Footprint

Grupo CLIOPE

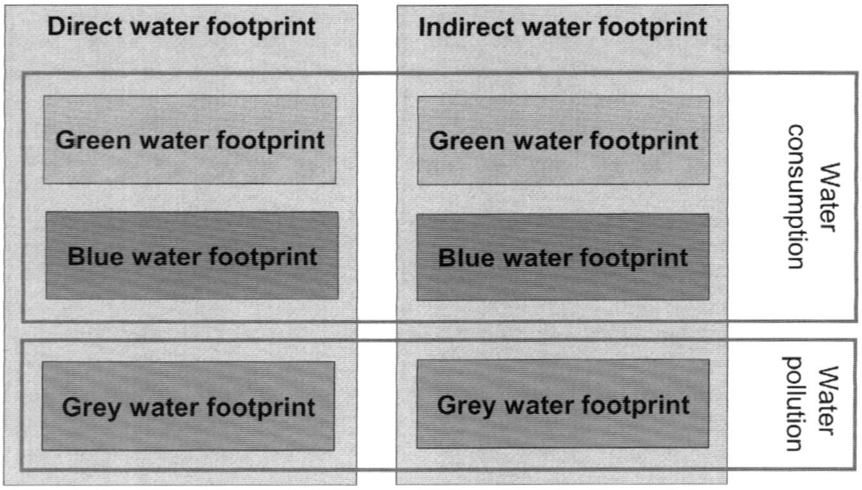

Hoekstra et al, 2011

www.cliope.org

Case study:

Prickly pear cactus for ethanol production in Northwest Argentina

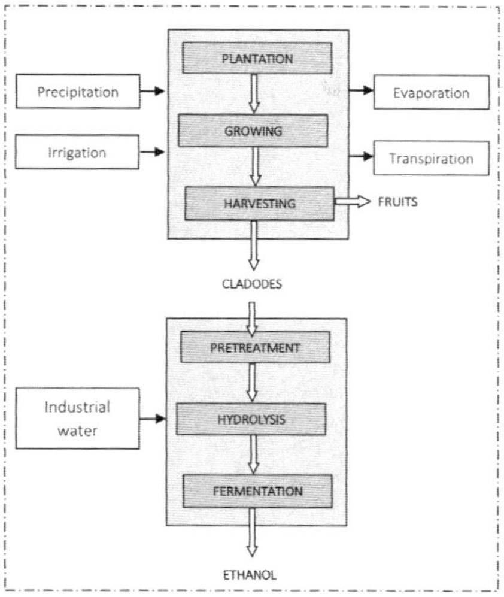

Grupo CLIOPE

www.cliope.org

Grupo CLIOPE

Water Footprint

WATER FOOTPRINT ACCOUNTING

$$WF_{green} = \sum \frac{CWU_{green}}{Y} \qquad WF_{blue} = \sum \frac{CWU_{blue}}{Y}$$

WATER SCARCITY ASSESSMENT

$$WS_{green(x,t)} = \frac{\sum WF_{green(x,t)}}{WA_{green(x,t)}} \qquad WS_{blue(x,t)} = \frac{\sum WF_{blue(x,t)}}{WA_{blue(x,t)}}$$

www.cliope.org

Water footprint for 1 MJ of energy

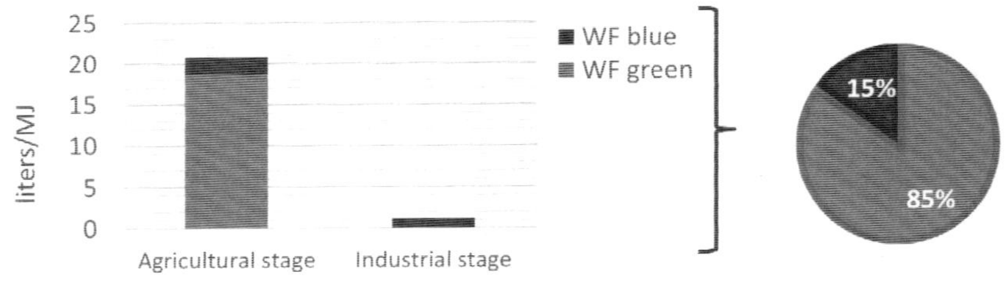

www.cliope.org

Grupo CLIOPE

Water Scarcity for 1 MJ of energy

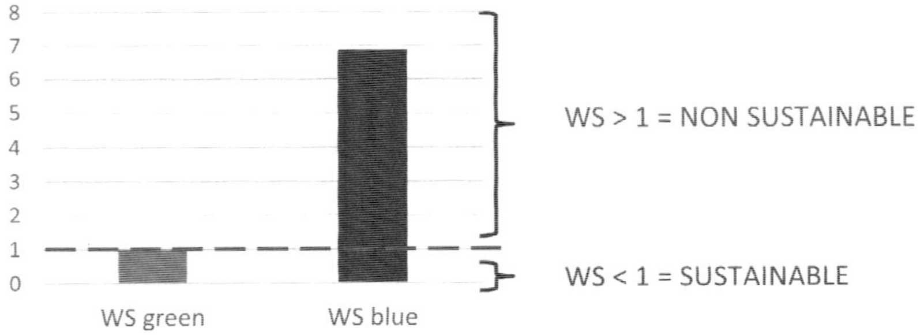

www.cliope.org

Grupo CLIOPE

Water Footprint and Water Scarcity for 1 ton of biomass for energy

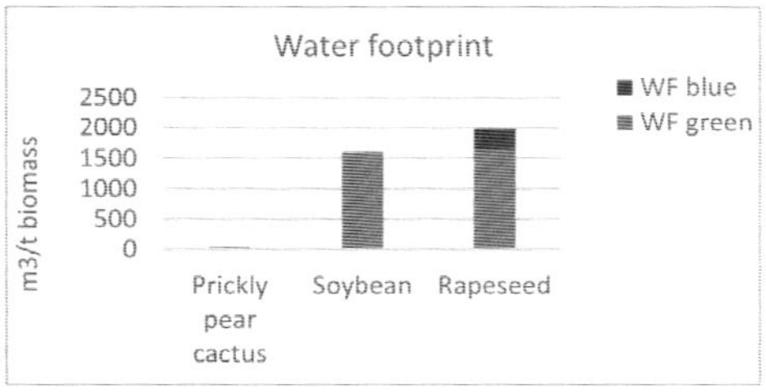

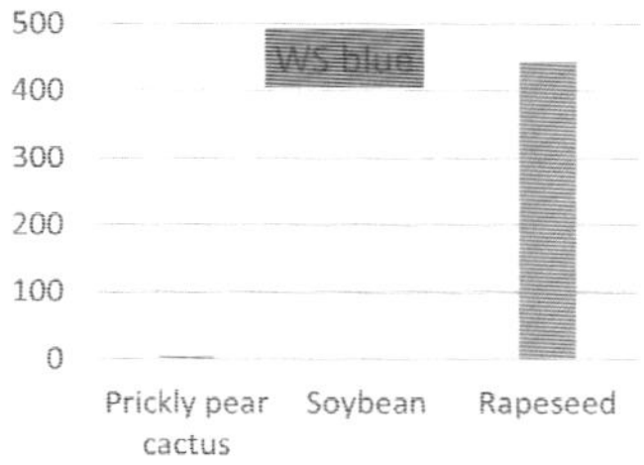

- Prickly pear cactus: arid region
- Rapeseed: arid region
- Soybean: humid region

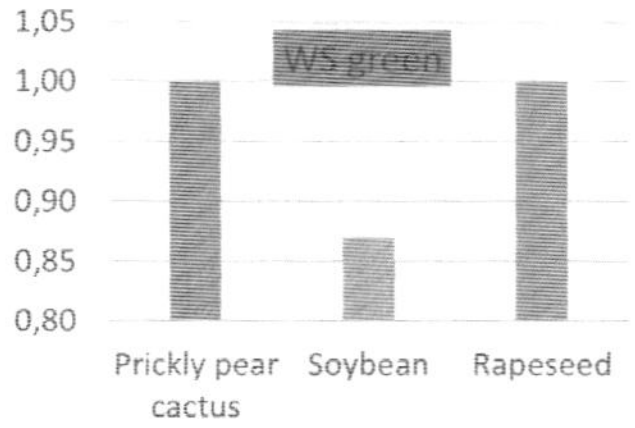

www.cliope.org

Prickly pear cactus for ethanol production in arid region:

✓ Less water consumption than C3 crops in arid and humid regions.

✓ Less impacts on water use than C3 crops in arid regions.

X Greater impacts on water use tan C3 crops inn humid regions.

Grupo CLIOPE *www.cliope.org*

Thanks for your attention
Grupo CLIOPE
roxana.ppp@gmail.com
+54 261 5244693
www.cliope.org
Coronel Rodriguez 273 - 5500 Mendoza - Argentina

LCA and Optimization of *Jatropha curcas* as a Sustainable Biodiesel Feedstock in Argentina

Antonio González, Alexa Beaver, M. Soledad Diaz
Planta Piloto de Ingeniería Química (PLAPIQUI), CONICET - Universidad Nacional del Sur, Bahía Blanca, Argentina

FULBRIGHT

PLAPIQUI U N S

Motivation

- Biofuels as environmentally-friendly alternative to fossil fuels?
- Global Biofuels Sustainability Requirements
- Global biodiesel market
- Biodiesel production in Argentina
  - 70% of total production exported
  - 90% soy-based

Biodiesel Outlook 2023

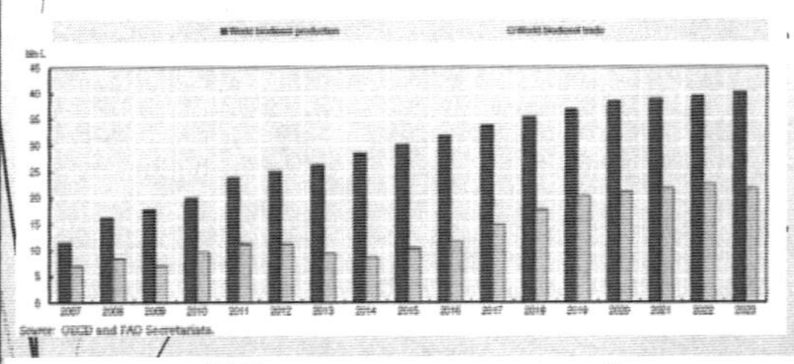

Source: OECD and FAO Secretariats.

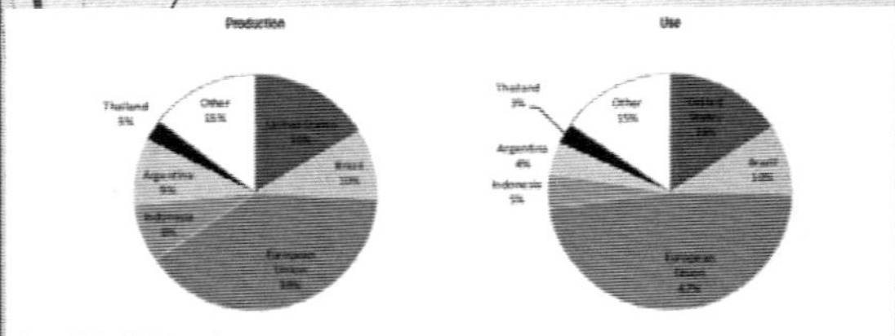

Source: OECD and FAO Secretariats.

- Global biodiesel production expected to reach 40 billon liters in 2023,+54% from 2013
- Argentina's share of global production will grow from 6.4% to 9%

Argentina is expected to maintain an export-oriented biodiesel industry

Jatropha as an alternative feedstock to soy

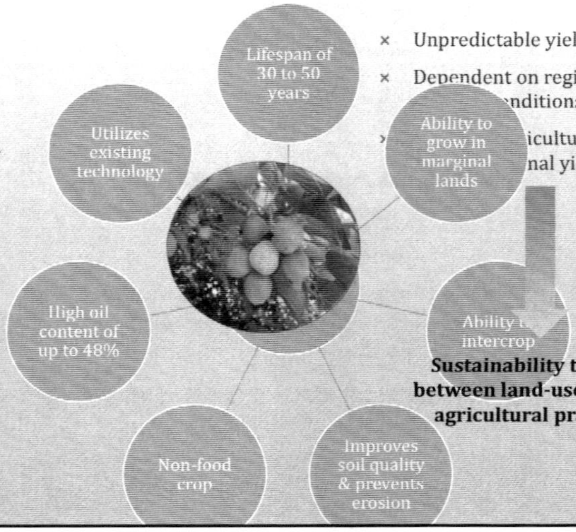

- × Unpredictable yields
- × Dependent on regional soil and ...nditions
- × ...icultural inputs to ...nal yields

Sustainability trade-off between land-use type and agricultural practices?

Global Jatropha Projects

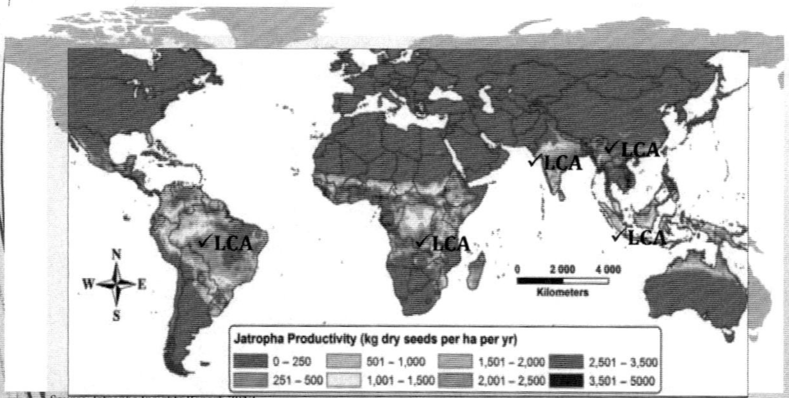

Source: Jatropha Insights Report, 2012
Source: Trabucco *et al*, 2010

Objective

- Assess the environmental impact of *Jatropha*-based biodiesel production in Argentina
- Design and optimization of biodiesel supply chain in Argentina (economic objective function)

Goal and Scope Definition

- Assess the environmental impact of Jatropha-based biodiesel production in Argentina
- Focus on use of country-specific data
- System boundaries: cultivation, transport, oil extraction, and biodiesel production ("cradle-to-gate")
- Functional unit: 1ton of Jatropha biodiesel produced
- Uncertainties of inventory data not evaluated
- Two Jatropha cultivation scenarios
  - Formosa province (actual)
  - Santiago del Estero province (theoretical)

PLAPIQUI U N S

System Boundary

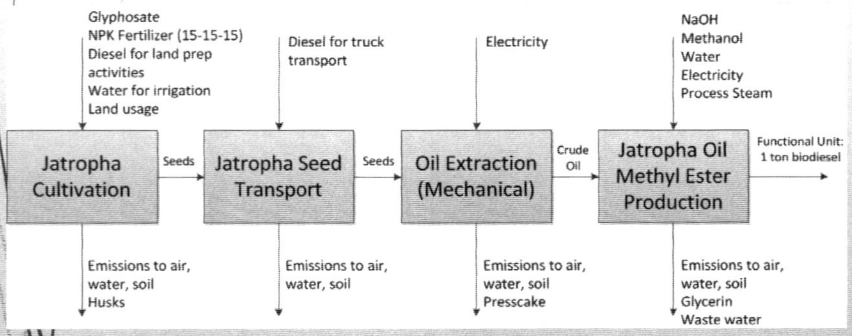

Formosa Plantation

- Location: Pilcomayo, Formosa
- First planted in 2008
- Plant spacing: 4 x 2 (1250 plants/ha)
- Annual precipitation: 1250 mm
- Average Min Temp: 17°C
- Average Max Temp: 28°C
- No irrigation
- Fertilizer only applied during initial field preparation
- Minimal use of herbicides
- **Yield in 2010: 750 kg seeds/ha**
- **Yield in 2014: 1400 kg seeds/ha**

Santiago del Estero Plantation

- Location: Santiago del Estero (based on agro-climatic maps)
- Marginal Jatropha-suitable land
- Irrigation requirements: CLIMWAT 2.0 and CROPWAT
- Crop, soil and region specific data
- Fertilizer
- Irrigation
- **Yield in 2010: 750 kg seeds/ha**
- **Yield in 2014: 1400 kg seeds/ha**

Data Sources: FAO CROPWAT & CLIMWAT

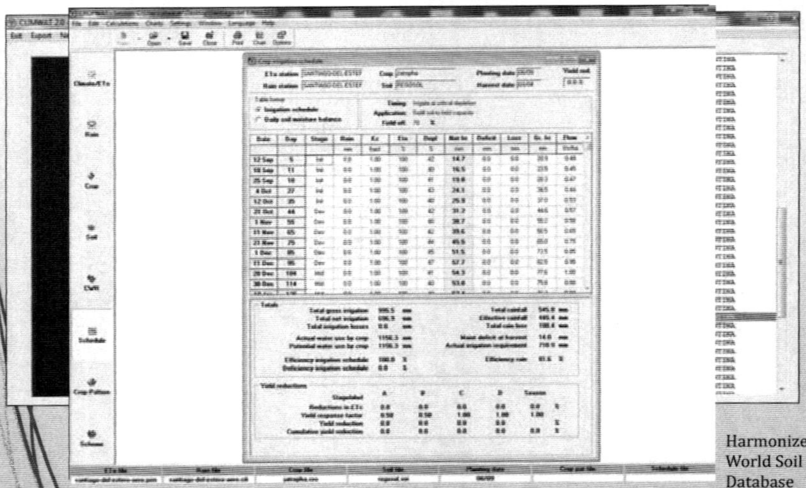

Harmonized World Soil Database

Inventory analysis: Cultivation & Transport

Input	Unit	Formosa	Santiago del Estero	Reference
Annual seed yield	kg ha^{-1}	1400	1400	Interview
Plant spacing	m^2	4 x 2	4 x 2	Conference proceedings
NPK Fertilizer	kg	3.4022E-3	0.0359	Achten et al., 2008; FACT Foundation, 2010
Glyphosate	kg	1.2138E-4	1.2138E-4	Panichelli et al., 2008
Diesel, field preparation	kg	0.00914	0.00914	Ecoinvent database
Diesel, irrigation	kg	-	4.46E-5	Ecoinvent database
Water, irrigation	m^3	-	4.127E-4	Calculation (CLIMWAT-CROPWAT)
Land, occupation	m^2·year	7.14	7.14	Calculation
		Permanent crop, non-irrigated, extensive	Permanent crop, irrigated, intensive	
Land, transformation to	m^2	0.357	0.357	Calculation
		Permanent crop, non-irrigated, extensive	Permanent crop, irrigated, intensive	
Land, transformation from	m^2	0.357	0.357	Calculation
		Agriculture	Shrubland	
Transport, truck >20t, EURO5	ton · km	0.735	0.752	Andersen et al., 2012

Inventory analysis: Extraction & Production

Input	Unit	Value	Reference
Oil Extraction			
Oil extraction efficiency	%	75	FAO, 2010
Seed oil content	%	32	Huerga et al., 2010b
Screw press capacity	kg seed hr^{-1}	200	Product spec
Screw press power	kW	11	Product spec
Electricity	kWh t^{-1} seed	55	Product spec
Biodiesel Production			
Transesterification efficiency	%	90	Berchmans and Hirata, 2008
Electricity	kWh t^{-1} crude oil	31	INTA
Process steam	MJ t^{-1} crude oil	619	INTA
Sodium hydroxide (NaOH)	kg t^{-1} crude oil	14	Berchmans and Hirata, 2008
Methanol	kg t^{-1} crude oil	240	Berchmans and Hirata, 2008

Life Cycle Assessment

- Software: Simapro v.8
- Ecoinvent database v.3
- Impact assessment method: ReCiPe, hierarchist perspective
- Attributional LCA
  - Comparison of two processes
  - Country-specific study
  - No economic analysis
  - No system expansion
- By-products
  - Presscake handled as organic waste to reflect current situation
  - Mass allocation used for glycerol

Results: Characterization of Cultivation Scenarios

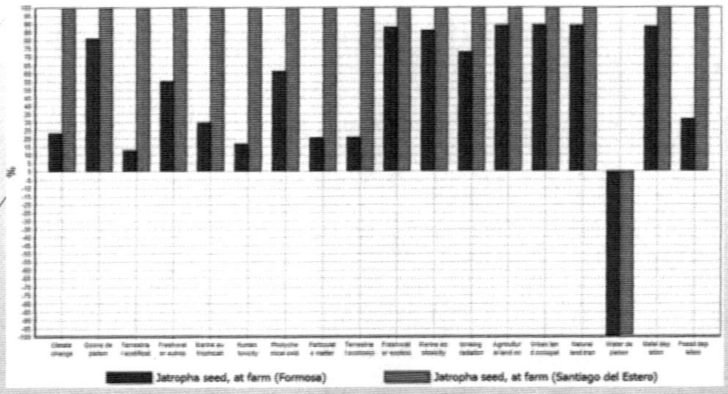

ReCiPe 18 midpoint indicators for Formosa and Sgo del Estero

Terrestrial Acidification 87% reduction wrt Sgo del Estero
Human Toxicity 83 % reduction
Particulate Matter Formation 80% reduction
Climate Change 76.7 % reduction

Inventory analysis: Cultivation & Transport

Input	Unit	Formosa	Santiago del Estero	Reference
Annual seed yield	kg ha^{-1}	1400	1400	Interview
Plant spacing	m^2	4 x 2	4 x 2	Conference proceedings
NPK Fertilizer	kg	3.4022E-3	0.0359	Achten et al., 2008; FACT Foundation, 2010
Glyphosate	kg	1.2138E-4	1.2138E-4	Panichelli et al., 2008
Diesel, field preparation	kg	0.00914	0.00914	Ecoinvent database
Diesel, irrigation	kg	-	4.46E-5	Ecoinvent database
Water, irrigation	m^3	-	4.127E-4	Calculation (CLIMWAT-CROPWAT)

Default Jatropha case irrigation: 0.0853 m3
Default Arg. Soy case: 13.1 m3

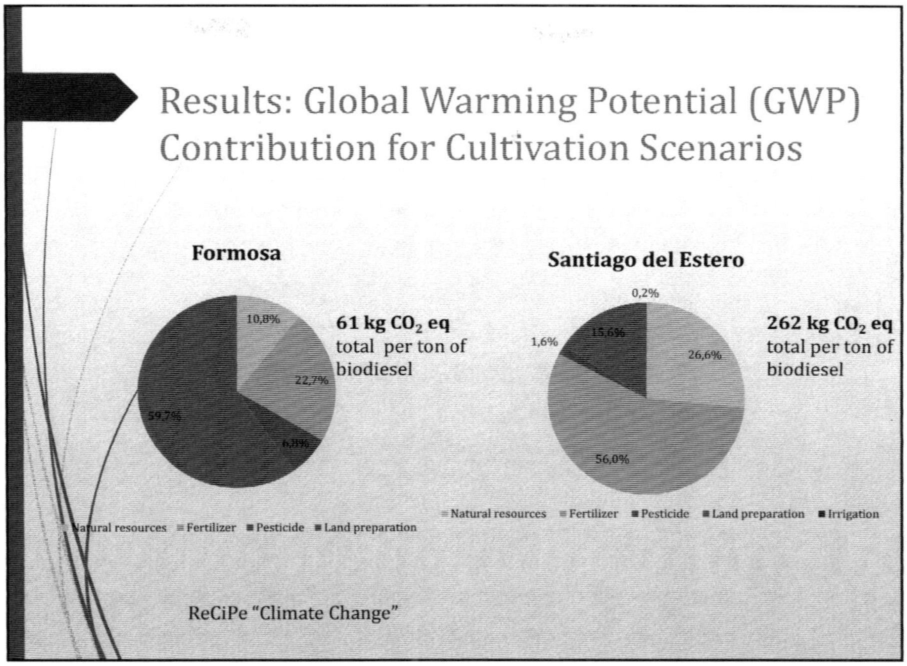
Results: Global Warming Potential (GWP) Contribution for Cultivation Scenarios
Formosa
10,8%
22,7%
6,8%
59,7%
61 kg CO_2 eq
total per ton of biodiesel
Natural resources
Fertilizer
Pesticide
Land preparation
Santiago del Estero
0,2%
1,6%
15,6%
26,6%
56,0%
262 kg CO_2 eq
total per ton of biodiesel
Natural resources
Fertilizer
Pesticide
Land preparation
Irrigation
ReCiPe "Climate Change"

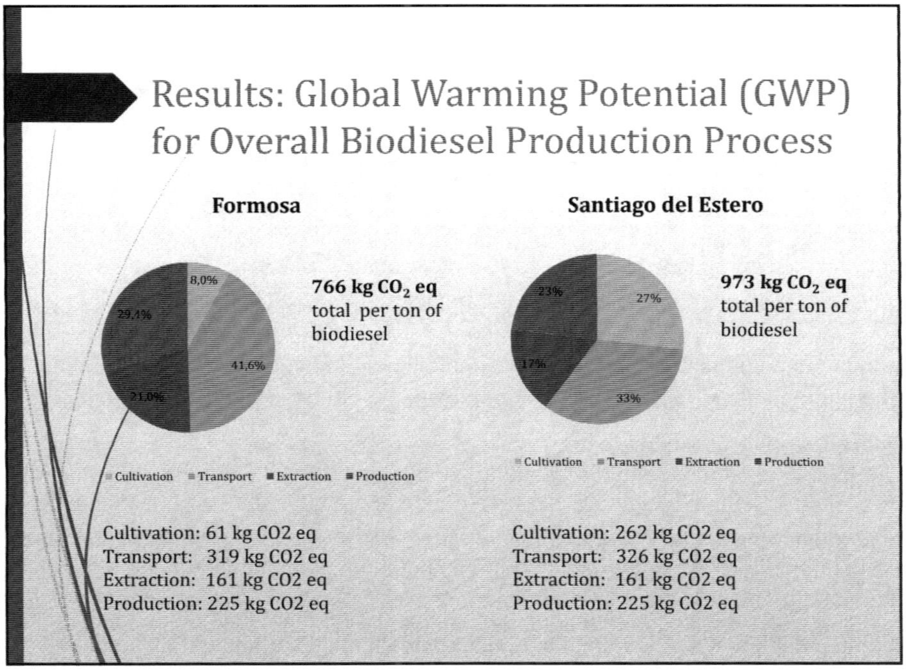
Results: Global Warming Potential (GWP) for Overall Biodiesel Production Process
Formosa
8,0%
41,6%
21,0%
29,4%
766 kg CO_2 eq
total per ton of biodiesel
Cultivation
Transport
Extraction
Production
Cultivation: 61 kg CO2 eq
Transport: 319 kg CO2 eq
Extraction: 161 kg CO2 eq
Production: 225 kg CO2 eq
Santiago del Estero
27%
33%
17%
23%
973 kg CO_2 eq
total per ton of biodiesel
Cultivation
Transport
Extraction
Production
Cultivation: 262 kg CO2 eq
Transport: 326 kg CO2 eq
Extraction: 161 kg CO2 eq
Production: 225 kg CO2 eq

Results: Damage Categories

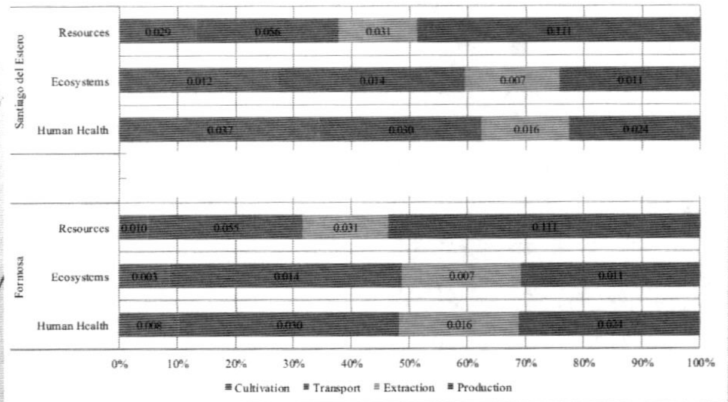

Santiago del Estero

Formosa

Results

Country	GWP (kg CO_2 eq per ton biodiesel produced)	Seed yield (kg per ha)
Argentina (this study)	766 or 973	1,400
Indonesia	741	5,000
Tanzania	672	3,024
China	894	1,485
India	1330	1,500

Biodiesel Supply Chain with land competition

Given:

- ✓ **Time horizon:** 7 years
- ✓ ***Multiperiod model.*** **(Each Period: 4 months)**
- ✓ *Existing capacity* **for oil and biodiesel plants & warehouses and blending**
- ✓ *Potential capacity* **for oil and biodiesel plants & warehouses and blending**
- ✓ **Maximal cultivated area and yield estimation from historical data**
- ✓ **23 regions**
- ✓ **Total available surface** for sewing **per region**
- ✓ **Fixed and variable** *investment and operation costs*
- ✓ *Economy of Scale* **for capital investment (small – medium – large sizes)**
- ✓ **Demand of Blends (B10)**

Determine:

- ✓ **Whether to** *install, expand or not* **oil and biodiesel plants & warehouses**
- ✓ *Optimal surface distribution* FOR **raw materials**
- ✓ *Flows in network* **for each time period**

Problem statement

Superstructure of "Technologies" for each region

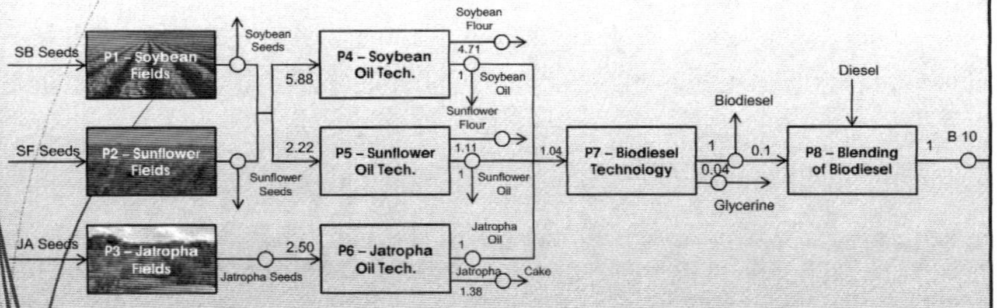

Supply chain model

(Andersen, Iturmendi, Espinosa, Diaz, 2012)

i → Products (final and intermediate)   l → Means of Transportation   p → Production Tech.
g → Regions   t → Time Set   z → Economy of Scale

1. Mass Balances

MASS BALANCES

$$SW_{igt-1} + \sum_{p \in SPP(i,p)} P_{igpt} + IP_{igt} + \sum_{l \in IL(i,l)} \sum_{g' \neq g} Q_{ilg'gt} = SW_{igt} + \sum_{p \in SPRM(i,p)} P_{igpt} + DP_{igt} + \sum_{l \in IL(i,l)} \sum_{g' \neq g} Q_{ilgg't} \quad \forall i,g,t$$

$$P_{igpt} = RHO_{ip} \sum_{i' \in MP(i',p)} P_{i'gpt} \quad \forall i,g,p,t$$

$$P_{igpt} = \sum_{t' \leq t} \eta_{ipg\,t-t'+1} SUPS_{ipgt'} \quad \forall i \in \{seeds\}, g,p,t$$

RAW MATERIAL CONSTRAINTS

$$SUPS_{SB\,g\,t} + SUPS_{SF\,g\,t} + \sum_{t' \leq t} SUPS_{JA\,g\,t'} \leq TOTSUP_{gt} \quad \forall g,t$$

$$\sum_{t' \leq t} SUPS_{JA\,g\,t'} \leq \left(TOTSUP_{gt} - SUPS_{SB\,gt} - SUPS_{SF\,gt}\right) P_{gt} \quad \forall g,t$$

SALES

$$DP_{igt} = DD_{igt} + DE_{igt} \quad \forall i,g,t$$

$$DD_{igt} \leq DDEM_{igt} \quad \forall i,g,t$$

$$DD_{igt} \geq DSAT_{igt}\, DDEM_{igt} \quad \forall i,g,t$$

$$DE_{igt} \leq DEOP_{i,g} \quad \forall i,g,t$$

$$\sum_{g} DE_{igt} \leq EXPMAX_{it} \quad \forall i,t$$

$$\sum_{g} DE_{igt} \geq EXPMIN_{it} \quad \forall i,t$$

$$DP_{igt} \leq DC_{i,g} (1+GR_i)^{t-1} \quad \forall i,g,t$$

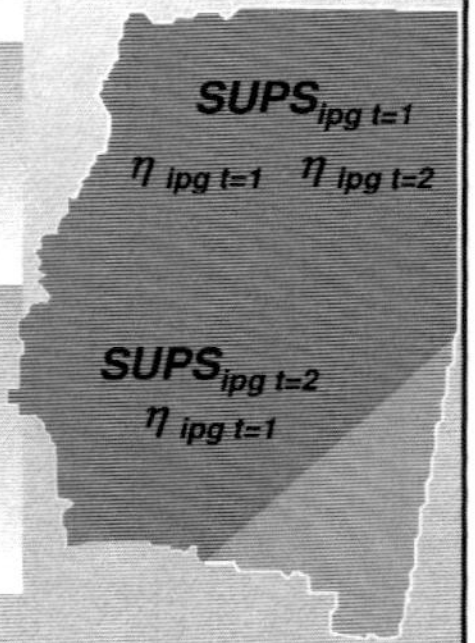

5

Supply chain model

i → Products (final and intermediate)   l → Means of Transportation   p → Production Tech.
g → Regions   t → Time Set   z → Economy of Scale

2. Capacity Eqs

PRODUCTION CAPACITY

$$\sum_{i \in MP(i,p)} PR_{igpt} \leq CP_{gpt} \quad \forall g,p,t$$

$$\sum_{i \in MP(i,p)} PR_{igpt} \geq CPSAT_{pt}\, CP_{gpt} \quad \forall g,p,t$$

$$CP_{gpt} = CP_{gpt-1} + \sum_{z} CEP_{gptz} \quad \forall g,p,t$$

ECONOMY OF SCALE

$$CEP_{gptz} \leq UBQP_{gpz}\, NP_{gptz} \quad \forall g,p,t,z$$

$$CEP_{gptz} \geq LBQP_{gpz}\, NP_{gptz} \quad \forall g,p,t,z$$

INTEGER VARIABLES

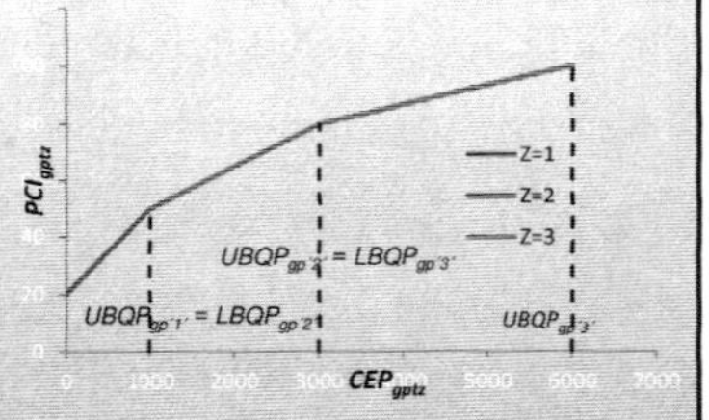

6

Supply chain model

3. Means of Transportation

Y(i,l,g,g´,t) → If Tpt of Prod I with mean of transportation L from Reg G to G´ in Year T
X(i,g,g´,t) → If Tpt of Prod I from Reg G to G´ in Year T

$$Q_{ilgg't} \leq QUP_{il}\, Y_{ilgg't} \quad \forall\ i,l,g,g',t$$

$$Q_{ilgg't} \geq QLO_{il}\, Y_{ilgg't} \quad \forall\ i,l,g,g',t$$

BOUNDS

$$X_{igg't} + X_{ig'gt} \leq 1 \quad \forall\ i,g,g',t$$

IF TRANSPORT G-G´ THEN NOT G´-G

$$\text{If } \{Y_{i\,\text{"TRUCK"}\,gg't} \vee Y_{i\,\text{"RAILWAY"}\,gg't} \vee Y_{i\,\text{"SHIP"}\,gg't}\} \Rightarrow X_{igg't} \quad \forall\ i,g,g',t$$

$$X_{igg't} \geq Y_{ilgg't} \quad \forall\ i,l,g,g',t$$

$$X_{igg't} \leq \sum_{l} Y_{ilgg't} \quad \forall\ i,g,g',t$$

Supply chain model

4. Economic Definitions

i → Products (final and intermediate) l → Means of Transportation p → Production Tech.
g → Regions t → Time Set z → Economy of Scale

OBJECTIVE FUNCTION ⟶ $$Max \quad NPV = \left(\sum_{t} \frac{CF_t}{(1+IR)^t} \right) + \frac{MV}{(1+IR)^{t_{END}}}$$

$$PCI_{gptz} = FICP_{gptz} NP_{gptz} + VICP_{gptz} CEP_{gptz} \quad \forall\ g,p,t,z$$

$$SCI_{igtz} = FICS_{igtz} NS_{igtz} + VICS_{igtz} CES_{igtz} \quad \forall\ i,g,t,z$$

CAPITAL INVESTMENT

$$POC_{gpt} = \sum_{i \in MP(i,p)} UPC_{igpt}\, PR_{igpt} \quad \forall\ g,p,t$$

$$SOC_{igt} = USC_{igt} \left(SW_{igt-1} + \frac{SW_{igt}}{2} \right) \quad \forall\ i,g,t$$

$$TOC_t = \sum_{i} \sum_{g} \sum_{g'} \sum_{t} Q_{igg't}\, DT_{gg'}\, F_{it} \quad \forall\ t$$

OPERATING COSTS

Input data

- ***Raw Materials*** → Soybean – Sunflower – Jatropha
- ***Fields*** → Total available for sewing
- ***SB/SF*** → historical data profiles
- ***JA*** → Agroecological aptitude of lands & free and suitable area
- ***Plants*** → SB, SF & JA Oil and BD Production existing capacity
- ***Warehouses*** → SB, SF & JA seeds, oil and BD existing capacity
- ***Distribution Centers***
- ***Multiperiod model*** time horiz 7 years, period length: 4 months
- ***Lower and Upper bounds in exports***
- ***Domestic Demand*** proportional to population in each region

JATROPHA

9

Results

# Equations	259,749
# Continuous Variables	387,481
# Discrete Variables	53,004
Solver	CPLEX 11.2

Processor → Intel Core 2 Duo P8600 2.40 GHz – 3GB RAM

	A	B	C
Objective Function	107,532	105,249	103,321
Gap [%]	0.01%	0.01%	0.02%
Resolution Time[sec]	1,265.03	1,915.57	3081.53
Truck	3.1%	25%	40%
Train	92.4%	65.4%	53.8%
Ship	4.5%	9.6%	6.2%

A → No transport constraint B →At least 25% by truck C → At least 40% by truck

Results

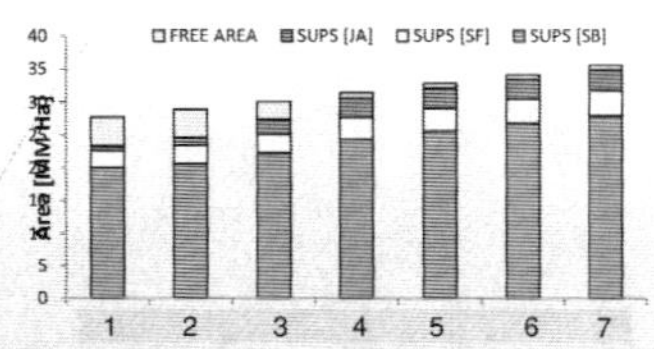

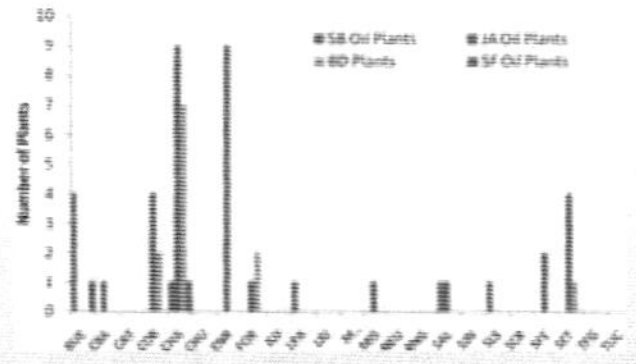

Sown Area – Optimal area distribution for soybean, sunflower and jatropha

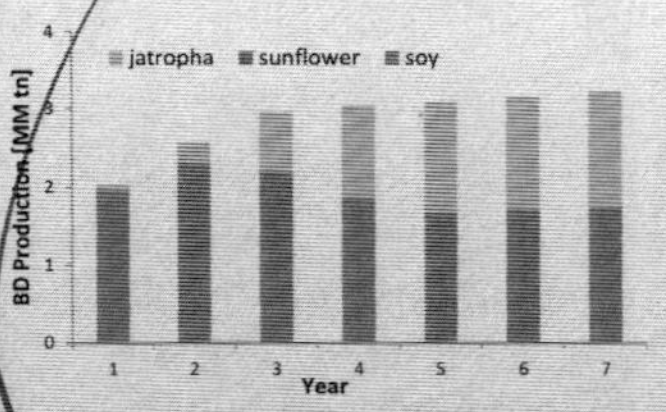

Biodiesel Source

Photo of 4th year, 1st period map;
Arrows: Biodiesel fluxes;
Green intensity proportional to soy cultivated áreas;
Blue dots: existing capacity;
Blue stars: new biodiesel plants

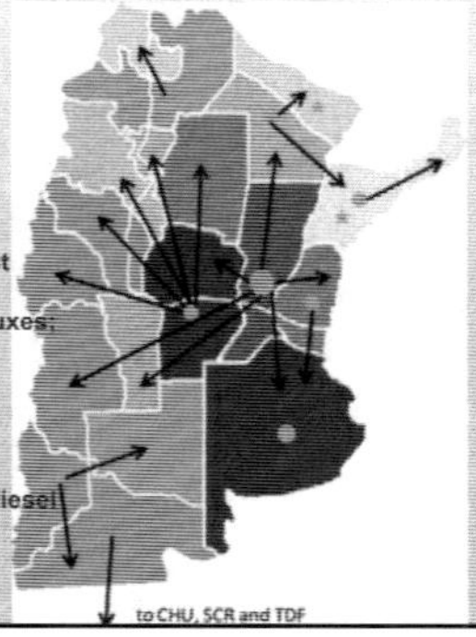

Conclusions

- Sustainability of *Jatropha* may be diminished when using marginal, yet suitable land, with agricultural inputs shown to have a greater negative environmental impact
- 81% and 75% reduction in kg CO_2 eq for Formosa and Santiago del Estero, respectively, when compared to EU RED default for diesel fuel
- 58% and 46% reduction in kg CO_2 eq for Formosa and Santiago del Estero, respectively, when compared to EU RED default for soybean biodiesel
- More field data needed both in Argentina and globally to better predict seed yields and minimize uncertainty and risk associated with *Jatropha* projects
- Incorporation of LCA results into existing biodiesel supply chain optimization model is part of current work

Thank you!

Corn Transforming Industry Environmental Analysis

Conference: RCN Conference on Pan American Biofuel and Bioenergy Sustainability

Year: 2016

Proceeding: 2nd RCN Conference on Pan American Biofuels & Bioenergy Sustainability

Group: Abstracts

Session: Session 8: Environmental LCA & Energy Balances of Biofuel Pathways/ anáLisis De Ciclo De Vida Ambientales y Balances De Energía De Las Cadenas De Biocombustibles

Time: Friday, September 16, 2016 - 12:00pm-12:20pm

Authors:

Hilbert, J. A. Sr., Instituto Nacional de Tecnología Agropecuaria
Carballo, S. M., INTA
Galbusera, S., Sec Medio Ambiente
Michard, N.
Dantur, M. A.
Galvan, M. J.
Acquaroli, S.

In the last five years a new generation of bioethanol plants in Argentina were born with multiple products production. From the technical point of view this plants are in line with new concepts linked to energy optimization and environment care. Evaluations methods employed in their study are in line with the bio refinery and bio economy concepts, which have different definitions according to the literature. Due to the strategic importance of this industry INTA through its rural engineering institute started a series of studies centered in the products analysis, the coproducts emission and the energy efficiencies. One case study is presented ACABIO Villa María Cordoba.The new study is based in a systemic approach looking at biofuels and bio products incorporating the whole transforming biomass chain identifying interactions between products. Complementary research analyzes the logistic and infrastructure aspects shared between the different products being produced. Local and international Environmental regulations and legal aspects define in which way this bio refinery industry is developing. Due the importance of the farm production stages of the crop a particular study was made regarding affected areas and yield variability within them using geographical information systems. Allocation factors were calculated according to the three most common methodologies used in the literature; mass balance, energy content and market price of each coproduct. Differences between them were significate in all the cases being considered in this paper.Due to EU emition reduction requirements and limits a specific study was made calculating the amount an emition reductions of the biofuels produced considering the default value defined at the directive 2009/28/CE. Overall contribution done by farm, industry and transport including overseas freight to Rotterdam grs.CO_2eq/Mj were calculated. The incorporation of the carbon dioxide capture and purification plant improve the overall results of all the rest of the products. In the second year ISO 14025 guidelines were introduced and product category rules (PCR) and a specific EPD is proposed.

Michigan Technological University

UADY
UNIVERSIDAD AUTÓNOMA DE YUCATÁN

Resource Assessment and Carbon Footprint Analysis of the Pyrolysis of Municipal Solid Waste (MSW) in Mexico

Ulises R. Gracida-Alvarez[1], Laura R. Nitz[1], Lauren M. Keenan[1], Julio C. Sacramento-Rivero[2], & David R. Shonnard[1,3]

[1]Department of Chemical Engineering, Michigan Technological University. Houghton, MI.
[2]Facultad de Ingeniería Química, Universidad Autónoma de Yucatán, Merida, Mexico.
[3]Sustainable Futures Institute, Michigan Technological University. Houghton, MI.

September 16, 2016

2nd RCN Conference on Pan American Biofuels & Bioenergy Sustainability

Outline

- Introduction
- Resource assessment for MSW in Mexico
- Estimation of potential yields for fuels
- Preliminary life cycle assessment (LCA) - (C footprint)
- Conclusion: Identify sustainable MSW pathways to guide research

2

Introduction

- In many countries, MSW is diverted to landfills.
- Landfills are major sources of CH_4 emissions.
- Area for landfills is becoming scarce.

- Avoid landfill emissions with biofuel production.
- Lower emissions from transportation fuels.
- Enhance energy security with additional sources of biofuel feedstocks.

Could using MSW as a feedstock improve waste disposal while providing a sustainable renewable fuel?

Study case: Mexico's production of fuels through thermochemical conversion

3

Objectives

- Determine the amount of MSW in Mexico that is available as feedstock (not recycled already)
- Estimate the yields to fuels and chemicals from the available MSW feedstock
- Estimate the contribution of these fuels and chemicals to the national demand
- Perform a preliminary (LCA - C footprint) of each proposed MSW to fuel pathway

4

Resource assessment for MSW

- Data Sources: Three most populated cities
  - Mexico city (5862 people/km^2)
  - Guadalajara (9876 people/km^2)
  - Monterrey (1270 people/km^2)

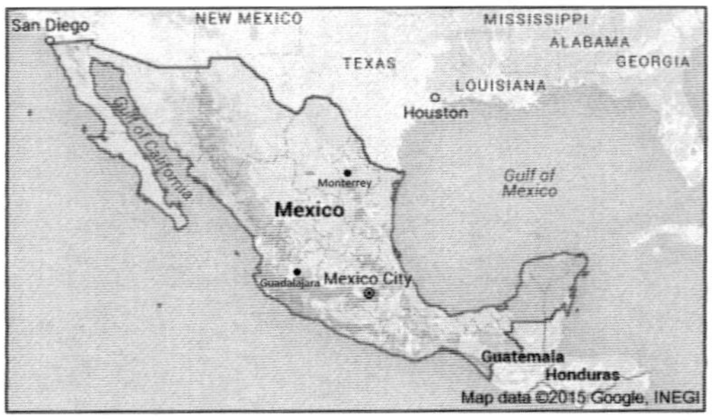

- Data Sources: National MSW statistics
- Methods
  - Literature review (Peer reviewed papers, national statistics)
  - Standardization of categories
  - Quantification

INEGI. (2012). Generación estimada de residuos sólidos por entidad federativa.
Durán-Moreno, A. et al (2013) Revista Internacional de Contaminación Ambiental, 29, 39-46.
Bernache-Pérez, G. et al (2001) Waste Management & Research, 19, 413-424.
OPS. (2003). Evaluación general de los servicios de manejo de residuos sólidos municipales.

Resource assessment for MSW

Available MSW not Recycled

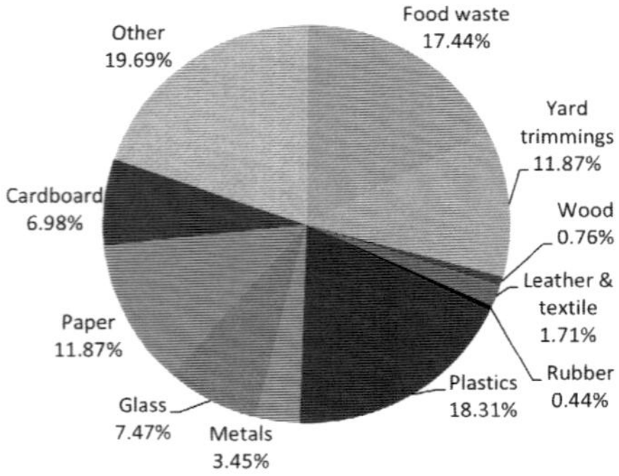

Total: 24.25 million tonnes per year

Usable sources:

- Paper
- Cardboard
- Plastics
- Wood
- Yard Trimmings
- Rubber

Available Plastic MSW

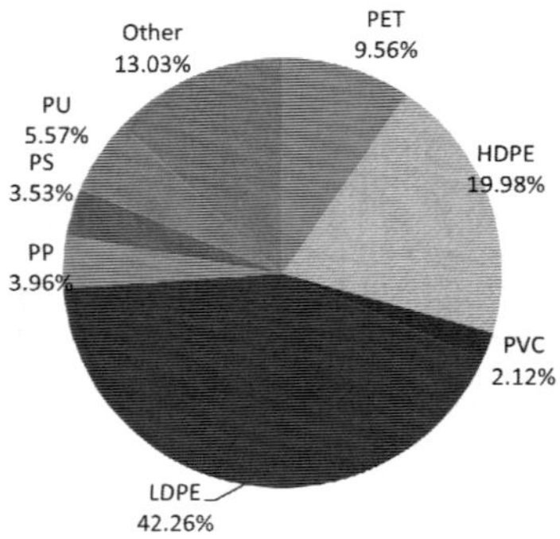

Total: 4.44 million tonnes per year

Usable sources:

- HDPE
- LDPE
- PP

Resource assessment results

- Available (not recycled) feedstock for further processing (dry weight basis)

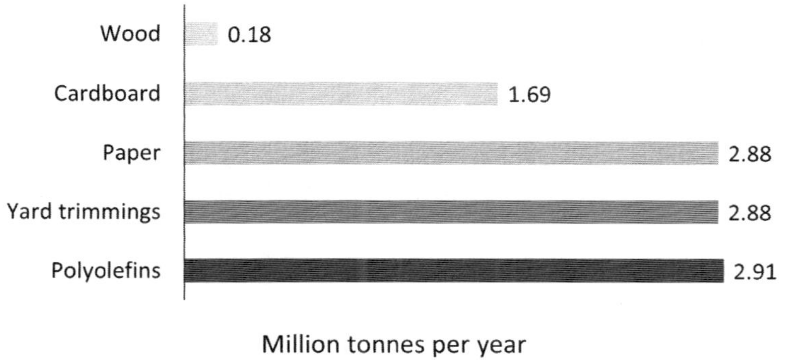

Total amount: **10.54 million tonnes per year**

Estimation of Yields Methods

- Proposed pathway for organic waste

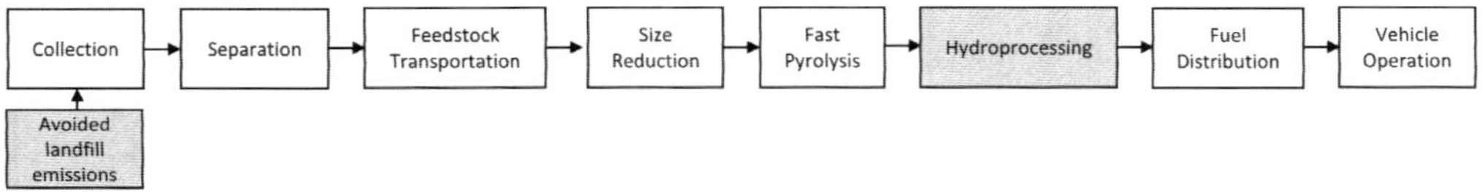

- Proposed pathway for polyolefin waste

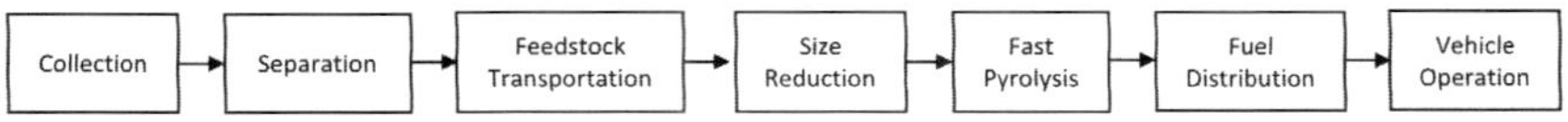

- Methods
  - Literature review
  - Yields estimation
  - Comparison with national consumption

Estimation of yields: Organic MSW

- Production of liquid fuels from organic waste (Hsu, 2011)

Ash and Moisture free Forest residues (kg)	128,366
Diesel (kg)	17,418
Gasoline (kg)	12,973
Total fuel (kg)	30,390

Yield: 0.237 kg fuel/kg organic waste

- For this study

Organic MSW (tonnes)	6,205,871.48
Estimated fuel production (tonnes)	1,469,234.59

Hsu, D. D. (2011) Life Cycle Assessment of Gasoline and Diesel Produced via Fast Pyrolysis and Hydroprocessing. NREL/TP-6A20-49341.

Estimation of yields: Polyolefins

- Polyolefin pyrolysis yields (Perugini *et al*, 2005)

Plastic feedstock (kg)	1.00
Gas fraction (kg)	0.147
Heavy fraction (kg)	0.448
Light fraction (kg)	0.265

Yield: 0.86 kg fuel/kg polyolefin waste

- So for the Plastic pathway

Polyolefin waste (tonnes)	2,933,698
Total fuel produced (tonnes)	2,522,980

Perugini ,F. et al (2005). Environmental Progress and Sustainable Energy, 24, 137-154.

Estimation of yields: Fuels from MSW

- **Contribution to the national consumption**

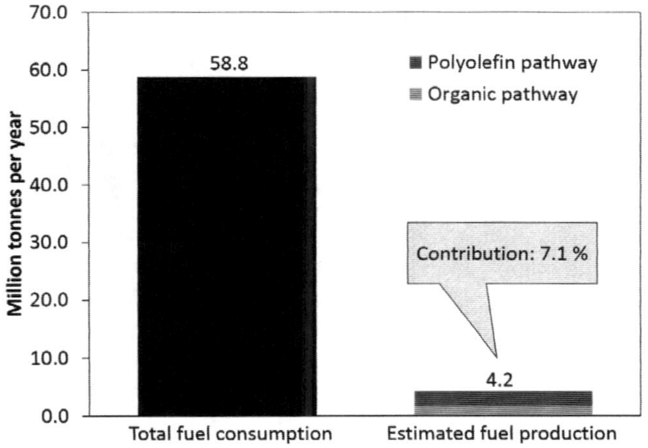

EIA (2015). International Energy Statistics.

Environmental assessment

- A Carbon footprint analysis was used to estimate the Global Warming Potential (GWP) for different process stages in CO_2 equivalent emissions.

Stage	Source	Assumptions
Collection	Fitzgerald et al, 2012	Factors for computing upstream emissions from provided combustion emissions. Diesel: 9.0 %, Electricity: 10.0 %, Natural Gas: 17.0 %
Separation	Fitzgerald et al, 2012	
Transportation	Simapro 8.0 database	Transportation distance of 108 km
Pyrolysis	Organic waste: Hsu, 2011 Polyolefin waste: Perugini et al, 2005	Upstream and combustion emissions obtained from ecoinvent™ in SimaPro 8.0
Hydroprocessing	Hsu, 2011	Only considered for organic waste
Fuel distribution	Simapro 8.0 database	Transportation distance of 108 km
Vehicle Operation	Organic waste: Hsu, 2011 Polyolefin waste: CO_2 produced from combustion	Stoichiometric CO_2 formation from representative compounds of each fraction.
Landfill	Organic waste: LCFS HSAD, 2012 Polyolefin waste: No emissions considered	Landfill gas production ($CH4:CO_2$ - 50:50) Active organic MSW carbon in landfill: 63.3 %
Carbon Accounting		All MSW C is accounted for; biogenic + fossil C along the pathways

C Footprint Analysis: Organic MSW

- Global warming potential for organic waste pyrolysis (g CO_2 eq/MJ)

Collection	7.3
Separation	2.2
Transportation	2.0
Size Reduction	19.8
Fast Pyrolysis	36.7
Hydroprocessing	45.2
Fuel distribution	0.1
Vehicle operation	72.3
Avoided Landfill emissions	-195.2
Net emissions	-9.5

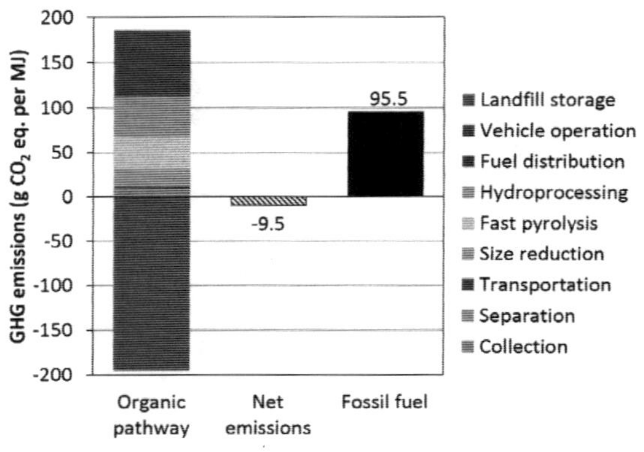

Fitzgerald ,G. C. et al (2012) Resources, Conservation and Recycling, 69, 50-56.
Hsu, D. D. (2011) Life Cycle Assessment of Gasoline and Diesel Produced via Fast Pyrolysis and Hydroprocessing. NREL/TP-6A20-49341.
LCFS HSAD, 2012, Proposed Low Carbon Fuel Standard (LCFS) Pathway for Production of Biomethane from High Solids Anaerobic Digestion (AD) of Organic (Food and Green) Waste. California Environmental Protection Agency.

C Footprint Analysis: Plastic MSW

- Global warming potential for polyolefin waste pyrolysis (g CO_2 eq/MJ)

Collection	1.1
Separation	0.3
Transportation	0.3
Size Reduction	3.2
Fast Pyrolysis	10.6
Fuel distribution	0.2
Vehicle operation	72.2
Avoided Landfill emissions	0.0
Net emissions	88.0

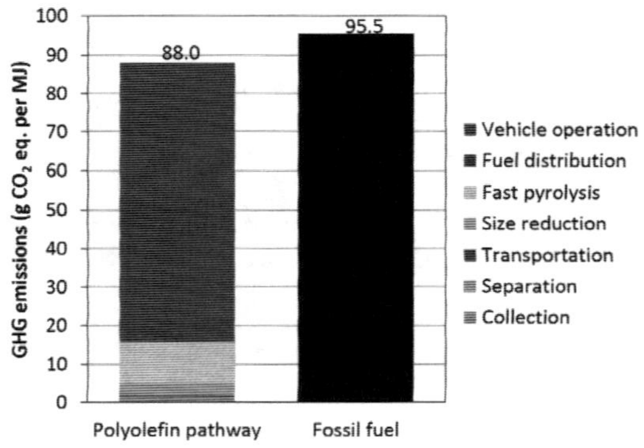

Fitzgerald ,G. C. et al (2012) Resources, Conservation and Recycling, 69, 50-56.
Perugini ,F. et al (2005). Environmental Progress and Sustainable Energy, 24, 137-154.
LCFS HSAD, 2012, Proposed Low Carbon Fuel Standard (LCFS) Pathway for Production of Biomethane from High Solids Anaerobic Digestion (AD) of Organic (Food and Green) Waste.

Fossil Energy Demand: Biomass Feedstocks

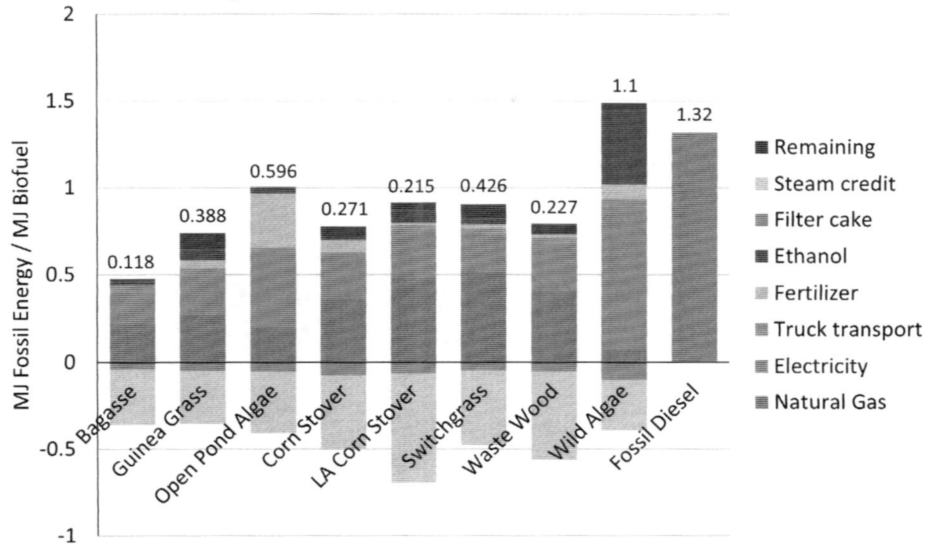

Mihalek, Matt, 2014, "Life Cycle Assessments (LCAs) of PyGasoline and PyDiesel From Different Regional Feedstocks: Corn Stover, Switchgrass, Sugar Cane Bagasse, Waste Wood, and Forest Residues. Phase 1: Preliminary LCAs for DOE Integrated Biorefinery (IBR) Project"

15

Cumulative Energy Demand: Biomass Feedstocks

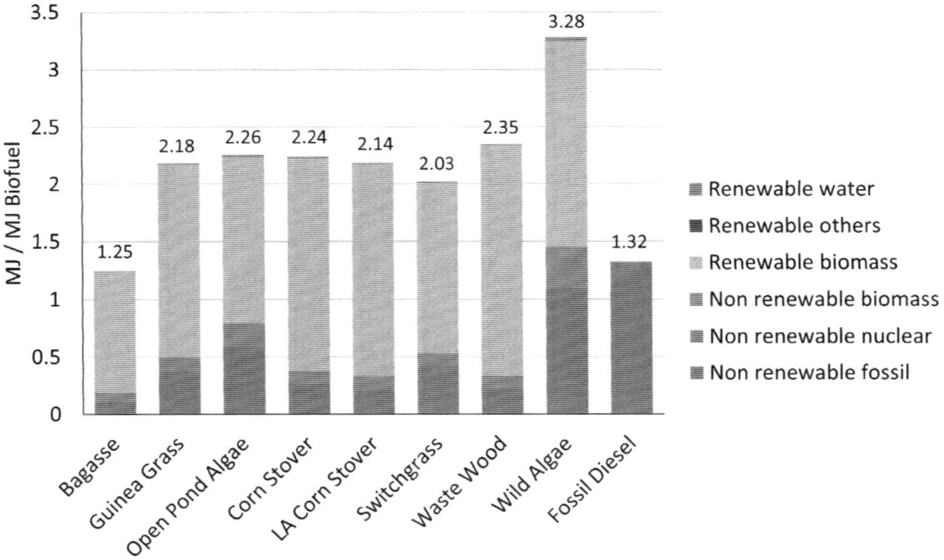

Mihalek, Matt, 2014, "Life Cycle Assessments (LCAs) of PyGasoline and PyDiesel From Different Regional Feedstocks: Corn Stover, Switchgrass, Sugar Cane Bagasse, Waste Wood, and Forest Residues. Phase 1: Preliminary LCAs for DOE Integrated Biorefinery (IBR) Project"

16

Conclusion

- There are 10.5 million tonnes/year of MSW available for thermochemical conversion in Mexico.

- The amount of fuels that can be produced accounts for 4.2 million tons/year which could satisfy 7.1 % of Mexico's annual fuel consumption.

- The C footprint for the pyrolysis-based biofuels is estimated to be -9.5 and 88.0 g CO_2 eq/MJ for organic and polyolefin plastics.

- Fossil Energy Demand (FED) is smaller for pyrolysis-based advanced hydrocarbon biofuel compared to fossil fuel, however Cumulative Energy Demand (CED) is higher except for bagasse feedstock. Renewable fuels require higher total energy, but most is renewable.

17

Acknowledgements

- To the Mexican Council of Science and Technology (CONACYT, award no. 383220) and the Graduate School of Michigan Technological University for the awarded PhD scholarships.

- To the Richard and Bonnie Robbins Endowment for its financial support.

18

Contact information

Ulises R. Gracida-Alvarez:	urgracid@mtu.edu
Laura R. Nitz:	lrnitz@mtu.edu
Lauren M. Keenan:	lmkeenan@mtu.edu
Julio C. Sacramento-Rivero:	julio.sacramento@correo.uady.mx
David R. Shonnard:	drshonna@mtu.edu

Gracida-Alvarez et al. (2016) **DOI: 10.1021/acssuschemeng.6b01143**

2nd RCN Conference on Pan American Biofuels & Bioenergy Sustainability

AUTHOR INDEX

AUTHOR INDEX